Environmental Technology

The Editors

Prof. D.R. Khanna (Ph.D., FIAES, FNC, FASEA, FZSI, FSESc, FMANU, FICER) is Professor in the Department of Zoology and Environmental Science, Gurukula Kangri Vishwavidyalaya, Haridwar. He is well known Limnologist and has published about more than 100 research papers on limnology and fish and fisheries. Prof. Khanna has written/ edited more than 32 books in limnology/ecology. He is the editor of various peer reviewed Journals & Member of Advisory Board of Several Journals. He has been awarded ASEA Excellence Award, 2000 and NATCON Impetus Award, 2004, International Environment Conservation Award, 2010 in Mauritius and Life time achievement award in 2011.

Prof. A.K. Chopra (Ph.D., FIAES, FASEA, FZSI, FMANU, FICER) is Professor in the Department of Zoology and Environmental Science, Gurukula Kangri Vishwavidyalaya, Haridwar. He is well known scientist in the field of environmental microbiology and pollution and has published about more than 150 research papers. Prof. Chopra has written/edited more than number of books respective field. He is the Editor-in-Chief Journal of Applied and natural science foundation and editor of various peer reviewed Journals & Member of Advisory Board of Several Journals. He has been awarded ASEA Excellence Award, 2005 and MANU - International Environment Conservator Award, 2011.

Dr. Gagan Matta (Ph.D., FASEA, FICER, FASNF, FMANU): did his M.Sc. and Ph.D. (Environmental Science) working for last more than 7 years in the field of Environmental Science and Hydrology. At present he is Assistant Professor in the Department of Zoology and Environmental Science, Gurukula Kangri Vishwavidyalaya, Haridwar. He has published more than 21 research paper with different aspects of aquatic science and environmental science. He is member of several academic societies and author/editor of 9 books. Dr. Matta is awardee if various national and international awards including International young scientist award in 2010 in Mauritius and Best Researcher Award in 2012 in Sri Lanka.

Dr. Rakesh Bhutiani (FASEA, FICER) did his M.Sc. and Ph.D. (Environmental Science) in the year 2000 and 2004 respectively from Gurukula Kangri Vishwavidyalaya. At present he is Assistant Professor in the Department of Zoology and Environmental Science, Gurukula Kangri Vishwavidyalaya, Haridwar. Currently he is doing research in the field of Limnological Modelling. Dr. Bhutiani is editor/associate editor of various books and also authored no. of books. He is editor of Environmental Conservation Journal & life member of several academic societies. He was awarded two junior Young Scientist Award & has published more than 45 research papers and articles.

Dr. Vikas Singh is a Young Researcher working in the field of Environmental Pollution, Monitoring and Biotechnology in Department of Zoology and Environmental Science, Gurukula Kangri University, Haridwar. He earned a B.Sc. from M.J.P. Rohilkhand University, Bareilly and a M.Sc. & Ph.D. from Gurukula Kangri University, Haridwar. He has qualified NET (Agriculture) in 2010 & received research fellowship under the scheme of BSR (UGC) for Meritorius Research Students. He has authored/edited 3 books and is Assistant Editor in peer reviewed journals like "Environment Conservation Journal" and "Essence". He has published more then a dozen paper in the field of Environmental Monitoring, Pollution & Bioremediation in National and International Journals of repute. He is fellow of "ASEA" and "ICMANU" societies. He earned Young Scientist Award two times in 2011 & 2012 and one Appreciation Award for contributing in Ist WCMANU-2011. He is also the President of Rotract Club, Jwalapur, Haridwar & give his precious time to the society for spreading Environment education & Awareness among the people.

Environmental Technology

Editors
Prof. D.R. Khanna
Prof. A.K. Chopra
Dr. Gagan Matta
Dr. R. Bhutiani
Dr. Vikas Singh

Associate Editor
Dr. Ajendra Kumar

2013
Daya Publishing House®
A Division of
Astral International Pvt. Ltd.
New Delhi – 110 002

Published by   :   **Daya Publishing House®**
A Division of
Astral International Pvt. Ltd.
– ISO 9001:2008 Certified Company
4760-61/23, Ansari Road, Darya
Ganj, New Delhi-110 002
Ph. 011-43549197, 23278134
E-mail: info@astralint.com Website:
www.astralint.com

Laser Typesetting   :   **Classic Computer Services**
Delhi - 110 035

Printed at   :   **Salasar Imaging Systems**
Delhi - 110 035

PRINTED IN INDIA

Preface

In the last few decades increasing numbers of laws have been enacted in an effort to stop and reverse environmental impacts of a previous era when there were few such laws. Technologies for removal and control of environmental pollutants have also shown an extraordinary growth, driven in many cases by the new legal requirements.

Technology, the term we generally give to the practical use of scientific discoveries to solve everyday problems. Technology and science are therefore closely associated so the "environmental technology" is a field of applied science. Basic idea of environmental technology is to search for explanation about the working, based on the principle of cause and effect, which is another way of saying that there, is a logical reasons or cause for every change in our environment and to apply it in different places with similar kind of problems.

Scientists study the factors that cause pollution of water and air, such as how various chemicals react in the atmosphere. Engineers use this information to develop technological solutions by designing more effective pollution control equipment. Technicians understand the basic scientific principles at work in pollution control equipment in order to operate, monitor, sample, or regulate it.

The overall goal of this book is to develop awareness of the many facets of technology in environmental conservation and preservation and sustainable development while following certain public policies. Environmental technology is a relatively new field that also has interesting and exciting connections with other non-science academic disciplines such as political science and sociology. You are encouraged to use the information available in the book in an integrated fashion.

This book is designed to provide the basic idea of new technological approaches developing for the conservation of environment and sustainable development including providing a logical flow of information and allowing for teaching as well as research to improve the quality of environment.

Editors and contributors of this book hope that this book provides the foundation for new researches to initiate the strides being made to protect the environment.

Editors

Contents

2013, Environmental Technology

Editors: D.R. Khanna, A.K. Chopra, Gagan Matta, R. Bhutiani & Vikas Singh

Published by: DAYA PUBLISHING HOUSE, NEW DELHI

Pages *1–14*

Chapter 1

Sustainable and Renewable Energy from DSSC: A Facile Method and Cost Effective Materials

Mohan Pal[1,2], Kushagra Bhatheja[1], Arun Singh[2], Sameer S.D. Mishra[1] and K.K. Saini[1]

[1]*National Physical Laboratory, Dr. K.S. Krishnan Road, New Delhi – 110 012*
[2]*Department of Physics, Jamia Millia Islamia, New Delhi – 110 025*

ABSTRACT

Solar energy can be converted into useful form of energy by two ways; solar photovoltaic (solar cell) and solar hydrogen (fuel cell). Solar hydrogen production is a form of artificial photosynthesis, where hydrogen is produced by decomposition of water with the help of solar photons on the surface of suitable catalyst which is subsequently converted into electrical energy by fuel cell. In solar photovoltaic, solar energy is converted directly into the electrical energy. There is renewed interest in the solar photovoltaic with the invention of dye sensitized solar cell (DSSC), popularly known as Gratzel cell, over the conventional p-n junction solar cell. This device employs wide band gap semiconductor active electrode. Out of several wide band gap materials, TiO_2 is non-toxic environmental friendly alternative. Conventional TiO_2 based PV devices use ruthenium metal based photo-sensitizer, which is rare element. Cost effectiveness of the device cannot be expected from this material. To popularize this environmental friendly sustainable source of energy we have to switch over to economic and abundant raw materials for the device fabrication which are environmental friendly too.

In this paper critical operating parameters of DSSC with new directions are discussed. We have developed new photo-sensitizer with low cost materials using facile method, suitable for large scale production. We have investigated its ability for photo-electron generation for useful power. Basic investigations of the material are carried out to tailor the useful properties. Technique is suitable for large scale production of DSSC.

Keywords: Solar cell, DSSC, Photo-sensitizer, Solar energy.

Introduction

Energy is essential for the developed of society and to upkeep the living standards. Major part of our energy requirements come from fossil fuels, nuclear and hydro power sources. According to world energy report 2008, we get around 80 per cent of our energy from-conventional fossil fuels like oil (34 per cent), coal (24 per cent) and natural gas (22 per cent). The modern renewable energy source (Solar, Hydro, Wind, Biomass, Geothermal and Marine energy) contribute 7 per cent whereas Nuclear energy contribution is around 6 per cent to the total energy requirement. Today we are facing threats from fossil fuels to their geographical location and exhaustion, there is uncertainty from the hydro- and wind power due to climatic conditions and safety and security is a major challenge due to recent incidents in different countries of the world. Fossil fuels are also prone to emission of greenhouse gases and hence a threat to our environment which poses serious health and weather related issues. We can no longer afford this destruction of our environment. As per Kyoto Protocol there are not much restrictions on India with regard to carbon emissions but with the development rate these emissions will increase and we shall not be permitted to pump more and more carbon into the atmosphere. Therefore, we fulfill our growing energy demands by renewable sources. Keeping this in mind India has already started solar mission program to harvest more and more solar energy to fulfill societal needs.

Tremendous energy from the Sun, much more than our requirement, is available on the Earth's surface. By using solar photovoltaic (solar cell) and solar hydrogen (fuel cell) we can convert the solar energy into useful form to cater the public demand. Solar cell device converts solar energy directly into electrical energy. Operation of solar cell relies on capture of Incident Photon by the Cell material which result in charge separation and finally the charge collection. Conversion of solar energy (photons) into electrical energy by silicon solar cell was demonstrated long back in 1954. But it could not gain momentum primarily because of the high cost of solar grade silicon (SGS). However silicon is abundant but its purification process requires lot of energy. Basically SGS is 2nd grade to device grade silicon, so Si - PV industry nurtured on the byproduct of silicon industry, where we cannot expect its growth more than the silicon industry. The next generation amorphous Si cell (a–Si) appeared in 1976, right after the first oil shock, when a great amount of hope and expectation was set upon PV energy. Amorphous silicon (a–Si) has a high absorption rate but its development has been blocked by the Staebler–Wronski [Kolodziej A 2004] light induced degradation effect.

Most other PV technologies have the same working principle as c–Si and search is mostly an empirical within a large number of semiconductors that resulted in a few promising materials. Polycrystalline CdTe was one of the first proposed PV thin-film materials, CdTe solar cells are typically hetero-junctions with CdS being the n-type component. CdTe has the adequate Eg (energy band gap) of 1.45 eV and a high optical absorption coefficient, but its toxicity is a troubling issue.

Chalcopyrite solar cells were based on $CuInSe_2$ (CIS) and $CuInGaSe_2$ (CIGS) are another thin film leading PV technologies. Of these, CIGS cells appear to have the best future potential due to higher efficiencies, confirmed maximum efficiency of 19.2 per cent and lower manufacturing energy consumption, which may even be the lowest for any PV technology, but they have toxicity related issues which may turn out to be a problem for mass production.

III–V semiconductor normally based on GaAs and InGaP based solar cells are very efficient but expensive devices, there efficiency can be further improved with multi-junction technology. But non-abundant materials will be more expensive and so will significantly increase the price of final product.

Molecular and polymer organic solar cells are simple PV devices that are made by organic semiconductors "sandwiched" between two electrodes. These cells are characterized by high optical absorption coefficients and low manufacturing costs. A great deal of attention has been given to these cells in recent times, as they are expected to play a key role in the future flexible PV market, particularly now that the 5 per cent efficiency barrier has been overcome.

Dye Sensitized Solar Cell (DSSC) is a successful attempt to create an anthropological analogous concept to photosynthesis [Smestad G.et al 1994]. The photoreceptor and the charge carrier are different elements. This separation of functions leads to lower purity demands on the raw materials side, and consequently makes DSCs a low cost alternative. State of the art DSCs, utilizing ruthenium metal based dye, achieve ~ 11 per cent energy conversion efficiency [Gratzel. M. and Mcevoy A.J. 1997]. But ruthenium is a rare element, so the price is expected to escalate with the pick up of this technology. Devices based on dyes without rare element does result in the desired efficiency. Secondly the organic dyes used in DSSCs are prone to degradation under solar irradiation, therefore have limited life span.

Alternate photosensitizers include chalcogenide quantum dots (CdS, CdSe, Bi_2S_3, CdTe, PbS etc.). These options appear promising because of low cost, ability to tailor the band edge absorption, easy fabrication and long life. But they have yet to prove their credibility for optimum device operation. Titanium oxide based DSSC type device with chalcogenide nano-particle sensitizer is popularly known as quantum dot sensitized solar cell (QDSC)[Kamat P. V. 2008, Nozik, A. 2002]. There are three routes to sensitize the TiO_2 electrode with quantum dots for light harvesting; (i) Chemical bath deposition process (CBD)[Diguna J. L. *et al.*, 2007, Niitsoo O. *et al.*, 2006], (ii) electrochemical deposition (ECD)[Robel I. *et al.*, 2006] and (iii) Selective ion absorption and reaction (SILAR) [Nicolau Y. F. *et al.*, 1985 and 1990]. I have adopted SILAR process to synthesize and assemble PbS quantum dot array on the TiO_2 film surface. I am presenting light harvesting properties of these quantum dots in this article as per the following layout:

1. Dye sensitized solar cell (DSSC) - General structure and its working.
2. Experimental Section –Sol-preparation, TiO₂ film coating, assembling of PbS nano dots, characterization.
3. Results and discussion.
4. Conclusions.

Dye Sensitized Solar Cell

Structure

In the case of the original Grätzel design, the cell has three primary parts. On the top is a transparent anode made of Indium tin oxide (ITO) deposited on the back of a (typically glass) plate. On the back of the conductive plate is a thin layer of titanium dioxide (TiO_2), which forms into a highly porous structure with an extremely high surface area. TiO_2 only absorbs a small fraction of the solar photons (those in the UV). The plate is then immersed in a mixture of a photosensitive ruthenium-polypyridine dye (also called molecular sensitizers) and a solvent. After soaking the film in the dye solution, a thin layer of the dye is left covalently bonded to the surface of the TiO_2 film/layer.

A separate backing is made with a thin layer of the iodide electrolyte spread over a conductive sheet, typically platinum metal. The front and back parts are then joined and sealed together to prevent the electrolyte from leaking. Structure of DSSC is shown in Figure 1.1.

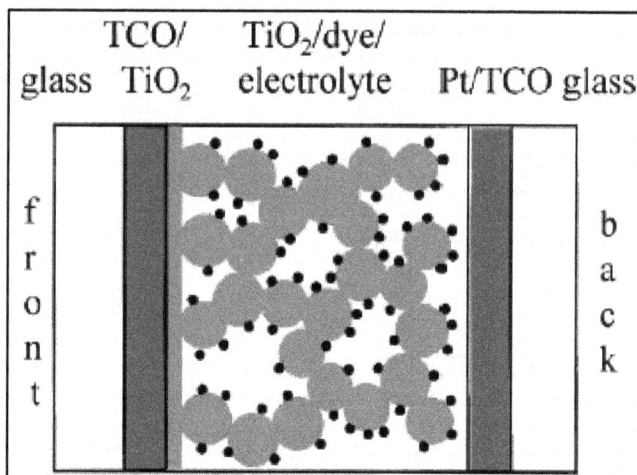

Figure 1.1: Structure of Dye Sensitized Solar Cell

Although they use number of "advanced" materials, these are inexpensive compared to the silicon, needed for normal cells because they require no expensive manufacturing steps. But slow electron transport and the poor absorption of low energy photons are the two main limiting factors in the performance of the nanoparticle based Dye-sensitized solar cell (DSSC). Incorporating highly-ordered TiO_2, nanoparticle arrays is a promising approach to improve the performance of DSSC.

The TiO$_2$ Nanoparticles have a high surface area to volume ratio. Furthermore, the nanoparticle structure enhance the electron percolation pathways, light conversion and ion diffusion at the semiconductor –electrolyte interface [Mor G.K et al., 2006].

Operating Principal of Nanocrystalline Dye Sensitized Solar Cell (DSSC)

In DSSC, dye (photosensitizer) molecule is excited by the absorption of a photon. The excited electron is then injected into the conduction band of the semiconductor. There is no hole formation in the valance band of semiconductor. The dye molecule is regenerated by accepting an electron from the electrolyte, which subsequently transports the generated hole to the back (counter) electrode as shown in Figure 1.2 and described in the following steps.

Step–1

Dye adsorbed on the semiconductor film (TiO$_2$), absorbs the incident photon of light. Dye is excited from the ground state(S) to the excited state(S*).

$$S + h \rightarrow S*$$

Step–2

Now these excited electrons are injected into the conduction band of theTiO$_2$, this process is done very fast ~10^{-15} seconds.

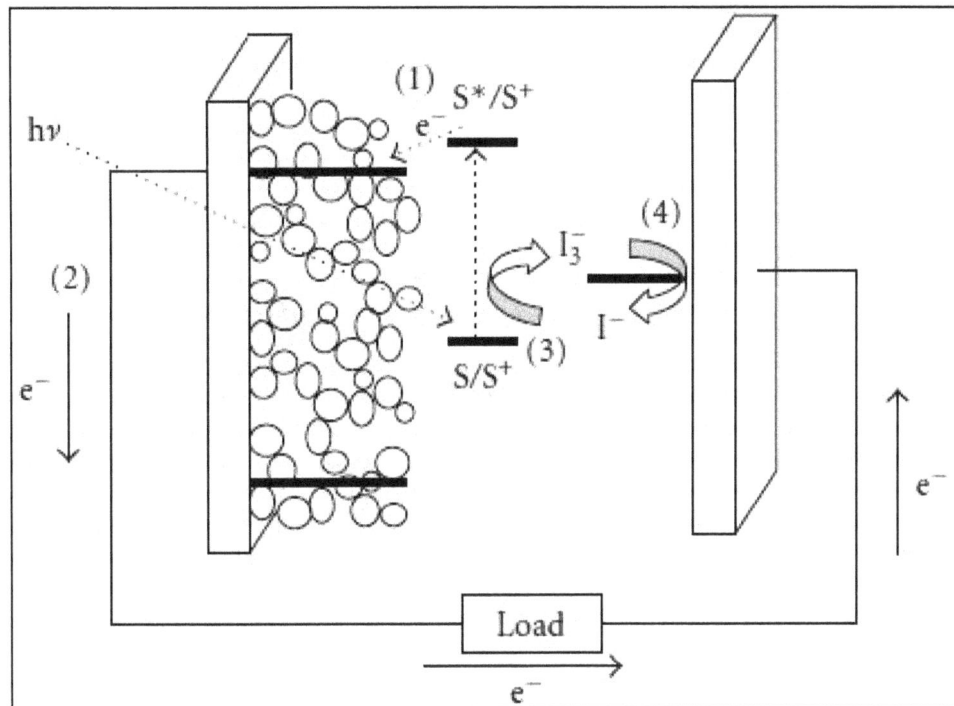

Figure 1.2: Schematic Structure of N nanocrystalline DSSC, with its Operating Principle

$$S^* \rightarrow S^+ + e^- (TiO_2)$$

In this process, a positive charge creates on the photosensitizer *i.e.* oxidized state of dye molecules.

Step–3

By diffusion, these injected electrons are transported through TiO_2 film, towards the back contact (made by SnO_2: F coated glass substrate) of the cell and reach the counter electrode, through the external circuit.

Step–4

Dye cations (S^+) accepts electrons, from a redox couple in the electrolyte solution, I^-/I_3^- is the most typical redox couple, regenerating the ground state (S).

$$S^+ + e^- \rightarrow S$$

In this process, I^- is oxidized to the oxidized state, I_3^-.

Step–5

The oxidized I_3^-, diffuse towards the counter electrode and is reproduced to I^- ions.

$$I_3^- + 2e^- \rightarrow 3I^-$$

This process is controlled by a kinetic balance [Haque S. A., Cho B. M., *et al.,* 2005] between the Dye excited state and electrolyte and other recombination process.

Overall electric powder is generated without any chemical transformation.

In DSSC, voltage is measured by, energy gap between the Fermi level (near the conduction band of TiO_2) of the semiconductor electrode and the redox potential level of the mediator in the electrolyte.

Experimental

Coating of TiO$_2$ Film on FTO Substrate

Cleaning of Glass Slides and Maintaining of Proper Environment

To have proper coating of TiO_2 film on the fluorine doped SnO_2 (FTO) glass substrate the substrate need to be properly cleaned and even the environment, where the coating would be done, has to be dust and humid free. Since absolutely dust and humid free environment cannot be obtained practically, environment with minimum possible humidity and dust is prepared. A de-humidifier machine, that minimizes the amount of humidity in the environment, is used. Thin film of TiO_2 was coated on the thoroughly cleaned glass substrates. Substrates were first cleaned with soap solution using cotton and then kept in chromic acid for three hours. Chromic acid solution is prepared from H_2SO_4 and propanol (alcohol), with propanol as the parent ingredient as it is a very good cleansing agent. After removing FTO glass substrate from the chromic acid solution they were rinsed with plenty of water to remove any traces of chromic acid. Cleaned substrates were handled carefully to avoid stain or finger prints on them.

Preparation of TiO$_2$ Precursor Sol

A 0.5 M TiO$_2$ sol was prepared by the partial hydrolysis and poly-condensation of titanium tetra-butoxide with water using isopropyl alcohol (IPA) as a solvent and HNO$_3$ as a catalyst. Titanium tetrabutoxide and water have been taken in 1:2 molar ratios. The hydrolysis and poly-condensation of titanium tetrabutoxide proceeds according to the following scheme:

$$Ti\,(OC_4H_9)_4 + 2H_2O \rightarrow Ti\,(OC_4H_9)2(OH)_2 + 2C_4H_9OH$$
$$Ti\,(OC_4H9)_2(OH)_2 + Ti\,(OC_4H_9)_2(OH)_2 \rightarrow Ti_2O\,(OC_4H_9)_4(OH)_2 + H_2O$$

This reaction stops with the inclusion of two water molecules:

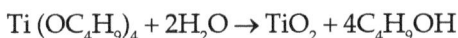

$$Ti\,(OC_4H_9)_4 + 2H_2O \rightarrow TiO_2 + 4C_4H_9OH$$

All the chemicals used were of AR (Analytical Reagent) grade and were used as procured *i.e.* without further purification. The mixture was stirred vigorously with a magnetic stirrer for 1 h. This results in a coloured transparent solution. The solution was kept overnight before film deposition [Sharma Sunil Dutta. *et al.*, 2006].

Deposition of TiO$_2$ Film on FTO Glass Substrate

Cleaned substrates were fixed in the dip coater and were allowed to dip in the precursor solution taken in the specified container. After a brief set-in period (1-2 minutes) they were pulled out vertically upward with a constant speed of 18 cm/min so as to obtain a uniform film of TiO$_2$. Coated substrates were allowed to hang in the coater to allow the coated film to dry and drip the excess solution at the lower edge of the substrate. Dried films were heated at 100°C for 30 min in an electric oven before annealing at 450°C for 1 hour in an electric furnace in under atmospheric conditions. Thick films were prepared by repeating the coating cycle several times [Sharma Sunil Dutta *et al.*, 2008].

Assembling PbS Quantum Dots on the TiO$_2$ Film Surface

For assembling the PbS quantum dots on TiO$_2$ film we use SILAR (Successive ionic Layer adsorption and reaction) method. Four beakers of 100ml capacity were taken, marked as I, II, III and IV and kept in line on the table sequentially. These beakers were filled with different solutions/solvents as; beaker No. I - 0.02 M Pb(NO$_3$)$_2$ solution in methanol, beaker No. II - pure methanol; beaker No. III - 0.02 M Na$_2$S in methanol and beaker No. IV - pure methanol. TiO$_2$ thin film coated FTO glass substrate was sequentially and successively immersed in each solution for one minute with one minute gap in between. One dip in all the four beakers completes one cycle[Lee HyoJoong *et al.*, 2009]. Several such cycles were repeated to obtain varying densities of PbS quantum dots onto the TiO$_2$ film. After drying at room temperature these slides were sintered at 300°C in an electric furnace under atmospheric conditions for one hour.

Characterization

Samples were characterized by XRD, SEM, UV-Vis spectrophotometery and CV techniques. XRD studies were carried out in 20-80° range on Simmon D-500 x-ray diffractometer which uses Cu Kα radiations (λ = 1.5406Å), SEM investigation were performed on LEO 400 scanning electron microscope, spectrometric studies on Shimadzu 1601 UV-Vis spectrophotometer and CV studied were performed on Eco-Chemie Netherlands, Model 302N electrochemical workstation.

Results and Discussion

Figure 1.3(A) shows XRD pattern of TiO_2 film deposited on FTO glass substrate and annealed at different temperatures 300°C, 400°C and 500°C. Crystallization of the material takes place after annealing at 500°C, therefore further investigations are carried out on these samples. Figure 1.3 (B) TiO_2 film with array of PbS quantum dots.

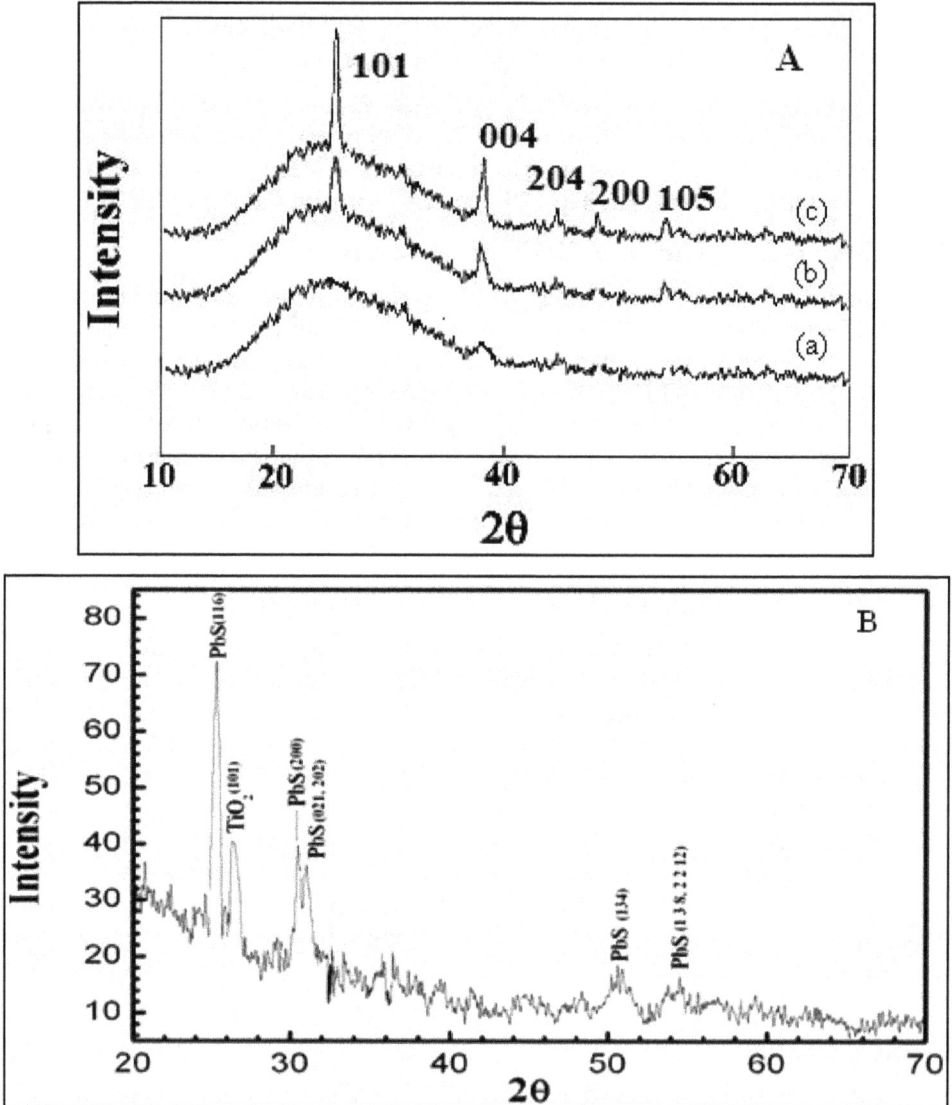

Figure 1.3: XRD Pattern of (A) TiO_2 Film Deposited on FTO Glass Substrate and (B) PbS Quantum Dot Assembled TiO_2 (TiO_2 + PbS) Electrode

XRD peaks were observed at 2q values 25.6, 37.6, 44.9, 47.3 and 54.4 degree in case of TiO_2 film without PbS nano-particles, these peaks were assigned as (101), (004), (204), (200) and (105) corresponding to anatase phase of TiO_2. In case of TiO_2 film with a layer of PbS nano-particles XRD peaks at 2q values; 25.6, 27.0, 30.4, 31.0, 51.1 and 54.4 degrees. These peaks corresponds to (khl) values PbS(116), TiO_2(101), PbS(202), PbS(021, 202), PbS(134) and PbS(1 3 8, 2 2 12) respectively as per the JCPDS card No. 80-1144.

Selected area SEM micrographs of TiO_2 films and PbS TiO_2 films are shown in Figure 1.4(a) and 1.4(b). This figure is clearly shows the formation of PbS nano-particles on the TiO_2 film surface. The surface morphology of TiO_2 film is very smooth whereas rough film surface morphology is evident after formation of PbS nano-particles. SEM image analysis at high resolution shows that the PbS nano-particles are mini rod shaped and uniformly distributed over the entire TiO_2 film surface. Average dimensions of these mini rods are ~ 50nm length and 14-20nm diameter.

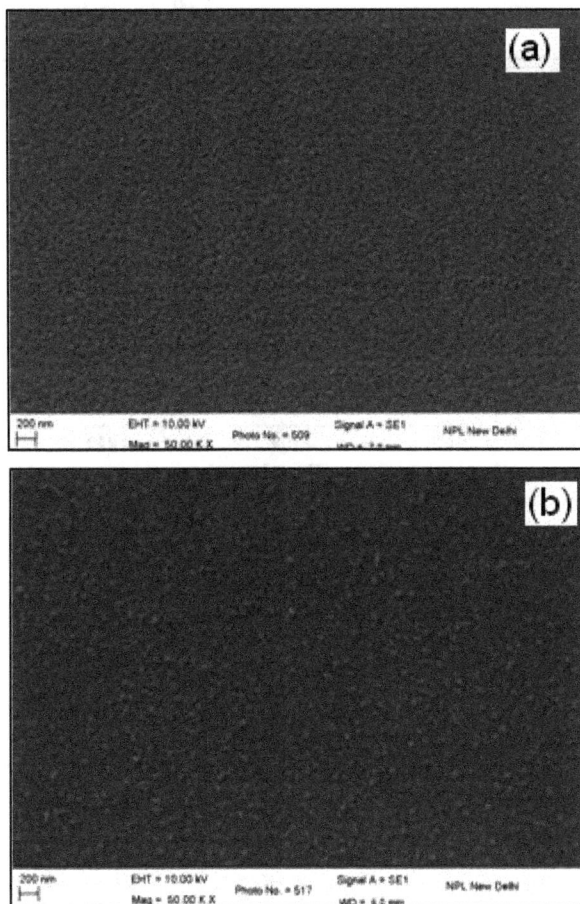

Figure 1.4: Represents the SEM of (a) Pure Tio$_2$ Film on FTO Glass Substrate and (b) PbS Assembled TiO$_2$ Film on FTO Substrate

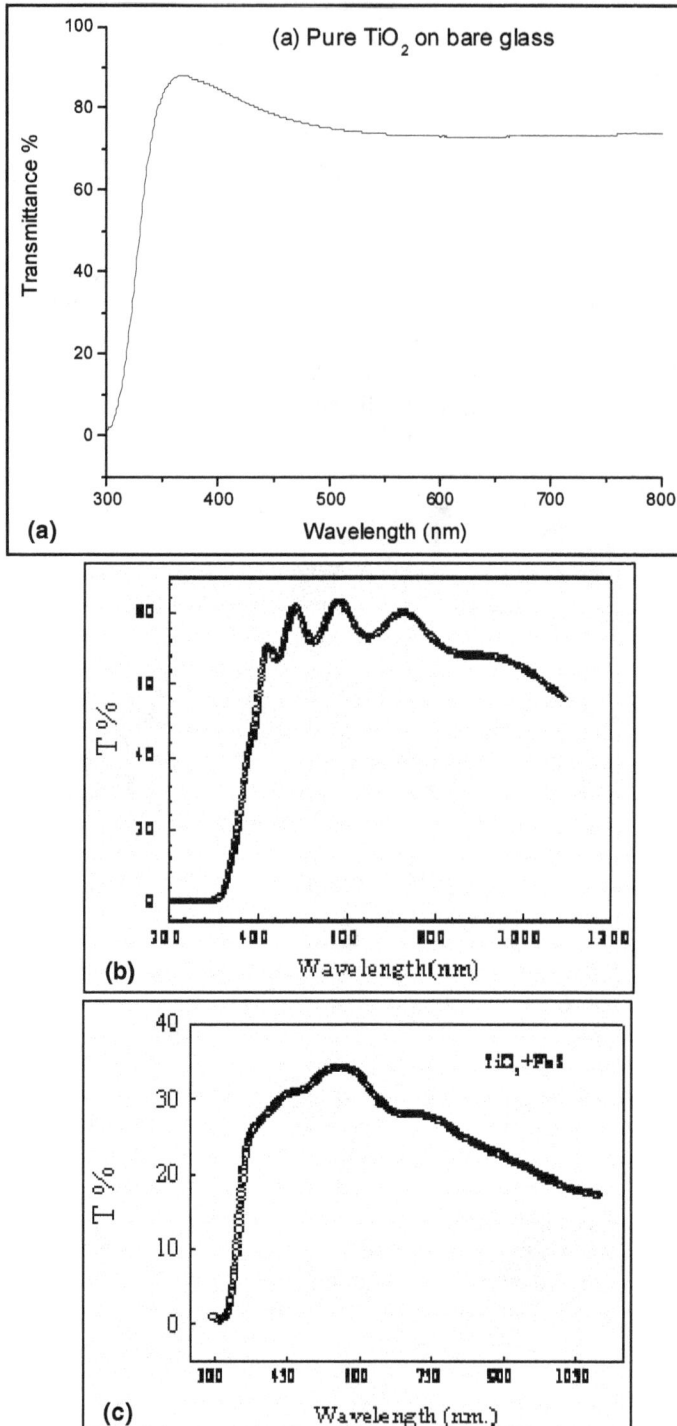

Figure 1.5: UV-Vis Transmission Spectra of
(a) Pure TiO$_2$ Film Coated on Simple Glass; (b) Pure TiO$_2$ Film Coated on FTO glass;
(c) (TiO$_2$ + PbS) Nano-Particle Coated Electrode

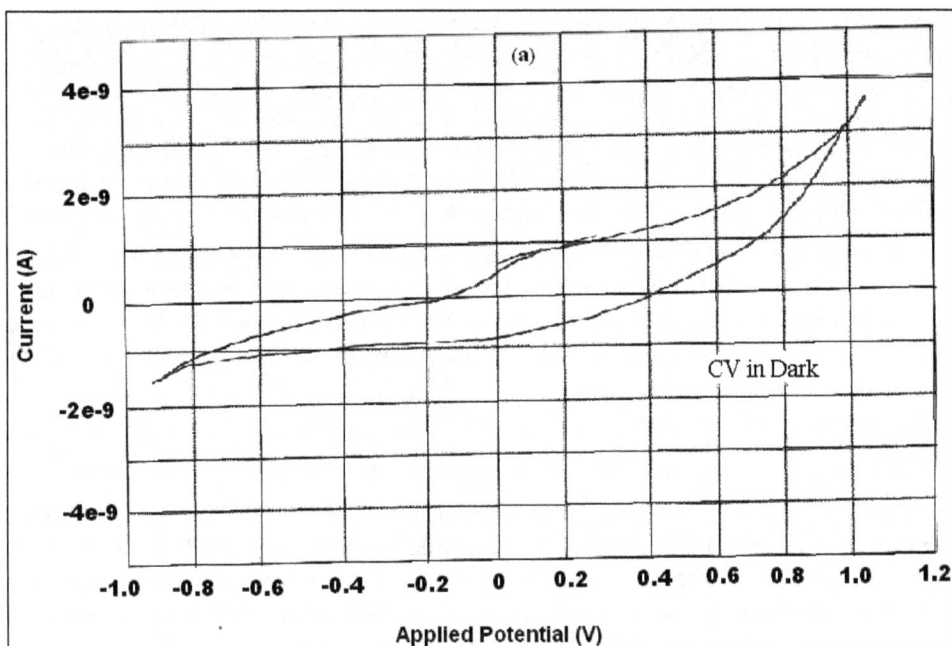

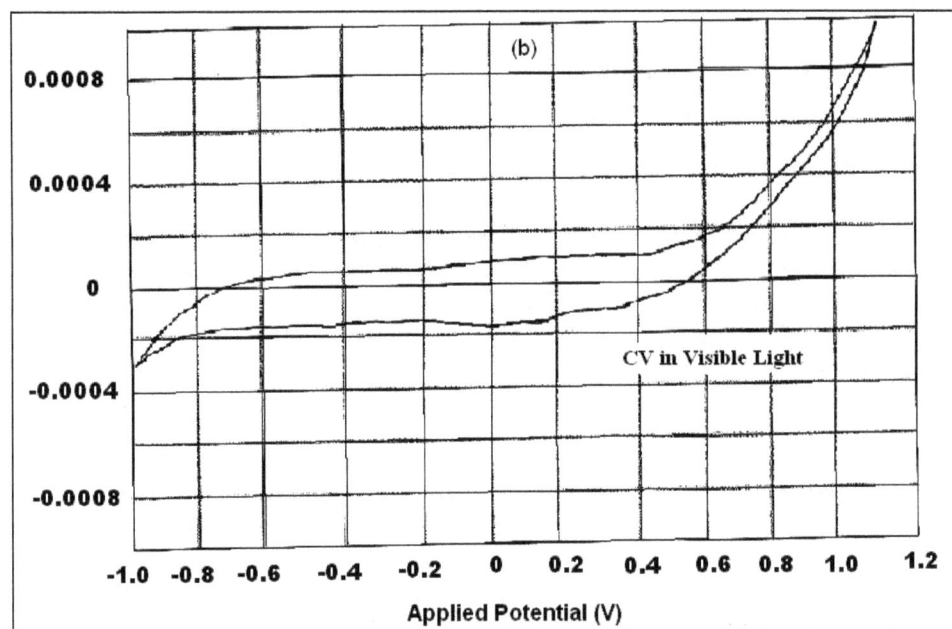

Figure 1.6: Cyclic Voltamograms of (TiO$_2$ + PbS) Electrode in I$^-$/I$_3^-$ Redox Electrolyte Under (a) Dark and (b) Illuminated Conditions

A smooth curve in the visible range with average 70–75 per cent transmission is observed in TiO$_2$ film samples without PbS nano-particles (Figure 1.5a). Sharp drop in transmittance at ~ 400nm corresponds to band edge absorption of the anatase phase of TiO$_2$. Light absorption properties of the sample are drastically modified with the deposition of PbS nano-particles. Average transmission in visible range is reduced ~ 35 per cent *i.e.* visible transmittance of the PbS decorated TiO$_2$ films is reduced to half as compared to pure TiO$_2$ film (Figure 1.5 b), which confirms the visible sensitization of the TiO$_2$ electrode by PbS nano-particles.

Photoelectron transfer from PbS nano-particles to the external electrode via TiO$_2$ conduction band are studied by linear sweep cyclic voltametery in the range -1.0V to 1.2V in three electrode geometery using I^-/I_3^- redox electrolyte, platinum counter electrode and Ag/AgNO$_3$ as reference electrode. Voltametric sweep is recorded with an increment of 20mV/sec in dark as well as under visible light illuminated conditions. Illumination was done with 100W visible light tungsten filament lamp. CV scans under as well as illuminated conditions are shown in Figure 1.6. An increase in CV current of about three orders is observed after illumination, which indicates that PbS nano-particles are efficient visible absorbers in visible range and can work as photo-sensitizer for DSSC. Increase in visible absorbance from 30 to 50 per cent and corresponding increase in the CV current is observed with increase in SILAR cycles upto five, after that a saturation behaviour is observed as there is no significant increase in CV current for further PbS deposition.

Light absorption properties of PbS quantum dots has recently been studied by [Ratan Debnath *et al.*, 2010, Pattantyus-Abraham Andras G. *et al.*, 2010 and YUE Dong *et al.*, 2011]. They have adopted different technique, where PbS quantum dots were synthesized as a colloidal solution which were subsequently transferred onto the substrate. Quantum dot size has been controlled by chemical techniques. [Braga Antonio *et al.*, 2011] have synthesized PbS and CdS quantum dots by SILAR process (used by us), but they have used complex poly-sulphide precursors and reported encouraging results. Study of photoelectron transfer dynamics through legends, dielectric inter layer and solution has also been taken seriously and modeling studies are in progress [Darren A. *et al.*, 2011, Hyun Byung-Ryool *et al.*, 2010, Kevin T 2011]. New results are being reported constantly and this field is progressing steadily to reach its goal.

Conclusions

PbS nano-particles absorbs efficiently in the visible range. These nano-particles can be synthesized on to the TiO$_2$ films by a facile technique, SILAR. The technique yields uniform coverage of the titanium oxide films with PbS quantum dots. Photo generated electrons can be efficiently and transferred into the TiO$_2$ conduction band for registering external current and hence the material can effectively be used as an inorganic photo-sensitizer for dye sensitized solar cell (DSSC) replacing the conventional organic dye.

Acknowledgements

Authors are thankful to Director, National Physical Laboratory, New Delhi for his constant encouragement and interest in this work. One of the authors, Mr. Mohan

Pal, is also thankful to Council of Scientific and Industrial Research (CSIR), Govt. of India, New Delhi, for providing him financial assistance in the form of SRF.

References

Braga Antonio, Gim_enez Sixto, Concina Isabella, Vomiero Alberto, and Mora-Sero Ivan. "Panchromatic Sensitized Solar Cells Based on Metal Sulfide Quantum Dots Grown Directly on Nanostructured TiO_2 Electrodes", J. Phys. Chem. Lett. 2011, 2, 454–460.

Darren Achey, Shane Ardo, Hai-Long Xia, Maxime A. Siegler, and Gerald J. Meyer, "Sensitization of TiO_2 by the MLCT Excited State of Co^I Coordination Compounds". J. Phys. Chem. Lett. 2011, 2, 305–308.

Diguna J. L., Shen Q., Kobayashi J., Toyoda, T., "High Efficiency of CdSe Quantum-Dot-Sensitized TiO_2 Inverse Opal Solar Cells". Appl. Phys. Lett. 2007, 91, 023116

Gratzel. M., Mcevoy A.J., "Dye-sensitized photovoltaic devices", 14th Eur. Solar Energy Conf. Barcelona 1997, Proc of the conf.

Haque S. A., Palomares E., Cho B., Green A.N.M., Hirata N., Klug D.R. and Durrant James R., "Charge Separation versus Recombination in Dye-Sensitized Nanocrystalline Solar Cells: the Minimization of Kinetic Redundancy". J. American Chemical Society, 127(10) pp.3456-3462, (2005).

Hyun Byung-Ryool, Bartnik A. C., Lee Jin-Kyun, Imoto Hiroaki, Sun Liangfeng, Choi Joshua J., Chujo Yoshiki, I Hanrath Tobias, Ober Christopher K., and Wise F. W., "Role of Solvent Dielectric Properties on Charge Transfer from PbS Nanocrystals to Molecules". *Nano Lett.* 2010, 10, 318-323.

Kamat P. V., "Quantum Dot Solar Cells. Semiconductor Nanocrystals as Light Harvesters". *J. Phys. Chem. C* 2008, 112, 18737–18753.

Kevin Tvrdya, Frantsuzovc Pavel A., and Kamata Prashant V., "Photoinduced electron transfer from semiconductor quantum dots to metal oxide nanoparticles". PNAS January 4, 2011 vol. 108(1) 29–34.

Lee HyoJoong, Leventis Henry C., Moon Soo-Jin, Chen Peter, Ito Seigo, Haque Saif A., Torres Tomas, Nu¨esch Frank, Geiger Thomas, Zakeeruddin Shaik M., Gra¨tzel Michael, and Nazeeruddin Md. Khaja, "PbS and CdS Quantum Dot-Sensitized Solid-State Solar Cells: Old Concepts, New Results" Adv. Funct. Mater. 2009, 19, 2735–2742.

Mor Gopal K., Varghese Oomman K., Paulose Maggie, Shankar Karthik, Grimes Craig A., "A review on highly ordered, vertically oriented TiO_2 nanotube arrays: Fabrication, material properties, and solar energy applications". Solar Energy Mater. Sol. Cells 90 (2006) *pp 2011-2075.*

Nozik, A. J., "Quantum Dot Solar Cells". *Phys. E* 2002, *14*, 115–200.

Niitsoo O, Sarka SK., Pejoux C, Rühle S, Cahen D, Hodes G, "Chemical Bath Deposited CdS/CdSe-Sensitized Porous TiO_2 Solar Cells". J. Photochem. Photobiol. A 2006, 181, 306–313.

Nicolau Y. F., "Solution Deposition of Thin Solid Compound Films by a Successive Ionic-Layer Adsorption and Reaction Process". Appl. Surf. Sci. 1985, 22(3), 1061–1074.

Nicolau Y. F., Dupuy M., Brunel M., "ZnS, CdS, and Zn1-XCdXS Thin-Film Deposited by the Successive Ionic Layer Adsorption and Reaction Process". J. Electrochem. Soc. 1990, 137, 2915–2924.

Pattantyus-Abraham Andras G., Illan J. Kramer, Barkhouse Aaron R., Wang Xihua, Gerasimos Konstantatos, Ratan Debnath, Levina Larissa, Raabe Ines, Nazeeruddin Mohammad K., Gratzel Michael, and Sargent Edward H., "Depleted-Heterojunction Colloidal Quantum Dot Solar Cells". ACS Nano 2010. 4(6) 3374-80.

Ratan Debnath, Tang Jiang, Barkhouse D. Aaron, Wang Xihua, Pattantyus-Abraham Andras G., Lukasz Brzozowski, Larissa Levina, and Sargent Edward H., "Ambient-Processed Colloidal Quantum Dot Solar Cells via Individual Pre-Encapsulation of Nanoparticles". J. AM. CHEM. SOC. 2010, 132, 5952–5953.

Robel I., Subramanian, V., Kuno, M., Kamat, P. V., "Quantum Dot Solar Cells. Harvesting Light Energy with CdSe Nanocrystals Molecularly Linked to Mesoscopic TiO_2 Films". J. Am. Chem. Soc. 2006, 128, 2385–2393.

Sunil dutta Sharma, Singh Davinder, Saini K.K., Chanderkant, Sharma Vikas, Jain S.C., C.P.Sharma "Sol–gel-derived super-hydrophilic nickel doped TiO_2 film as active photo-catalyst" Applied Catalysis A: (General) 314 (2006) 40–46.

Sharma Sunil Dutta, Saini K.K., Kant Chander, Sharma C.P., Jain S.C., "Photodegradation of dye pollutant under UV light by nano-catalyst doped titania thin films" Applied Catalysis B: Environmental 84 (2008) 233–240.

Smestad G., Bignozzi C., Argazzi R., "Testing of dye sensitized TiO_2 solar cells-I: Experimental photocurrent output and conversion efficiencies". Solar Energy Mater 32 (1994) 259-272.

YUE Dong, ZHANG Jian-Wen, ZHANG Jing-Bo, LIN Yuan, "Preparation of PbS Quantum Dots Using Inorganic Sulfide as Precursor and Their Characterization". *Acta Phys. -Chim. Sin.* 2011, 27 (5), 1239-1243.

2013, Environmental Technology
Editors: **D.R. Khanna, A.K. Chopra, Gagan Matta, R. Bhutiani & Vikas Singh**
Published by: **DAYA PUBLISHING HOUSE, NEW DELHI**

Pages 15–23

Chapter 2

Evaluation of Anaerobic Biological Treatment for Wastewater of Ion Exchange (IX) Resin Manufacturing Facility

Rajesh Singh and V.K. Choubey

*Environmental Hydrology Division, Jal Vigyan Bhawan,
National Institute of Hydrology, Roorkee – 247 667*

ABSTRACT

Ion Exchange (IX) Resin manufacturing unit utilizes considerable amount of water during the course of resin production. The effluent produced is rich in organics and inorganic solids to the tune of 8000-10000 mg/L and 15000-20000 mg/L respectively. The industry under study was having full- fledged effluent treatment plant but was undergoing huge financial loss due to the new norms set by pollution control board. Intensive treatability study was performed to find out the possibility of using anaerobic biological treatment which has not been applied for such type of effluents so far. The study implied that 47-50 per cent COD reduction was possible with anaerobic biological treatment alone and 90-95 per cent COD reduction by anaerobic followed by aerobic biological treatment. A final process design as well as economic study based on the laboratory findings has been carried out.

Keywords: Resin, Wastewater, COD, BOD, UASBR.

Introduction

Ion exchange resins consist of a polymeric matrix and a functional group with a mobile ion which can be exchanged with other ions present in the solution to be treated. The most common synthetic structures are – cross liked polystyrene, cross liked polymethacrylate, phenoil-formaldehyde etc. The manufacture of ion exchange resins involve the preparation of a cross liked copolymer followed by sulfonation in the case of strong acid cation resins, or chloromethylation and the amination of the copolymer for anion resins (Dow, 2000). The production process involves use of variety of chemicals like styrene, divenyl benzene, ethylene glycol dimethacrylate, isobutyl alcohol, formaldehyde, methanol, ethylene dichloride, trimethyl amine, dimethyl amine, methacrylates, polyvinyl alcohol, oleum, etc. The unused solvents and chemicals make the effluent from the manufacturing process high in organics, total dissolved solids, and low in pH. Presence of variety of chemicals in the wastewater makes the treatment challenging and requires a cost effective as well as robust process.

Anaerobic digestion is used for treating the high strength organic wastewater. Since the late seventies, anaerobic digestion has experienced an outstanding growth in research and full scale application, particularly for the treatment of food and beverage industry effluent and to a lesser extent for municipal wastewater (Hulshoff Pol *et al.*, 1998; Yu *et al.*, 2004; Fountoulakis *et al.*, 2004; Filik Iscen *et al.*, 2007). Anaerobic digestion is a complex, natural, multi stage process. During the process, organic compounds are degraded through a variety of intermediates into methane and carbon dioxide, by the activity of a consortium of micro organisms. Interdependence of the bacteria is a key factor in the anaerobic digestion process (Parawira *et al.*, 2005). The upflow anaerobic sludge blanket process is one of the most commonly used wastewater treatment system, with several installations treating industrial wastewater (Techobanoglous *et al.*, 2004). The upflow anaerobic sludge blanket reactor (UASBR) is a reactor of upflow where the organic material on its way through the covering of sludge composed of a large population of anaerobic bacteria begins its biodegradation. The reactor is composed of three essential parts: a zone of digestion, a zone of sedimentation, and a separator of gas - solids-liquids. These are integrated into one column where the primary sedimentation process, the bio-digestion of the sludge and the secondary sedimentation is done simultaneously as a primary and secondary treatment of residual waters achieving efficiency in the removal of organic material up to 85 per cent (Sponza, 2001). Anaerobic digestion treatment is one of the technologies being considered to provide a solution to the treatment high strength organic wastewater and maximum amount of biodegradable fraction can be converted into useful energy end product in the form of biogas and fertilizer in the form of digestate (Fernandez *et al.*, 2001; Saravanan *et al.*, 2004; Song *et al.*, 2004).

The objective of the work was to study the anaerobic treatability and possibility of utilizing UASBR in order to optimize the operational cost, reduce green house gas emissions and convert waste into useful end products.

Materials and Methods

Pilot Scale Digester

The UASB reactor was constructed from MS-FRP sheet with 2 m Length, 1 m width, and 4 m height. The working volume of the reactor was 7 m^3 (Figure 2.1).

Sampling ports were provided for collection of samples. A centrifugal pump (Grundfos, Chiu) of capacity 5 m³/hr was used for feeding wastewater into the reactor. A buffer tank constructed of HDPE with 500 L capacity was utilized as buffer tank.

Seed and Inoculation

The reactor was initially seeded with inoculum from an anaerobically digested sludge of a sewage treatment plant. On subsequent days, Jaggery solution along with urea and DAP was added to obtain the desired mixed liquor suspended solids (MLSS) and mixed liquor volatile suspended solids (MLVSS) in the reactor. After achieving 4 per cent MLSS and 0.75 MLVSS/MLSS ratio, effluent injection at the rate of 0.5 m³/d started for acclimatization of the micro-organisms. The acclimation period in this study was 60 days.

Sampling and Analysis

The functioning of the reactor was monitored over a period of four months. The samples at feed, in the reactor, and treated effluent were taken on regular basis to monitor the performance of the reactor.

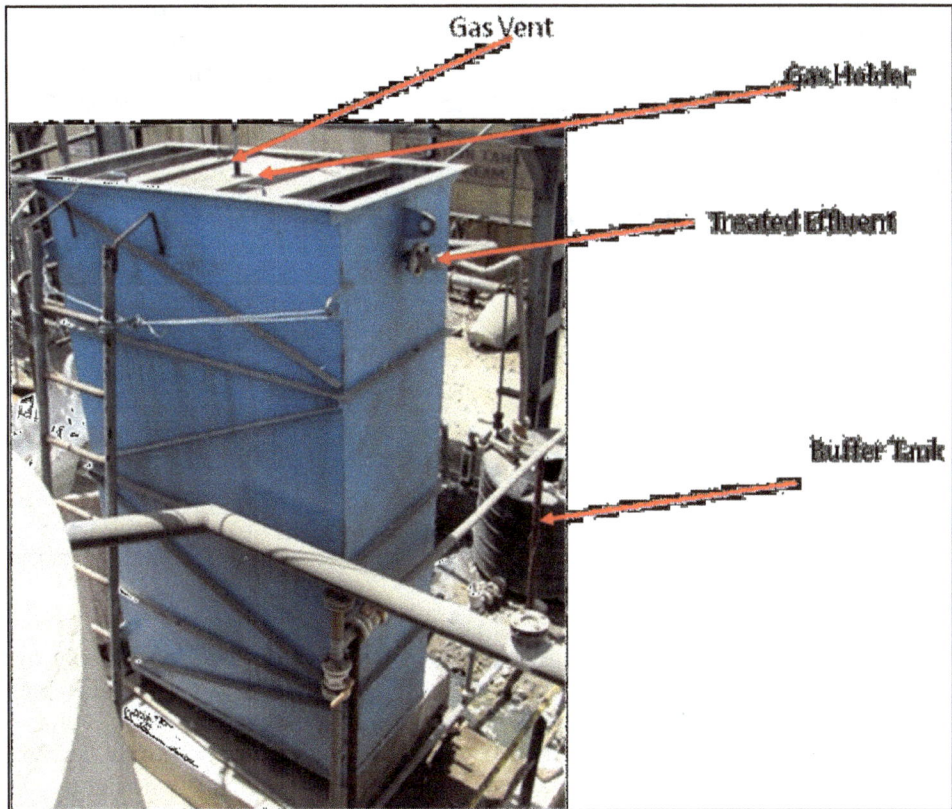

Figure 2.1: Pilot Upflow Anaerobic Sludge Blanket Reactor

Chemical oxygen demand (COD), total suspended solids (TSS), total dissolved solids (TDS), MLSS, and MLVSS were regularly performed for the untreated and treated effluent as well as sludge according to the standard methods (APHA, 1996).

Operating Conditions

The reactor was operated in fill, react, and withdrawal during initial period of operation. After stabilization of the process, the reactor was operated in continuous mode for a period of three months.

Result and Discussion

Characterization of Wastewater

The pilot plant was installed in the Ion exchange resin manufacturing facility located in Ankleshwar, Gujarat to understand the actual operational conditions and the associated difficulties. The manufacturing facility was already having a state of the art effluent treatment facility consisting of collection tank, solids contact clarifier, aerobic reactor based on membranes technology and hence, it was decided to install the pilot plant after clarifier (Figure 2.2) in order to get rid of suspended solids which are polymeric in nature and non biodegradable and replicate the future condition. Hence the samples were collected from the outlet of solids contact clarifier and the characterization is given in Table 2.1. It can be seen from the table that the effluent is very high in COD (7000-8000 mg/L), BOD (2500-3500 mg/L), and TDS (15000-25000 mg/L). Apart from these harsh parameters, the BOD/COD ratio (0.3-0.35) is not very much favourable for biological treatment.

Table 2.1: Wastewater Characteristics

Sl.No.	Parameters	Unit	Inlet
1.	pH	–	7.0-9.0
2.	Total dissolved solids	mg/L	15000–25000
3.	Total suspended solids	mg/L	100-200
4.	Volatile suspended solids	mg/L	50-150
5.	COD	mg/L	7000-8000
6.	BOD	mg/L	2500-3500
7.	Oil and grease	mg/L	< 30

UASBR Performance

Acclimatization of micro organisms for the wastewater was judged from the analysis of MLSS and MLVSS in the reactor and the organic loading rate (OLR) was increased or decreased based on the MLVSS in the reactor. During start up phase, ups and downs were observed in the MLVSS value. If sharp reduction in MLVSS value (25 per cent) was observed than wastewater feed was replaced with Jaggery in order to maintain 0.5 Food/Micro-organism (f/m) ratio. The process stabilized in 60 days and after that continuous increase in MLVSS was observed (Figure 2.2).

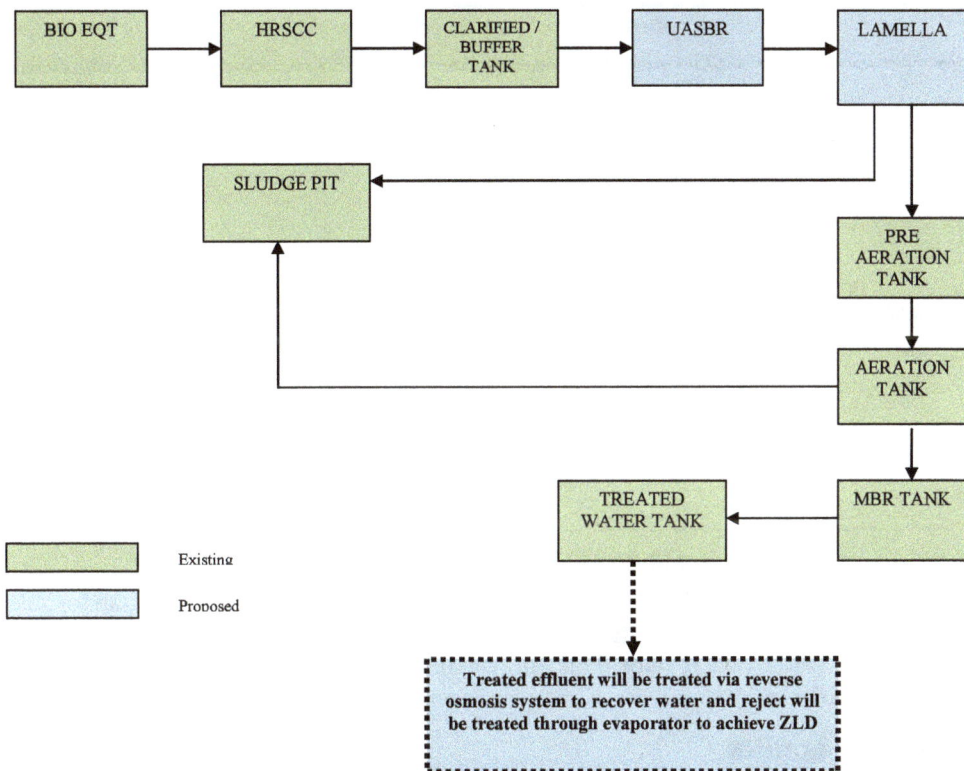

Figure 2.2: Schematic Diagram of Effluent Treatment Plant for Resin Manufacturing Facility

OLR was always maintained above 1 kg COD/m³/d by combination of organics supplied from process effluent and jaggery. Initial OLR from effluent was kept as low as 0.5 kg/m³/d and increased in a stepped manner to 4.5 kg/m³/d over a period of 75 days (Figure 2.3). The process was found stabilized at this point and the system operated for almost 1 month with this OLR with consistency.

Upflow velocity is regarded as one of the main parameter significantly affecting microbial ecology and characteristics of UASBR. It also helps in flushing the hazardous gases thereby keeping the system in healthy condition. The optimum upflow velocity for the wastewater and the system under study was found to be 0.5 m/h. In order to maintain the desired OLR and upflow velocity, the feed to UASBR is 4-5 times of the influent and hence the same was recycled back to buffer tank/UASBR feed tank. This also helps in minimizing the toxicity and shock load to UASBR.

The COD reduction in the initial phase of the start up was on the higher side due to higher percentage of COD from the jaggery which is easily biodegradable. COD reduction stabilized to 47-50 per cent (Figure 2.4) which was desired by the manufacturing facility and based pilot plant observation, full scale plant was observed.

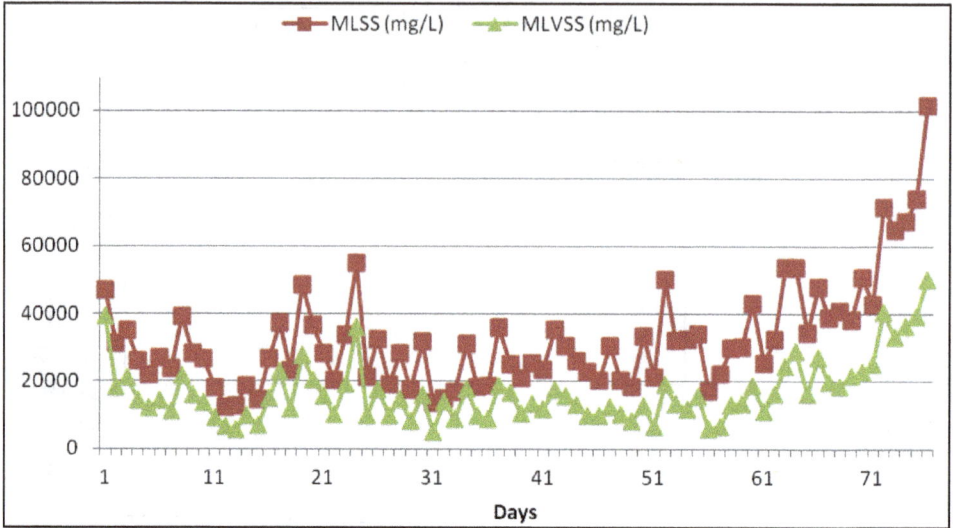

Figure 2.3: MLSS and MLVSS Trend

COD reduction of 90-95 per cent was achieved with the combination of anaerobic followed by aerobic reactor, whereas only 85-90 per cent COD reduction was observed with two stage aerobic reactor. This may be due to conversion of non biodegradable fraction into biodegradable fraction by anaerobic bacteria.

Cost Economics

The manufacturing unit was planning to increase the production of IX resins, which will lead to increase in flow as well as organic load to the ETP. The expansion of the manufacturing facility will lead to increase in the flow to ETP from 180 m^3/day to 240 m^3/d with marginal or no change in the COD and BOD. A cost comparison of UASBR against installation of an aerobic reactor was carried out which is presented in Table 2.2.

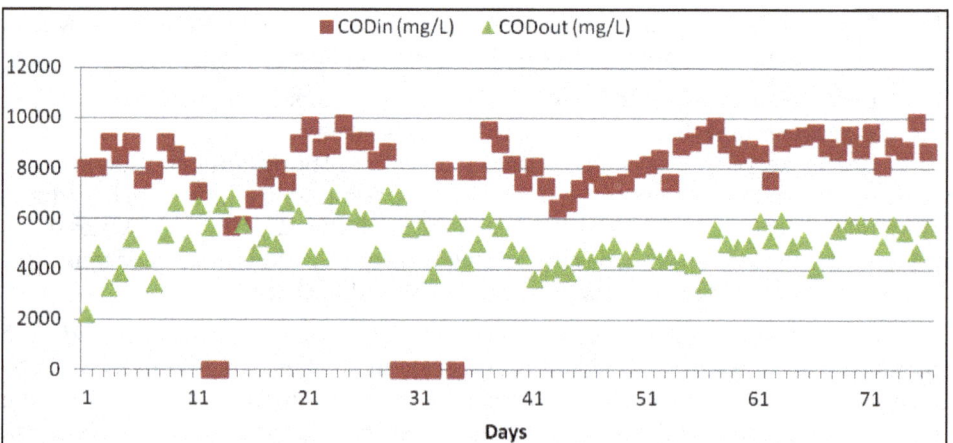

Figure 2.4: COD Profile Over the Trial Period

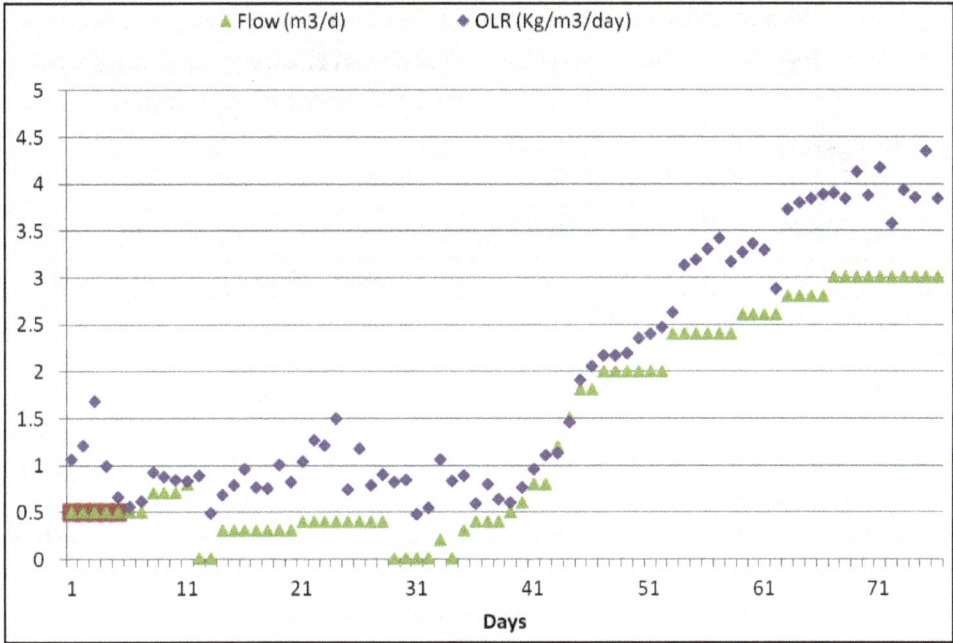

Figure 2.5: Flow and OLR Profile Over the Trial Period

Extra investment in case of UASBR	= (88.18 – 84.70) Lacs INR
	= 3.48 Lacs INR
Savings in operating cost	= (11.02 – 1.07) Lacs INR
	= 9.95 Lacs INR
Pay back period	= 3.48/9.95
	= 0.35 yr = 4.5 months

Apart from the above said advantage, savings in term of methane generation and power production can also be considered.

COD reduction in the reactor	= 3500 mg/L x 240 m^3/d/1000
	= 840 kg/d

Approx. Methane generation @ 0.40 m^3 CH$_4$/kg COD reduced

$$= 0.40 \times \text{COD reduction}$$
$$= 0.40 \times 840 = 336 \text{ m}^3/\text{d}$$
$$= 14 \text{ m}^3/\text{hr}$$

Power production	= (Vol. of methane x cal. Value x Engine eff.)/860
	= 14 mm^3/hr x 9500 Kcal/m^3 x 0.35/860
	= 54 KWH

Savings in term of power consumption @ Rs. 6 per unit

$$= 54 \text{ KWH} \times 24 \text{ hr} \times 365 \text{ days} \times 6 \text{ Rs.}$$

$$= 28.3 \text{ Lacs INR per annum}$$

The cost benefit analysis indicates long term advantage of UASBR over the widely used aerobic reactor.

Table 2.2: Cost Comparison for Aerobic Reactor Versus UASBR

Sl.No.	2 Stage Aerobic Reactor		Anaerobic Followed by Aerobic Reactor	
1	Mechanical Equipments			
	AT Blower(2x1010 m³/hr)	5,60,000	UASBR feed pump (60 m³/hr)	68,000
	Sludge pump(2 x 10 m³/hr)	50,000	Sludge recirculation pump (1 m³/hr)	50,000
	Sec. clarifier mechanism	1,80,000	Lamella clarifier	5,00,000
	Diffuser (100)	4,50,000	UASBR Internals + Gas flare system	40,00,000
	TOTAL	**12,40,000**		**46,18,000**
2	Civil Equipments @ 6000 INR/m3³			
	Aeration Tank (1145 m³)	68,70,000	UASBR tank (700 m³)	42,00,000
	Sec. Clarifier(60 m³)	3,60,000		
	TOTAL	**72,30,000**		**42,00,000**
	GRAND TOTAL	**84,70,000**		**88,18,000**
3	Operating Cost @ 4.5 INR/KWH (INR/ANNUM)			
	AT Blower	10,08,000	UASBR feed pump	1,41,000
	Sec. clarifier mechanism	52,000	Sludge pump	29,000
	Sludge recirculation pump	42,000		
	TOTAL	**11,02,000**		**1,70,000**

Conclusion

The following conclusions can be drawn from the present study-

1. UASBR can be successfully employed for the treatment of IX resin manufacturing facility wastewater.

2. The optimum COD removal efficiency of reactor found to be 50 per cent corresponding to optimum HRT and organic loading rate of 3 days and 4.5 kg/m³/d respectively. The removal efficiency is expected to improve over the operational period.

3. The stabilization/acclimatization period for the system is 60 days.

4. Sludge recycling mode was found to be effective technique for process stabilization.

5. Cost benefit analysis indicates UASBR a better choice if compared to extended aerobic reactors.

6. UASBR is also helpful in reducing green house gas emissions.

References

1. DOW Chemical Company, 2000. DOW Liquid separation - Fundamentals of Ion Exchange.

2. Hulshoff Pol L. W., Lens P., Stams A. J. M., Lettinga G. (1998). Anaerobic treatment of sulfate rich wastewaters: microbial and process technological aspects. Biodegradation, 9, 213-224.

3. Filik Iscen C., Ilhan S., Yildirim M. E., 2007. Treatment of cake production wastewater in Upflow Anaerobic Packed bed Reactors. International Journal of Natural and Engineering Sciences, 1(3), 75-80.

4. Yu Y., Park B., Hwang S., 2004. Co-digestion of lignocellulosics with glucose using thermophilic acidogens. Biochemical Engineering Journal, 18(3), 225-229.

5. Fountoulakis M, Drillia P., Stamatelatou k., Lyberatos G., 2004. Toxic effect of pharmaceuticals on methanogenesis. Water Science and Technology, 50(5), 335-340.

6. Parawira W., Mutto M., Zvauya R., Mattasson B., 2005. Comperative performance of a UASB reactor and an anaerobic packed bed reactor when treating potato waste leachate. Renewable energy, 31(6), 893-903.

7. Techobanoglous G., Burton F., and Stensel H. D., 2004. Wastewater Engineering: Treatment and Reuse. McGraw-Hill, New York.

8. Sponza D. T., 2001. Anaerobic granule formation and tetrachloroethylene (TCE) removal in an upflow anaerobic sludge blanket reactor. Enzyme and Microbial Technology, 29(6-7), 417-427.

9. Fernandez B., Porrier R., Chammy R., 2001. Effect of inoculums-substarte ration on the start up of solid waste anaerobic digesters. Water Science and Technology, 44(4), 103-108.

10. Saravanane R., Sivasankaran M. A., Sundararaman S., Sivacoumar R., 2004, Anaerobic sustainability for integration of sugar mill waste and municipal sewage. Journal of Environmental Science and Engineering, 46(2), 116-122.

11. Song Y. C., Kwon S. J., Woo J. H., 2004. Mesophilic and thermiphilic temperature co-phase anaerobic digestion compared with single stage mesophilic and thermophilic digestion of sewage sludge. Water Reasearch, 38(7), 1653-1662.

12. APHA, AWWA, WPCF, 1995. Standar methods for the examination of water and wastewater. APHA, Washington DC, USA.

2013, Environmental Technology
Editors: D.R. Khanna, A.K. Chopra, Gagan Matta, R. Bhutiani & Vikas Singh
Published by: DAYA PUBLISHING HOUSE, NEW DELHI

Pages 25–30

Chapter 3

Extracting CO_2 from Atmosphere by Constructing Green Building

G.C. Mishra, Prashant Johl and Sumit Gulati

Department of Civil Engineering,
Lingaya's University, Faridabad,

ABSTRACT

In the wake of overall Industrial and social development of a country the need of energy especially electricity will continue to grow. The use of fossil fuels has become inevitable in the modern times for the production of electricity, transport systems and to meet to the various industrial requirements. This is a major contributory factor for the rate of increase in the Green House Gases namely carbon dioxide in the atmosphere. Increase in the carbon dioxide content in the atmosphere is a major cause of concern. Several control measures have been evolved and innovated to decrease the carbon dioxide content in the atmosphere. In our study a model has been devised to capture the atmospheric carbon dioxide and injecting it into dried up oil wells or in the deep oceans or converting it to some useful compounds for various purposes. In order to capture atmospheric CO_2 a model has been designed in which certain specific chemical will be used to absorb CO_2 from atmosphere. A carbon sink being used in a building at certain height is the most promising approach to capture and reduce CO_2. Such a building would work by letting air pass through the especially designed structure which will be on roof of the building and then catching the CO_2 via a "sorbant" material, such as sodium hydroxide. This procedure does not involve the use of any expensive catalysts and the use of sophisticated equipments and is hence expected to be cost effective. This includes a simple process, when air will flow through fin like structure in the upper parts of the system which contain sodium hydroxide will absorb the CO_2, at another end of the system by increasing the temperature it will discharge the CO_2. After extracting CO_2 it can be dumped at suitable locations *i.e.* inside earth or Deep Ocean. Extracted CO_2 can be converted into useful product. Each house can abstract large quantity of carbon dioxide annually which in turn mitigate the climate change in sustainable manner for holistic growth and development of the society.

Introduction

Carbon dioxide is a naturally occurring chemical compound composed of two oxygen atoms covalently bonded to a single carbon atom. It is a gas at standard temperature and pressure and exists in Earth's atmosphere in this state, as a trace gas at a concentration of 0.039 per cent by volume. As part of the carbon cycle known as photosynthesis, plants, algae, and cyanobacteria absorb carbon dioxide, light, and water to produce carbohydrate energy for themselves and oxygen as a waste product. But in darkness photosynthesis cannot occur, and during the resultant respiration small amounts of carbon dioxide are produced. Carbon dioxide also is a by-product of combustion; emitted from volcanoes, hot springs, and geysers; and freed from carbonate rocks by dissolution.

Carbon dioxide in earth's atmosphere is considered a trace gas currently occurring at an average concentration of about 390 parts per million by volume or 591 parts per million by mass.[1] The total mass of atmospheric carbon dioxide is 3.16×10^{15} kg (about 3,000 gigatonnes). Its concentration varies seasonally (see graph at right) and also considerably on a regional basis, especially near the ground. In urban areas concentrations are generally higher and indoors they can reach 10 times background levels. Carbon dioxide is a greenhouse gas. It is estimated that volcanoes release about 130–230 million tones (145–255 million tons) of CO_2 into the atmosphere each year. Carbon dioxide is also produced by hot springs such as those at the Bossoleto site near Rapolano Terme in Tuscany, Italy. Here, in a bowl-shaped depression of about 100 m diameter, local concentrations of CO_2 rise to above 75 per cent overnight, sufficient to kill insects and small animals, but it warms rapidly when sunlight and the gas is dispersed by convection during the day. Concentration of is CO_2 shown in below curve:

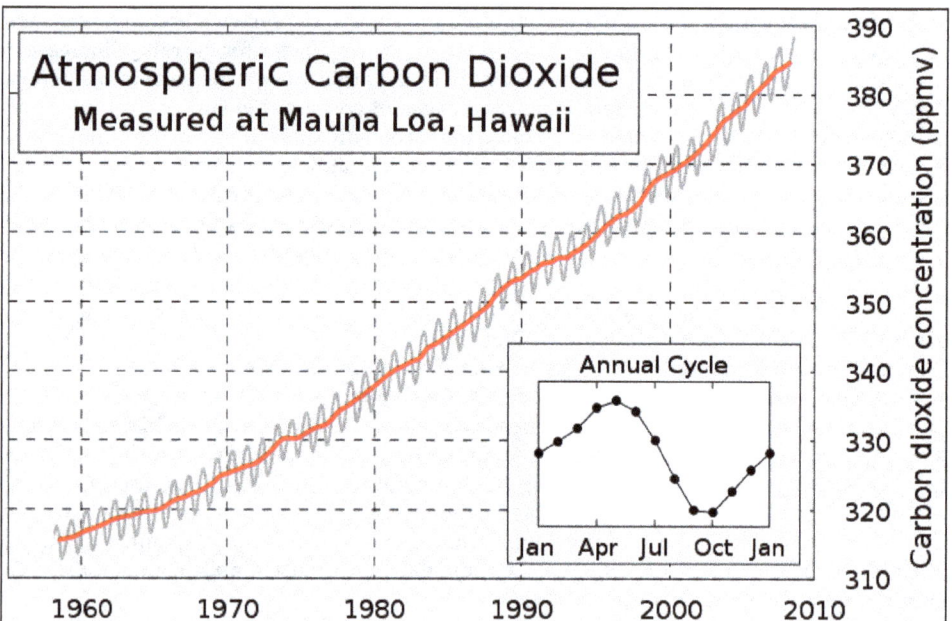

Figure 3.1

Impact of CO_2 Emission on Environment

☆ Rises in sea level of as much as three feet across the globe, flooding land where millions of organisms including human beings now live.

☆ Changes in rainfall pattern across vast areas where agricultural activities are carried out.

☆ The melting of many glaciers and a rise in snow elevations, affecting water supplies across the globe.

☆ Storms including hurricanes of increasing intensity and frequency.

☆ Loss of Biodiversity/Extinction of animal and plant species as the pace of change in habitat driven by global warming outstrips their ability to adjust.

Objective

Use of Carbon dioxide absorbent chemicals in building structures which will extract atmospheric carbon dioxide, by which there will be reduction in the concentration of carbon dioxide from atmosphere in order to mitigate climate change.

Methodology

For Buildings

Construction of these building will be same as the normal building. The carbon extraction setup will be fixed at the top of the building structure. In this structure specially designed fins will be created which will allow the flow of air inside that structure which will contain liquid sodium hydroxide in flowing state in order to maximize the surface area for better absorption of carbon dioxide to give a liquid solution of sodium carbonate. Then this liquid will be piped away to special facility made nearby big buildings where the carbon dioxide will be extracted from aqua solution of sodium carbonate, as sodium carbonate looses carbon dioxide and again converted into sodium hydroxide by increasing small temperature. This process doesn't need change of sodium hydroxide as sodium hydroxide is regenerated or recycled at the end to continue the absorption cycle. The purified air can be sent inside the building as this air is carbon dioxide free this will be of low temperature and can be used for cooling the building.

For Industries

This setup can be constructed at the industries where carbon emission is high. Smoke coming out of these industries is firstly passed through Electrostatic precipitators which can easily remove fine particulate matter such as dust and smoke from the air stream, then it will be passed through the carbon dioxide extraction setup which will be same as for building, size will vary as per requirements and in this way carbon-dioxide free air can be released into the atmosphere.

Extracted CO_2 will be Used As

☆ Approximately 115 million metric tons of CO_2 used annually by the global chemical industries.

★ CO_2 can be recycled back into valuable hydrocarbon products such as methanol, synthetic ethanol and diesel fuel.

★ For Food and Beverage products as carbon dioxide is used in beverage carbonation.

★ For Water/Wastewater Treatment as Industrial and municipal wastewater must be neutralized before being discharged to the environment. Carbon dioxide replaces harsher acids for the alkaline neutralization process. It's safer and cheaper than sulphuric-acid systems, improves controllability, and there's less downtime and no labor to handle chemicals. It is also less corrosive, and easier to handle and store.

★ For Metal Fabrication as a shielding gas during welding.

★ For Plant Growth carbon dioxide systems greatly improve growth and quality of plants in the greenhouse. Increasing concentrations of the gas results in larger, healthier and faster-growing plants and lower operating costs.

★ Saving the Forest.Carbon dioxide is used to make precipitated calcium carbonate (PCC), which is used to reduce the use of virgin wood fiber in paper making.

★ Energy Source.Storage of carbon dioxide at its triple point (the temperature-pressure combination at which carbon dioxide can exist simultaneously as a solid, liquid or gas) is being tested as a means of providing closed-loop refrigeration in order to shift electrical-energy demand to off-peak consumption hours. Under test in Japan, the process offers the potential to customers to shift electrical load while maintaining temperatures as low as minus 60°F (-51°C).

★ Firefighting.Carbon dioxide smothers fires without damaging or contaminating materials and is used for fighting fires when water is ineffective, undesirable or unavailable.

★ Carbon dioxide can be used in "green" plastic production.

★ Using existing oil drilling technology, channels thousands of meters deep would be bored into the sea bed. The carbon dioxide gas would be injected into it, permeating the surrounding porous rock.

Advantage

★ Save environment.

★ Carbon dioxide fixation.

★ Simple technology.

★ One time investment and it required cheap maintenance.

★ Can generate revenue after some time.

★ It can be installed at old big buildings and industries.

★ Generate employment

Result and Discussion

This technology will help to accelerate the industrial growth, as carbon dioxide will be reduced from environment. Extracted carbon dioxide may find application for various industrial purposes as well as a cooling agent for preservation of cryogenic materials. Initially concept of this CO$_2$ elimination technique was conceived by some researchers in order to mitigate the green house gases specially CO$_2$ which is an issue of concern for climate change. Since last couple of decades efforts are on by eminent scientist and researchers to tackle this twin problem of global warming and climate change. Since urbanization is taking place at very fast pace in most of the countries of the world especially the developing countries are also moving very fast for holistic industrialization in order to achieve development at par with developed countries. In the process of industrialization they are also contributing large quantity of green house gases into the environment which is issue of concern, therefore in this study an effort is being made to find out solution to reduce carbon content of atmospheric gases by creating a structure at top of high rise residential buildings and industrial installations in which certain chemical which work as an adsorbent will be utilized to convert a useful product which will find multiple application as a raw material for various product as well as ingredient/supplementary substances to generate some product for various utilization such as in food beverage, cooling agent etc. Growth and development is need of the human civilization but unsustainable development need correction at early stage for holistic development of society. This research will find application not only in the new building structures or new industrial installations but also this could find application in existing high rise civil structures to extract excessive carbon dioxide from atmosphere. however the feasibility of this research is yet to be decided which needs further research in terms of its efficiency, cost effectiveness and workability, therefore further research required to study in detail to achieve sustainable pattern of development and mitigate the climate change by using this technology which was earlier devised in form of a tree structure which was very costly, time consuming and require extra space for construction and installation.

References

Biosphere 2000: Protecting our Global Environment, Donald G. Kaufman, Kendall Hunt, 2000.

Genthon, G., Barnola, J. M., Raynaud, D., Lorius, C., Jouzel, J., Barkov, N. I., Korotkevich, Y. S., Kotlyakov, V. M. (1987). "Vostok ice core: climatic response to CO$_2$ and orbital forcing changes over the last climatic cycle".Nature **329** (6138): 414.

http: //www.gizmag.com/CO$_2$-used-to-impregnate-plastics/17455

http: //www.tomcoequipment.com/water-treatment/direct-co2-gas-injection.html

http: //news.bbc.co.uk/2/hi/6374967.stm

http: //www.specialtyminerals.com/our-minerals/what-is-pcc/

http: //firefighter.co.in/fire_equipments_co2-flooding-system.asp http: //www.popsci.com/environment/article/2009-06/installing-plastic-trees-help-environment

Hydrocarbons for the 21st Century - The work of the Loker Hydrocarbon Research Institute, Published in Chemical and Engineering News

Petit, J.R., *et al.,* 2001, Vostok Ice Core Data for 420, 000 Years, IGBP PAGES/World Data Center for Paleoclimatology Data Contribution Series #2001-076. NOAA/ NGDC Paleoclimatology Program, Boulder CO, USA.

United Nations Environmental Program.

IUPAC, *Compendium of Chemical Terminology*, 2nd ed. (the "Gold Book") (1997). Online corrected version: (2006–) "electrostatic precipitator

2013, Environmental Technology
Editors: D.R. Khanna, A.K. Chopra, Gagan Matta, R. Bhutiani & Vikas Singh
Published by: DAYA PUBLISHING HOUSE, NEW DELHI

Pages 31–42

Chapter 4

Applications of Nanotechnology in Water Treatment: Recent Advancements and Future Possibilities

*Indranil Saha¹, Sayan Bhattacharya², Kaushik Gupta¹,
Krishna Biswas¹, Sushanta Debnath¹ and Uday Chand Ghosh¹*

¹*Department of Chemistry, Presidency University
(formerly Presidency College),
85/1, College Street, Kolkata – 700 073*
²*Department of Environmental Science, University of Calcutta
51/2, Hazra Road, Kolkata – 700 019*

ABSTRACT

Nano-materials have gained special attention in water pollution mitigation researches since last decade because these kinds of materials have unique properties like high surface-to-volume ratio, enhanced magnetic property, special catalytic properties etc. Nanoltration techniques are now widely used to remove cations, natural organic matter, biological contaminants, organic pollutants, nitrates and arsenic from groundwater and surface water. Nano-membranes are used to treat contaminated water by filtration or separation techniques. Nanosorbents are widely used as separation media in water purication to remove inorganic and organic pollutants from contaminated water. During the last decade, titanium dioxide (TiO_2) nanoparticles have emerged as promising photocatalysts for water purication. Consequently, different workers had adopted the methods such as chemical precipitation, sol–gel, vapour deposition, solvo thermal, solid state reaction etc. for the synthesis of some nanostructured mixed oxides, which can be effectively used for groundwater treatment.

Nanoparticle agglomerates of mixed oxides such as iron–cerium, iron–manganese, iron–zirconium, iron–titanium, iron–chromium, cerium-manganese etc. have been synthesized and characterized for pollutants removal (*i.e.* arsenic, fluoride etc.) from aqueous solutions. Among the available different technologies, adsorption is one of the best due to its easy handling, low cost and high efficiency. The difficulty encountered when nano-materials are used for the treatment of contaminated water is the separation of material from its colloidal suspension by simple filtration, but the advantage is the generation of low sludge volume which diminishes disposal problem. The environmental fate and toxicity of a material are critical issues in materials selection and design for water purication. The information so far collected is rather insufficient for judging the environmental fate, transport and toxicity of nanomaterials. The success of the techniques in field conditions is a factor of interdisciplinary works, which need successful collaboration of chemistry, material science and geology and bio-sciences.

Keywords: Nanomaterials, Pollutants, Sorption, Filtration, Adsorption.

Introduction

Water is a ubiquitous substance. Man has always been dependent on water for his daily needs primarily for drinking purposes. As such, water is the most immediate and critical limiting factor to both human and environmental well-being. Sustainable supplies of clean water are vital to the world's health, environment and economy. At present we are facing a tremendous crunch in meeting rising demands of potable water as the available supplies of freshwater are decreasing due to extended droughts, population growth, decline in water quality particularly of groundwater due to increasing groundwater and surface water pollution, unabated flooding and increasing demands from a variety of competing users. Water being a prime natural resource, a basic human need and a precious national asset, its use needs appropriate planning, development and management. Increasing population coupled with overexploitation of surface and groundwater over the past few decades has resulted in water scarcity in various parts of the world. Wastewater is increasing significantly and in the absence of proper measures for treatment and management, the existing freshwater reserves are being polluted. Increased urbanization is driving an increase in per capita water consumption in towns and cities. Hence there is a need to recognize the requirement to manage existing water reserves in order to avoid future water strain. Today availability of safe drinking water is a concern.

For almost all the water needs of the country, groundwater is by far the most important water resource. Worldwide, according to a United Nations Environment Study (UNEP) study over 2 B people depend on aquifers for their drinking water. 40 per cent of the world's food is produced by irrigated agriculture that relies largely on groundwater. Groundwater constitutes about 95 per cent of the freshwater on our planet (discounting that locked in the polar ice caps), making it fundamental to human life and economic development. However, the ever increasing scarcity of groundwater coupled with diminishing water quality has posed a serious threat to the population especially the rural community and has forced everyone to look at treatment of groundwater because clean water is fast becoming an endangered commodity. The unabated use has taken a serious toll on the availability of groundwater resources and as such the world is facing a severe crunch in the availability of groundwater. So we have no other option to move from "groundwater development" to "groundwater management" which means that we have to move towards optimal usage of groundwater which would be sustainable in the long run. Today the onus is on everybody to provide safe drinking water and for that water treatment processes need to be developed that are easy to implement, cost effective and sustainable in the long run.

India is a vast country having diversified geological, climatological and topographic set-up, giving rise to divergent groundwater situations in different parts of the country. Unsustainable uses of resources and indiscriminate applications of pesticides, fertilizers, industrial pollutants are continuously disturbing the status of purity of groundwater. Shallow aquifers generally suffer from agrochemicals, domestic and industrial waste pollution. Major water pollutants include microbes (like intestinal pathogens and viruses), nutrients (like phosphates and nitrates), heavy metals and metalloids (like arsenic, lead, mercury), organic chemicals (like DDT, lubricants, industrial solvents), oil, sediments and heat. Virtually all industrial and goods-producing activities generate pollutants as unwanted by-products. Heavy metals can contaminate the aquifer and subsequently can bioaccumulate in the tissues of humans and other organisms. For example, more than 100 million people are living in the arsenic affected districts of India and Bangladesh. 9 districts out of 19 in West Bengal, 78 blocks and around 3150 villages are affected with arsenic-contaminated groundwater (Chakraborti *et al.*, 2002). Groundwater is regularly used for agricultural and household purposes in these areas. As rainwater is insufficient to support the water demand of the increasing population and intensive agricultural system of West Bengal, thousands of shallow tube-wells were installed for irrigation in last 40-45 years. A vast majority of these tube wells have been installed privately with locally available expertise, without any check of the quality and yield of the water that originates from them. The use of arsenic contaminated groundwater for irrigation purpose in crop fields elevates arsenic concentration in surface soil and in the plants grown in those areas (Meharg and Rahman, 2003). Many millions of cubic meters of undergroundwater are used for agricultural irrigation. Much of this groundwater is contaminated with arsenic, which is deposited in the soil in contact with the irrigation water throughout the year. Groundwater systems are very vulnerable freshwater resources and prone to contamination. Pollutants can take years to reach the aquifers, but, once it reaches the water source, it is very difficult and costly to remove the pollutants. More than 80 per cent of sewage in developing countries is discharged without proper treatment which can pollute the river systems, lakes and coastal water bodies (UNESCO, 2009).

In the present context the recent advancement of nanoscale science and engineering is opening up a hitherto unknown and novel gateway to the development and deployment of water purification processes which are in tune with the above mentioned parameters. Nanoscience is the study of phenomenon and manipulation of materials at atomic, molecular and macromolecular scales, where properties differ significantly from those at a larger scale. Nanotechnology is the design, characterization, production and applications of structures, devices and systems by controlling shape and size at nanometer scale. In recent years, a great deal of attention has been focused onto the applicability of nanostructured materials as adsorbents or catalysts in order to remove toxic and harmful substances from wastewater. Nano-materials had gained special attention since last decade because the materials of such kind posses unique properties than the bulk materials. Like different nano materials, single and multi metal or doped metal oxides are also subject of much interest since that materials posses high surface-to-volume ratio, enhanced magnetic property, special catalytic properties etc. Consequently, different methods *viz.* chemical precipitation, sol-gel, vapour deposition, solvo thermal, solid state reaction etc were adopted for the synthesis of specified oxides by various workers (Zhang, 2003). Nano-enabled technologies for water treatment are already on the market. Nanofiltration currently seeming to be the most mature and eco-friendly technology and many more are on their way of development and applications. The environmental fate and toxicity of any material are critical issues in choice of materials for water purification. Nanotechnology while being questionably better than other techniques used in water treatment, the knowledge about the environmental fate, transport and toxicity of nanomaterials is still inadequate. The high surface area and surface reactivity compared with granular forms enable the nanoparticles to remediate more material at a higher rate and with a lower generation of hazardous by-products. Advances in nanoscale science and engineering suggest that many of the current problems involving water quality could be resolved or greatly diminished by using nanosorbents, nanocatalysts, bioactive nanoparticles, nanostructured catalytic membranes, nanotubes, magnetic nanoparticles, granules, flake, high surface area metal particle supramolecular assemblies with characteristic length scales of 9-10 nm including clusters, micromolecules, nanoparticles and colloids have a significant impact on water quality in natural environment. The defining factor which characterizes the capability of nanoparticles as a versatile water remediation tool includes their very small particle sizes (1–100 nm) in comparison to a typical bacterial cell which has a diameter on the order of 1 m (1000 nm). Hence nanoparticles can be transported effectively by the groundwater flow. They can also remain in suspension for sufficient time in order to launch an in situ treatment sphere. As a result, nanoparticles can be anchored onto a solid matrix such as a conventional water treatment material like activated carbon and/or zeolite for enhanced water treatment.

Nanomaterials in Wastewater Treatment: Opportunities and Challenges

Nanomaterials are fast emerging as potent candidates for water treatment in place of conventional technologies which, notwithstanding their efficacy, are often very expensive and time consuming. This would be in particular, immensely beneficial for developing nations like India and Bangladesh where cost of implementation of any new removal process could become an important criterion in determining its success. Qualitatively speaking nanomaterials can be substituted for conventional materials that require more raw materials, are more energy intensive to produce or are known to be environmentally harmful. Employing green chemistry principles for the production of nanoparticles can lead to a great reduction in waste generation, less hazardous chemical syntheses, and an inherently safer chemistry in general. However, to substantiate these claims more quantitative data is required and whether replacing traditional materials with nanoparticles does indeed result in lower energy and material consumption and prevention of unwanted or unanticipated side effects is still open to debate. There is also a wide debate about the safety of nanoparticles and their potential impact on the environment. There is fervent hope that nanotechnology can play a significant role in providing clean water to the developing countries in an efficient, cheap and sustainable way. On the other hand, the potential adverse effects of nanoparticles cannot be overlooked either. For instance the catalytic activity of a nanoparticle can be advantageous when used for the degradation of pollutants, but can trigger a toxic response when taken up by a cell. So this Janus face of nanotechnology can prove to be a hurdle in its widespread adoption. However as mentioned before nanotechnology can step in a big way in lowering the cost and hence become more effective than current techniques for the removal of contaminants from water in the long run. In this perspective nanoparticles can be used as potent sorbents as separation media, as catalysts for photochemical destruction of contaminants; nanosized zerovalent iron used for the removal of metals and organic compounds from water and nanofiltration membranes.

Mechanisms of Adsorption of Pollutants from Wastewater

Nanosorbents

Two vital properties make nanoparticles highly lucrative as sorbents. On a mass basis, they have much larger surface areas compared to macro particles. They can also be enhanced with various reactor groups to increase their chemical affinity towards target compounds. These properties are increasingly being exploited by workers to develop highly selective and efficient sorbents for removal of organic and inorganic pollutants from contaminated water. Many materials have properties dependent on size. Hematite particles with a diameter of 7 nm, for example, adsorbed Cu ions at lower pH values than particles of 25 or 88nm diameter, indicating the enhanced surface reactivity for iron oxides particles with decreasing diameter. Peng *et al.* (2005) have developed a novel sorbent with high surface area (189 m^2/g) consisting of cerium oxide supported on carbon nanotubes (CeO_2-CNTs). They showed that the CeO_2-CNT particles are effective sorbents for As(V). This goes to show that how chemically modified nanomaterials can enhance the adsorption capacity of a traditional substance. Deliyanni *et al.* (2003) have also synthesized and characterized a novel As(V) sorbent consisting of akaganeite [β-FeOOH] nanocrystals. For the removal of metals and other inorganic ions, mainly nanosized metal oxides. Equilibrium adsorption of As(III) and As(V) by nanocrystalline TiO_2 occurred within 4 hours and the adsorption followed pseudo-second-order kinetics in experiments conducted by Pena *et al.* (2005). Bang *et al.* (2005) reported that equilibrium was reached in 63 minutes for adsorption of As(V) adsorption and 240 minutes for adsorption of As(III). Manna *et al.* (2004) investigated the removal of As(III) using a synthesized crystalline hydrous titanium dioxide. They found that 70 per cent of As(III) adsorption occurred within the first 30 minutes of contact time. Nano-agglomerates of mixed oxides such as iron–cerium, iron–manganese, iron–zirconium, iron–titanium, iron–chromium, cerium-manganese etc. have been synthesized and successfully employed for pollutant removal (*i.e.* arsenic, fluoride etc.) from aqueous solutions. Metals such as zinc and tin possess similar reduction capabilities of iron (Boronina 1995). Like iron, these metals are converted to metal oxides in the decontamination process. Other metals have been combined with iron as well to produce similar results. Both iron-nickel and iron-copper bimetallic particles have been demonstrated to degrade trichloroethane and trichloroethene (Lien 2001). Another example is iron-platinum particles, which possess similar capabilities in degrading chlorinated benzenes (Lien 2001). The photo-oxidation of As(III) to As(V) in the presence of TiO_2 and light and subsequent adsorption into TiO_2 has also been reported. Bissen *et al.* (2001) have showed that photo-oxidation of As(III) to As(V) occurs within minutes. No reverse reaction of As(V) to As(III) was observed, and while As(III) was oxidized by UV light in the absence of TiO_2, the reaction was way too slow to be feasible in water treatment. The reaction rate did not depend on the pH of the solution. Pena *et al.* (2005) reported that rapid photo-oxidation of As(III) to As(V) occurred in the presence of sunlight, nanocrystalline TiO_2, and oxygen. In natural groundwater, Pena *et al.*, believed that oxidation of As(III) to As(V) and subsequent adsorption of As(V) onto TiO_2 would completely eradicate arsenic at slightly acidic pH values.

Nanofiltration

Membrane processes such as nanofiltration (NF) are emerging as key contributors to water purification (US Bureau of Reclamation, 2003). Nanofiltration membranes (NF membranes) are deployed in water treatment for drinking water or wastewater treatment. It is a low pressure membrane process that separates materials in the 0.001-0.1 micrometer size. NF membranes are pressure-driven membranes with properties between those of reverse osmosis and ultra filtration membranes and have pore sizes between 0.2 and 4 nm. NF membranes have been shown to remove turbidity, microorganisms and inorganic ions such as Ca and Na. They are used for softening of groundwater (reduction in water hardness), for removal of dissolved organic matter and trace pollutants from surface water, for wastewater treatment (removal of organic and inorganic pollutants and organic carbon) and for pretreatment in seawater desalination. Van der Bruggen and Vandercasteele (2003) have studied the use of nanofiltration to remove cations, natural organic matter, biological contaminants, organic pollutants, nitrates and arsenic from groundwater and surface water. Favre-Reguillon *et al.* (2003) found that nanofiltration can be used to remove minute quantities of U(VI) from seawater. Mohsen *et al.* (2003) have evaluated the use of nanofiltration to desalinate water. They found that nanofiltration in combination with reverse osmosis could effectively render brackish water potable. An improvement in water quality was shown by Peltier *et al.* (2003) for a large water distribution system using nanofiltration. Carbon nanotubes filters are also gaining prominence in water treatment processes. Srivastava *et al.* (2004) recently reported the successful fabrication of carbon nanotube filters. These new filtration membranes consist of hollow cylinders with radially aligned carbon nanotube walls. They showed that the filters were effective at removing bacteria (*Escherichia coli* and *Staphylococus aureus*)

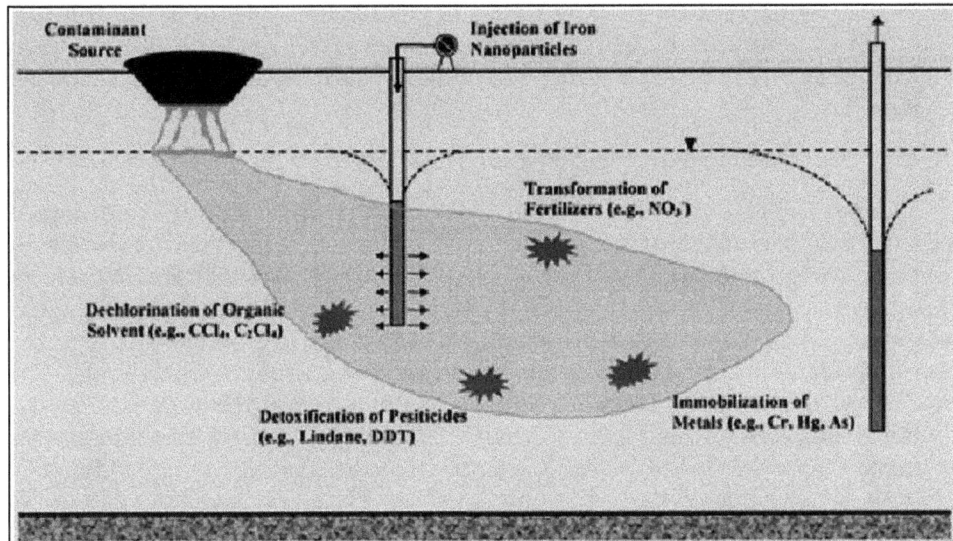

Figure 4.1: Nanoscale Iron Particles for *in situ* Remediation

from contaminated water. The carbon nanotube filters are readily cleaned by ultrasonication and autoclaving.

Nanoceramic filters are a mixture of nanoalumina fiber and micro glass with high positive charge and can retain negatively charged particles. Nanoceramic filters have high efficiency for removing virus and bacteria. They have high capacity for particulates and less clogging and can chemisorb dissolved heavy metals (Shah and Ahmad, 2010).

Nanoscale Zerovalent Iron

The use of Nanoscale Zerovalent Iron (nZVI) for groundwater purification has been the most widely investigated environmental nanotechnological technique. It has been established that nanoscale metallic iron is very effective in destroying a wide variety of common contaminants such as chlorinated methanes, brominated methanes, trihalomethanes, chlorinated ethenes, chlorinated benzenes, other polychlorinated hydrocarbons, pesticides and dyes. The basis for the reaction is the corrosion of zerovalent iron in the environment:

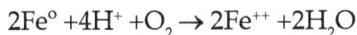

$$2Fe^\circ + 4H^+ + O_2 \rightarrow 2Fe^{++} + 2H_2O$$

$$Fe^\circ + 2H_2O \rightarrow Fe^{++} + H_2 + 2OH^-$$

It has been found that nZVI can reduce not only organic contaminants but also inorganic anions like nitrate, perchlorate, selenate, arsenate, arsenite and chromate. The reaction rates for nZVI are several times faster and also the sorption capacity is much higher compared to normal granular iron. nZVI is also capable in removing dissolved metals from solution, *e.g.* Pb and Ni. The metals are either reduced to zerovalent metals or lower oxidation states.

Typically, nZVI can be prepared by using sodium borohydride as the principle reductant. For example, $NaBH_4$ (0.2 M) is added into $FeCl_3.6H_2O$ (0.05 M) solution (1:1 volume ratio). Ferric iron is reduced by the borohydride according to the following reaction:

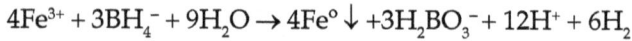

$$4Fe^{3+} + 3BH_4^- + 9H_2O \rightarrow 4Fe^\circ \downarrow + 3H_2BO_3^- + 12H^+ + 6H_2$$

Permeable Reactive Barrier (PRB) technology is a novel groundwater remediation method which enables physical, chemical or biological in situ treatment of contaminated groundwater by means of reactive materials. Granular ZVI in the form of reactive barriers has been used for many years at numerous sites all over the world for the remediation of organic and inorganic contaminants in groundwater. In recent years nZVI have gained ground as attractive candidates using this technology. The reactive materials are placed in underground trenches downstream of the contamination zone forcing it to flow through them and by doing so, the contaminants are treated without water excavation. Generally, this cost-effective removal technology causes less environmental harm than other methods. The arrangement of a permeable reactive barrier employing nZVI is illustrated in Figure 4.2.

Removal of Nanoparticles from Treated Water

The use of nanoparticles in environmental applications will invariably lead to the release of nanoparticles into the environment. Assessing their potential risks in the environment requires an understanding of their mobility, bioavailability, toxicity and persistence. Little is known about the possible exposure of aquatic and terrestrial life to nanoparticles in water and soil. The rapidly growing use of engineered nanoparticles in a variety of industrial scenarios and their potential for wastewater purification and drinking water treatment raise the inevitable question how these nanoparticles can be removed in the urban water cycle. Traditional methods for the removal of particulate matter during wastewater treatment that have been in vogue include sedimentation and filtration. However, due to the small sizes of nanoparticles the sedimentation velocities are relatively low and significant sedimentation will not occur as long as there is no formation of larger aggregates. Common technologies such as flocculation might be inappropriate to remove nanoparticles from water, which points to the need of finding new solutions to the problem. Till now, membrane filtration (*e.g.* nanofiltration and reverse osmosis) has been already applied for the removal of pathogens from water. Hence, this technique can also be used for the removal of nanoparticles. Most nanoparticles in technical applications today are functionalized in nature and therefore studies using virgin nanoparticles may not be relevant for assessing the behaviour of the actually used particles. Functionalization is often used to decrease agglomeration and therefore increase mobility of particles. Unfortunately little is known to date about the influence of functionalization on the behaviour of nanoparticles in the environment.

Conclusion

While nanotechnology is considered to be the new buzzword by many in the scientific community, information regarding the subject remains largely dispersed

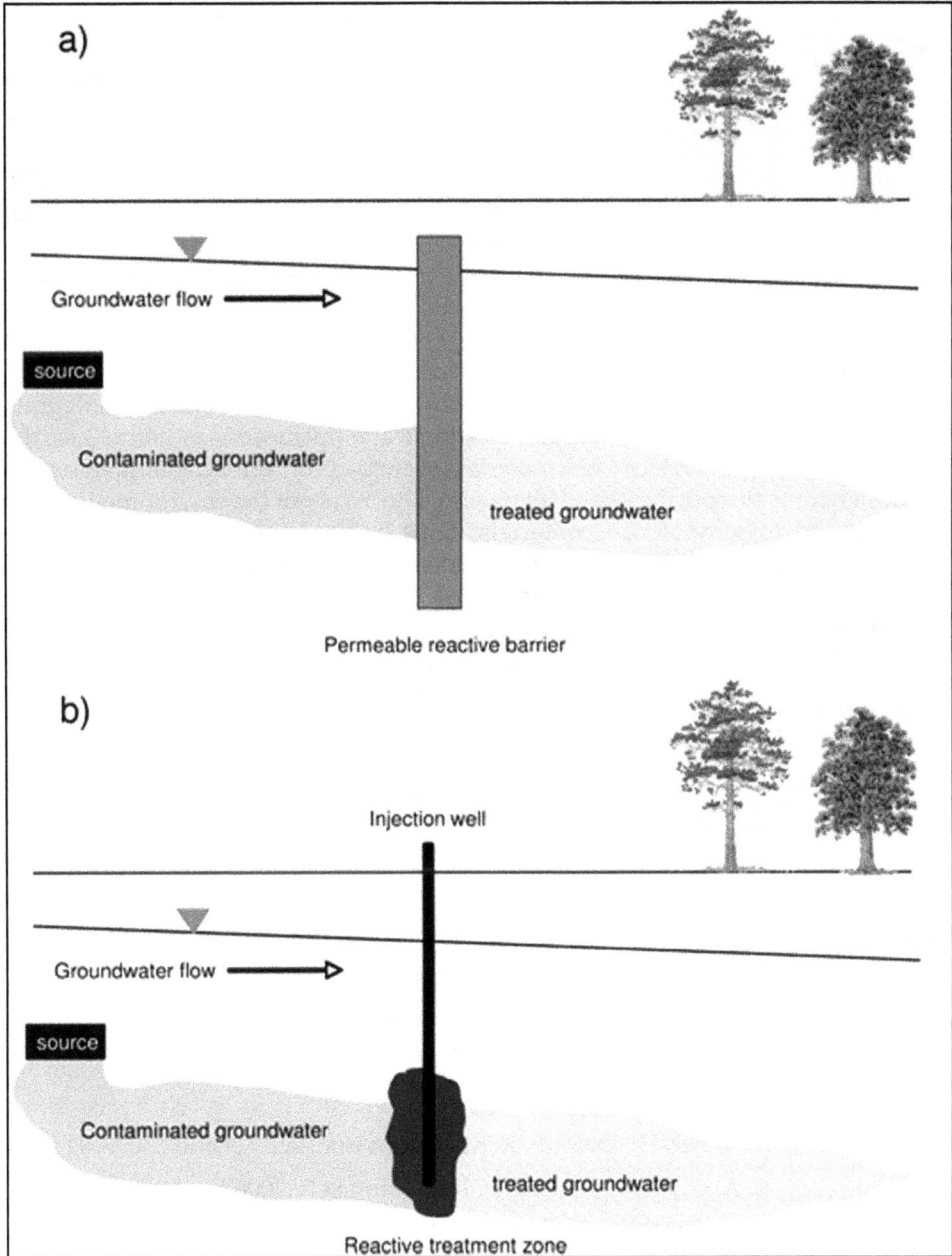

Figure 4.2: Three Forms of ZVI Application for Groundwater Remediation: (a) Conventional reactive barrier using granular ZVI; (b) Injection of nZVI to form an immobile reaction zone; (c) Injection of mobile nZVI.

and fragmented due to the relative novelty of the technology. But the increasing trends of researches which have been discussed so far have made it clear that nanotechnology holds an immense potential to be developed into a very potent water treatment tool of the 21st century. In fact nanomaterials and their various incarnations are the drivers for the nanotechnology revolution. Nanoparticles in particular will have important impacts on various fields of environmental technology and engineering not least in water treatment. However most of techniques for the treatment of wastewater involving nanotechnology so far have only been investigated in laboratory scale and not all of them are likely to be feasible alternatives for existing treatment technologies mainly perhaps due to economic reasons. This makes it difficult to predict what the future holds for us at this stage concerning this nascent technology. Also the incorporation of nanomaterials into existing water purification systems is another key challenge. Membrane processes such as RO, NF are becoming the standardised water purification techniques for public utilities and industry because they are flexible, scalable, modular and relatively easy to operate and maintain. Thus further laboratory investigations and pilot scale testing will be needed to integrate novel nanostructured membranes into existing water purification systems. Also the environmental fate and toxicity of a material are areas of concern in material selection and design for water purification. Not much is known about the environmental fate, transport and toxicity of nanomaterials. Thus it should be borne in mind that nanotechnology can become a double edged sword and each positive and desired property of nanomaterials could pose a risk to the environment. Thus a careful weighing up of the opportunities and risks of nanotechnology with respect to their impact on the environment is therefore needed. No systematic investigations regarding the stability of nanomaterials in natural and engineered environmental systems have been carried out till date to the best of our knowledge. On a positive note, due to their extremely high potential in combination with the high specificity, nanoparticles can be developed into ideal candidates for water treatment and may contribute to solving future challenges in the area of water treatment technologies. Thus nanotechnology holds a lot of promise in the remediation of groundwater and for this there is further scope in research and development.

References

A.A. Meharg, M.M. Rahman, M.M. Environ. Eng. Sci. 2003, 37 (2), 229-234.

A. Nel, T. Xia, L. Moadler, N. Li, Science 2006, 311, 622.

A. S. Madden, M. F. Hochella, T. P.Luxton, Geochim. Cosmochim. Acta 2006, 70, 4095.

B. A. Manning, J. R. Kiser, H. Kwon, S. R. Kanel, Environ. Sci. Technol. 2007, 41, 586.

B. D. Chithrani, A. A. Ghazani, W. C. W. Chan, Nano Lett. 2006, 6, 662.

Colvin, V.L., . The Nature Biotech., 2003, 10, 1166-1170.

D. Chakraborti, M.M. Rahman, K. Paul, U.K. Chowdhury, M.K. Sengupta, D. Lodh, C.R. Chanda, K.C. Saha, S.C. Mukherjee, Talanta, 2002, 58(4), 3–22.

D. E. Giammar, C. J. Maus, L. Y. Xie, Environ. Eng. Sci. 2007, 24, 85.

E. A. Deliyanni, E. N. Peleka, K. A. Matis, J. Hazard. Mater. 2007, 141, 176.

G. Jegadeesan, K. Mondal, S. B. Lalvani, Environ. Prog. 2005, 24, 289.

H. L. Lien, W. X. Zhang. Colloids and Surfaces. 2001, 191, 97-105.

Ichinose, N., Y. Ozaki and S. Kashu, 1992. Superfine particle technology. Springer, London, (Book).

Lecoanet H.F., J.Y. Bottero and M.R. Wiesner, 2004. Environ. Sci. Technol. 38(19), 5164–5169.

L. K. Limbach, Y. Li, R. N. Grass, T. J. Brunner, M. A. Hintermann, M. Muller, D. Gunther, W. J. Stark, Environ. Sci. Technol. 2005, 39, 9370.

Mamadou, S.D. and N. Savage, 2005. J. Nano. Res., 7: 325-330

Mohsen M.S., J.O. Jaber and M.D. Afonso, 2003. Desalination 157(1), 167–167.

M. Siegrist, A. Wiek, A. Helland, H. Kastenholz, Nat. Nanotechnol. 2007, 2, 67.

M. C. Roco, Environ. Sci. Technol. 2005, 39, 106A.

M. F. Hochella, Geochim. Cosmochim. Acta 2002, 66, 735.

M. R. Hoffmann, S. T. Martin, W. Choi, D. W. Bahnemann, Chem. Rev. 1995, 95, 69.

Manna, B., M. Dasgupta, and U.C. Ghosh. 2004. Journal of Water Supply: Research and Technology, 53(7): 483-495.

M. R. Wiesner, G. V. Lowry, P. Alvarez, D. Dionysiou, P. Biswas, Environ. Sci. Technol. 2006, 40, 4336

N. Hilal, H. Al-Zoubi, N. A. Darwish, A. W. Mohammad, M. Abu Arabi, Desalination 2004, 170, 281.

Peltier S., E. Cotte, D. Gatel, L. Herremans and J. Cavard, 2003. Water Sci. Tech. Water Supply 3, 193–200.

R. Jones, Nat. Nanotechnol. 2007, 2, 71.

Srivastava A., O.N. Srivastava, S. Talapatra, R. Vajtai and P.M. Ajayan, 2004. Nature Mater. 3(9): 610–614.

S. Pacheco, J. Tapia, M. Medina, R. Rodriguez, J. Non-Cryst. Solids 2006, 352, 5475.

S. R. Kanel, J. M. Greneche, H. Choi, Environ. Sci. Technol. 2006, 40, 2045.

S. R. Kanel, B. Manning, L. Charlet, H. Choi, Environ. Sci. Technol. 2005, 39, 1291.

S. M. Ponder, J. G. Darab, T. E. Mallouk, Environ. Sci. Technol. 2000, 34, 2564.

Stoimenov, P.K., R.L. Klinger, G.L. Marchin and K.J. Klabunde, 2002. Langmuir, 18: 6679-6686.

S. O. Obare, G. J. Meyer, J. Environ. Sci. Health 2004, 39, 2549.

T. Hillie, M. Munasinghe, M. Hlope, Y. Deraniyagala, Nanotechnology, water and development, Meridian Institute, 2006.

T. K. Boronina, G. Klabunde. Environ. Sci. Technol. 1995, 39, 1511-1517.

UNESCO, 2009. *World Water Assessment Programme. The United Nations World Water Development Report 3: Water in a changing world.*

US Bureau of Reclamation and Sandia National Laboratories, 2003. Desalination and water purification technology roadmap a report of the executive committee Water Purification.

US Environmental Protection Agency, 1998b. Microbial and disinfection by-product rules. Federal Register, 63: 69389-69476.

US Environmental Protection Agency, 1999. Alternative disinfectants and oxidants guidance manual. EPA Office of Water Report 815-R-99-014.

W.X. Zhang, 2003. J. Nanopart. Res. 5, 323–332.

World Health Organization, 1996. Guidelines for drinking-water quality. Geneva: WHO, Vol: 2.

Wang C. and W. Zhang, 1997. Environ. Sci. Technol. 31(7), 2154–2156.

X. Q. Li, W. X. Zhang, Langmuir 2006, 22, 4638.

X. Q. Li, D. W. Elliott, W. X. Zhang, Crit. Rev. Solid State Mater. Sci. 2006, 31, 111

2013, Environmental Technology   *Pages 43–50*
Editors: **D.R. Khanna, A.K. Chopra, Gagan Matta, R. Bhutiani & Vikas Singh**
Published by: **DAYA PUBLISHING HOUSE, NEW DELHI**

Chapter 5

GIS Mapping Posibility Using Google Trends for Disease Outbreaks

Silki Anand

International Institute of Health Management Research,
New Delhi

ABSTRACT

Google has been accepted as one of the key search engines widely used on the internet.It uses 'netbots' as Artificial Intelligence (AI) tool for fetching information in sub femto seconds. Google started 'Google Trends' to compare world's interest in chosen topics over time and geographical location. "Infodemiology" was the term introduced from information and epidemiology to track health demand and supply trends. Google Trends is a feature which allows users to graph the frequency of searches for a term, a string of multiple terms, or a phrase. The graphs are scaled to the average search traffic for the selected term and are also normalized on a relative, rather than absolute, basis. Since it offers possibility to track infectious disease activity faster than by traditional surveillance systems, it is proposed to monitor disease trends in developing countries and map it on suitable Geographical information system (GIS), such as Quantum GIS, for effective disease surveillance, control and management. Mapping on Quantum GIS, an open source mapping tool, can lead to improved understanding and management of disease outbreaks, including describing the level and distribution of disease and development of zones for disease control on specified time frame. This application can also be used for cross referencing information collected from legacy sources.

Keywords: *Google trends in health care, Disease surveillance, Geographical Information System (GIS) and Quantum GIS.*

Introduction

The Internet is an increasingly important source of health information for both patients and physicians (2). The large number of searches conducted through popular websites creates trend data which can be analyzed over time to identify disease outbreaks and supplement traditional surveillance methods (3). This area of inquiry falls under the realm of "infodemiology", the study of the distribution and determinants of information on the Internet (3).Google, a mainstream general search engine which also includes several additional tools and resources, represents an important gateway to online health information for patients and physicians (15). Trend data generated by the number of Google searches over time in a particular geographic region have recently been made available by Google Trends (http:// www.google.com/trends). The diseases prevalent such as Tuberculosis used as key words can be mapped on geographic location desired to provide trend data (Figure 5.1). Google Trends is a feature which allows users to graph the frequency of searches for a term, a string of multiple terms, or a phrase. The data used to generate these graphs are scaled to the average search traffic for the selected term and are also normalized on a relative, rather than absolute, basis (1).These graphs can also be manipulated to restrict results to specific time frames and geographic regions. The generation of this type of data may have valuable future implications in aiding surveillance of a broad range of diseases and mapping on geographical information systems (GIS). In addition to epidemic surveillance, health searches can also be conducted for endemic health concerns

The internet company Google recently launched an experimental tool for near real-time detection of influenza outbreaks by monitoring and analyzing health care–seeking behavior in the form of queries to its online search engine. The tool, Google Flu Trends, is a sophisticated Web-based tool for detection of regional outbreaks of influenza. It is so promising that the Centers for Disease Control and Prevention (CDC) is testing it. Preliminary testing suggests that Google Flu Trends can detect regional outbreaks of influenza 7–10 days before conventional CDC surveillance [13]. The CDC uses laboratory and clinical data to publish national and regional weekly statistics, typically with a 1–2 week lag in reporting. Real-time surveillance would alert public health care practitioners in the early phases of an outbreak, enabling them to promptly institute control measures and case finding and to ensure adequate access to treatment, thereby reducing morbidity and mortality [5,9]. With international concerns about emerging infectious diseases, bioterrorism, and pandemics, the need for a real-time surveillance system is at an all-time high [6]. The data generated would also be useful for public health care practice, clinical decision making, and research [10]. The main aim of this paper is to introduce the more generic Google Trends (GT) tool to health professionals, to show how they can track disease activity of interest to them. We describe GT, how the data are processed, potential uses, and the tool's strengths and limitations

Methods

The study is exploratory on using google trends for mapping diseases.

Tuberculosis   **1.00**

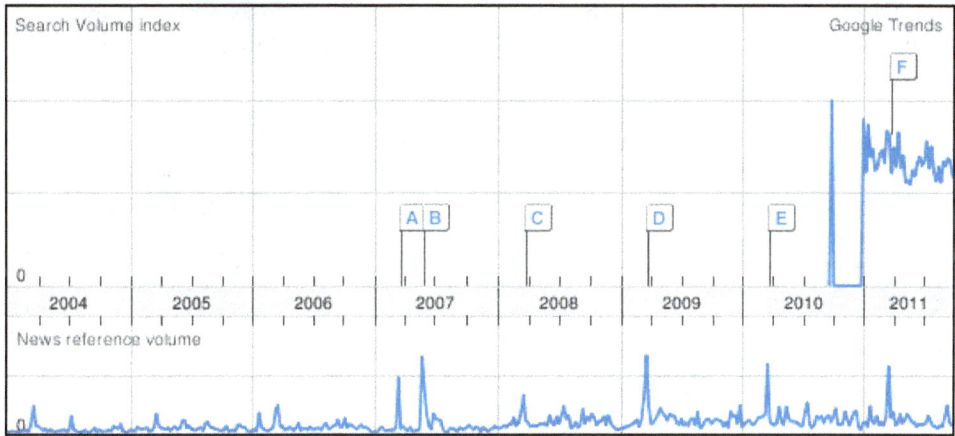

Rank by   tuberculosis ▾

A Global Tuberculosis Rates Level Off–FOX News - Mar 22 2007; B Near Misses Allowed Man With Tuberculosis to Fly–New York Times - May 31 2007; C Nigeria: World Tuberculosis Day - Containing Ravages of a Pandemic–AllAfrica.com - Mar 24 2008; D Let sunshine in to fight tuberculosis: WHO–Times of India - Mar 24 2009; E Global Fund seeks $20 billion to fight AIDS, tuberculosis, malaria in poor nations–CANOE - Mar 24 2010; F US Tuberculosis Cases Hit Record Low–CDC Says Newsday - Mar 24 2011.

Languages

1. English

Subregions

1. Kerala, India
2. Punjab, India
3. Tamil Nadu, India
4. Delhi, India
5. Andhra Pradesh, India
6. Rajasthan, India
7. Uttar Pradesh, India
8. Karnataka, India
9. Maharashtra, India
10. Madhya Pradesh, India

Cities

1. Trivandrum, India
2. Mangalore, India
3. Chandigarh, India
4. Chennai, India
5. Cochin, India
6. New Delhi, India
7. Coimbatore, India
8. Hyderabad, India
9. Bhopal, India
10. Mumbai, India

Figure 5.1: Tuberculosis Trend in India by Google Trends

Google Flu Trends

Google Flu Trends is available at http://www.google.com/flutrends/(Figure 5.2). There is a close relationship between the number of people searching for influenza related topics and those who have influenza symptoms. Naturally, all the people searching for influenza-related topics are not ill, but trends emerge when all influenza-related searches are added together. Google Flu Trends has strong correlations with retrospective surveillance data from the CDC and accurately estimated influenza levels 1–2 weeks earlier than published CDC reports (Figure 5.3). Google Flu Trends was developed from the more generic GT tool, which is available to all internet users

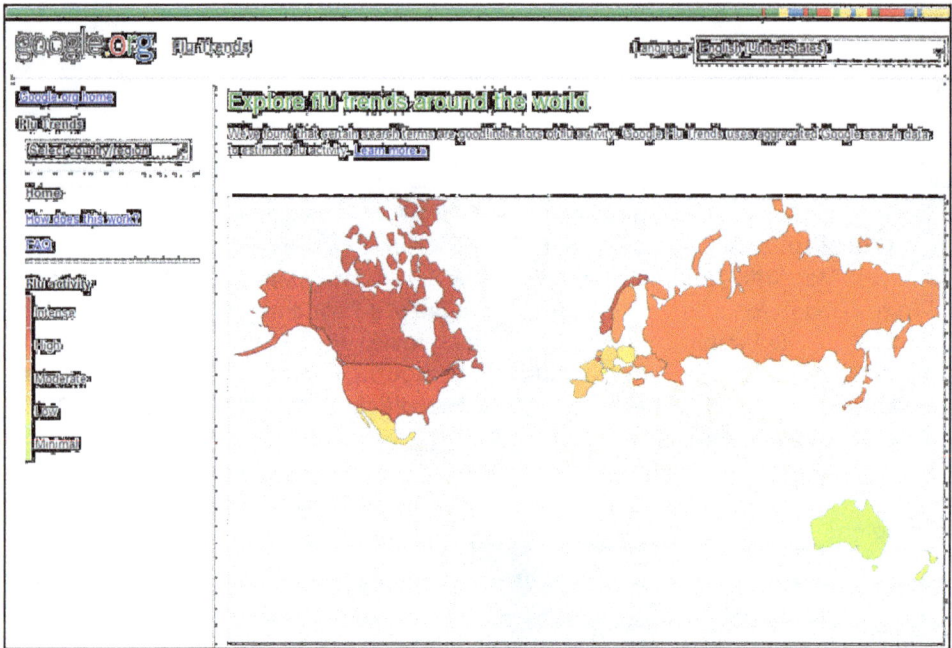

Figure 5.2: Google Flu Trends Web Tool Interface
(Available at http://www.google.com/flutrends/)

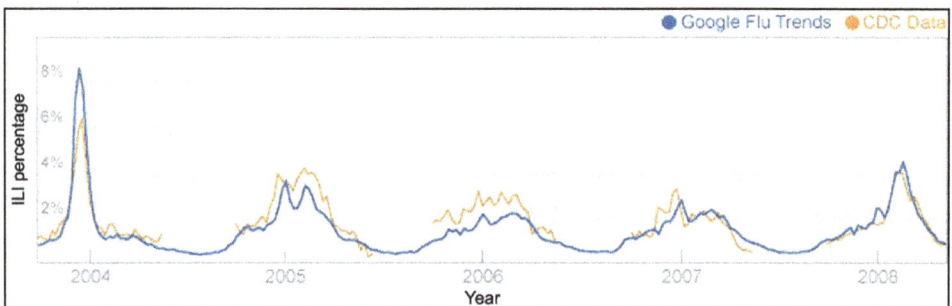

Figure 5.3: Correlation Between Google Flu Trends and Centers for Disease Control and Prevention (CDC) Surveillance Data for the US Mid-Atlantic Region from 2004 through 2008

at http://google.com/trends/. Users enter Web search queries to see the relative search volume of these queries–for example, queries for "flu" (Figure 5.4). GT analyzes a fraction of the total Google Web searches over a period of time and extrapolates the data to estimate the search volume. This information is displayed in a search volume index graph, which is currently updated daily. Beneath the search volume index graph is the news reference volume graph. This graph shows the frequency with which the Web search queries appeared in Google News stories. When a spike is detected in the news reference volume graph, GT labels the search volume index graph with a headline of a relevant but randomly selected Google News story published near the time of the spike. These headlines are shown to the right of the search volume index graph. The regions, cities, and languages with the highest search volume are displayed on the bottom of the page.

Scaling the Data

GT data are scaled in 2 ways: relative and fixed [4,8]. The difference between them is the time frame used to normalize the search volume. In relative mode, data are scaled using the average search volume over the time period selected. For example, the search volume index graph for "flu" is normalized using the extrapolated search volume for "flu" from January 2004 to the present. If the time frame is restricted–for example, restricted to 2008–then the data are scaled using the average search volume

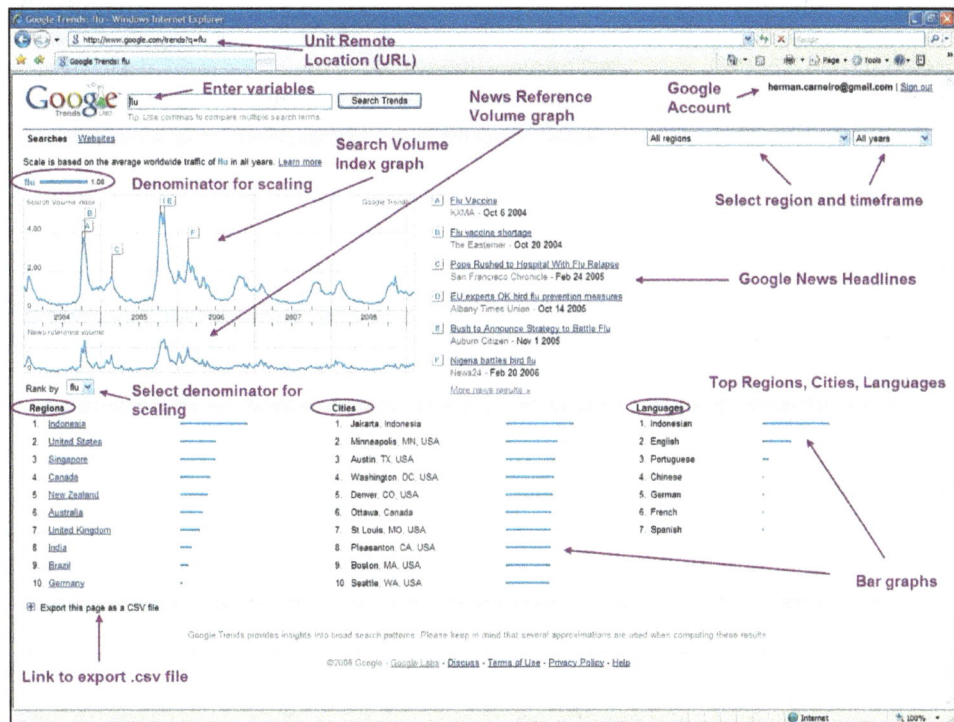

Figure 5.4: Google Trends Output for Web Search Queries for the Term "Flu" Worldwide

for "flu" in 2008 as the denominator. In fixed mode, the data are normalized using the extrapolated search volume at a fixed time point (January 2004), which is when GT data start. Because the denominator does not change, users can look at different time periods and can relate them to each other. A Web search query for the terms being investigated must have been done at the reference time point (January 2004); otherwise, no denominator exists for comparison. To view data scaled in fixed mode, the data must be downloaded in a comma-separated values (.csv) file and subsequently imported into a database or spreadsheet program. This feature is available only to users who are logged into their Google accounts. The news reference volume graph shows the raw number of Web search queries that appeared in Google News stories.

Data Normalization

An increase in the volume of a Web search query increases its own average over time and thus its denominator for future comparisons. This reduces the sensitivity in detection of changes in future search volume trends. GT controls for this by dividing by an unrelated, common Web search query. For example, the search volume for the term "fever" may be normalized by dividing it by the search volume for the unrelated and common term "baseball." Normalization also compensates for population sizes, making it possible to rank cities purely on the basis of search volume trends ranking the top regions, cities, and languages, GT takes a sample of all Web search queries and determines the areas or languages from which the most searches for the entered terms originate. Internet protocol addresses from server logs are used to establish the origin of Web search queries. Language information is based on the language version of Google used for Web searches. The algorithm then calculates the ratio of a variable's search volume from each city and the total search volume from those cities. The city name and bar chart

Alongside it represent this ratio. When these are close together, the ranking between the cities is less meaningful.

Results

Google Flu Trends uses a multitude of Web search queries that correlate well with physician visits for influenza-like symptoms to estimate current weekly levels of influenza activity at regional and state levels. GT users, on the other hand, can enter only up to 5 Web search queries, which raise questions about its ability to monitor disease trends effectively.

Discussion and Conclusions

GT is in the early phases of development, and its data "may contain inaccuracies for a number of reasons, including data sampling issues and approximation methods used to compute the results" [7]. Currently, GT search criteria are not standardized. Users may enter symptoms differently, depending on their level of education and their cultural and language backgrounds. For example, users may enter fever, pyrexia, chills, and rigors for the same symptom. Detailed analyses are required to find search query proxies that correlate well with specific diseases. To be most effective, GT requires large populations of Web search users, which means that GT is currently better suited to tracking disease activity in developed countries. Furthermore, GT is available only in a limited number of languages and region-specific versions. GT is an exciting tool with enormous potential, as shown by Google Flu Trends. It is a convenient, easy, and accessible source of search data. In its current stage of development, GT can be used in conjunction with traditional surveillance systems to improve the efficacy of disease surveillance. Google Flu Trends has detected influenza outbreaks 7–10 days before the traditional surveillance systems used by the CDC Still in its early phases of development, GT data may contain inaccuracies, which Google is working to resolve. Research is needed to find suitable Web search query proxies that correlate well to actual cases of diseases of interest. These proxies then can be used to establish specialized tools for infectious diseases, using Google Flu Trends as a blueprint, or to setup syndromic surveillance of Web search queries. Also, there are privacy issues involved in using Google Web search data. Google stores and uses data from personal Web searches for public research, often without the consent and knowledge of Internet users. In some cases, Google search data may be traced to individuals if they are signed into their accounts when they conduct online searches. Google assures users that personal search data remain safe and private. In conclusion, GT is currently better suited to track epidemics, diseases with high prevalences, and diseases in developed countries than other types of diseases, because it requires large populations of Web search users to be most effective. However, the world is changing, and society is becoming more dependent on the Internet, thus providing a wealth of information that reflects the "collective intelligence" of populations. Google Flu Trends, and possibly GT, make it possible to track infectious disease activity faster than by traditional surveillance systems. This unique and innovative technology takes us one step closer to true real-time outbreak surveillance.

References

1. About Google Trends, 2009. http://www.google.com/intl/en/trends/about.html#7 (accessed on 16 July 2009).

2. Dickerson S, Reinhart AM, Feeley TH, Bidani R, Rich E, Garg VK, Hershey CO, 2004. Patient Internet use for health information at three urban primary care clinics. J Am Med Inform Assoc 11, 499-504.

3. Eysenbach G, 2009. Infodemiology and infoveillance: framework for an emerging set of public health informatics methods to analyze search, communication and publication behavior on the Internet. J Med Internet Res 11, e11

4. Eysenbach G. Infodemiology: tracking flu-related searches on the Web for syndromic surveillance. AMIA Annu Symp Proc **2006**: 244–8

5. Ferguson NM, Cummings DA, Cauchemez S, *et al.*, Strategies for containing an emerging influenza pandemic in Southeast Asia. Nature **2005**, 437: 209–14.

6. Frenk J, Gomez-Dantes O. Globalization and the challenges to health systems. Health Aff (Millwood) 2002, 21: 160–5.

7. GoogleTrendsLabs. About Google Trends. Availableat: http://www.google.com/intl/en/trends/about.html#7. Accessed 1 October 2009

8. Johnson HA, Wagner MM, Hogan WR, *et al.*, Analysis of Web access logs for surveillance of influenza. Stud Health Technol Inform **2004**, 1107: 1202–6.

9. Longini IM Jr, Nizam A, Xu S, *et al.*, Containing pandemic influenza at the source. Science **2005**, 309: 1083–7.

10. Mandl KD, Overhage JM, Wagner MM, *et al.*, Implementing syndromic surveillance: a practical guide informed by the early experience. J Am

11. Med Inform Assoc **2004**, 11: 141–50.

12. National Electronic Disease Surveillance System Working Group. National Electronic Disease Surveillance System (NEDSS): a standards based approach to connect public health and clinical medicine. J Public Health Manag Pract **2001**, 7: 43–50.

13. New York Times. Google uses searches to track flu's spread. 11 November **2008**. Available at: http: //www.nytimes.com/2008/11/12/technology/internet/ 12flu.html?_rp1. Accessed 19 January 2009.

14. Polgreen PM, Chen Y, Pennock DM, Forrest ND. Using Internet searches for influenza surveillance. Clin Infect Dis **2008**, 47: 1443–8.

15. Wheeler D, 2006. Google as a pathology portal. Adv Anat Pathol 13, 275-276

2013, Environmental Technology

Pages 51–59

Editors: **D.R. Khanna, A.K. Chopra, Gagan Matta, R. Bhutiani & Vikas Singh**

Published by: **DAYA PUBLISHING HOUSE, NEW DELHI**

Chapter 6

Bioremediation as a Naturally Adopted Alternative Technology to Clean Dye Industry Messes

Kinshuk Soni and D.M. Kumawat

School of Studies in Environment Management,
Vikram University, Ujjain

ABSTRACT

In the present study textile dye industry effluent having diverse chemical composition ranging from inorganic compounds to polymers and several organics was subjected to a ecofriendly treatment technology *i.e.* bioremediation. The bioremediation treatment was carried out with different test organisms (bacteria and fungi). Degradation was achieved by metabolic pathways operating with in these organisms. These metabolic pathways are functioning within the cell or by enzyme secreted by the cell in the particular sample. These processes are the crux of environmental technologies. The studied results were promising and lowering of the pollution load in term of chemical oxygen demand (COD) was significant. Bioremediation was found to be cost effective alternative technology for remediation of textile dye industry messes. As textile dyeing effluent sample containing recalcitrant dyes and polluting water bodies of adjoining area due to their colour and by formation of toxic or carcinogenic intermediates such as aromatic amines from azo dye stuff. This technology provides very valuable support in cleaning up of environmental messes especially from dye industry. This technology constitutes the use of natural biota and their processes for pollution reduction.

Keywords: Textile messes, Dyes, COD reduction, Bacteria, Fungi.

Introduction

The world's ever increasing population and her progressive adoption of an industrial– based lifestyle has inevitably led to an increased anthropogenic impact on the biosphere. Textile effluents can seep into the aquifer and pollute the undergroundwater, or where it is discharged without proper treatment into water bodies. The pollutants cannot be confined within specific boundaries. Textile wastewater is generally characterized as highly colored, with high concentrations of BOD, COD, total organic carbon (TOC) and dissolved solids. Considering both volumes discharged and effluent composition, the wastewater generated by the textile industry is rated as the most polluting among all industrial sectors. Nature has demonstrated its capacity to disperse, degrade, absorb or otherwise dispose of unwanted residues in the natural sinks of the atmosphere, waterways, ocean and soil. It is realized however that this ability is not finite. The discharge of these waste residues into the environment eventually poison, damage or affect one or more species in the environment, with resultant changes in the ecological balance.

Bioremediation constitutes the use of natural biota and their processes for pollution reduction; it is a cost effective process and the end products are non-hazardous (Ahmedna *et al.*, 2004). Microbial communities are of primary importance in bioremediation contaminated soil and water, because microbes alter chemistry of organic and inorganic pollutant through reduction, accumulation, mobilization and immobilization. Heterotrophic fungi (*Aspergillus*, *Penicillium* and *Fusarium*) and Bacteria (*Bacillus megaterium* and *Pseudomonas fluorescens*) can remove both soluble and insoluble metal species from solution and are also able to leach metal cations from solid waste (White *et al.*, 1997).

Materials and Methods

Location

The study area characterized in this study was Bhairavgarh Textile Dye Industry, Bhairavgarh villager, near Ujjain city. A general survey of the locality at Bhairavgarh was conducted and different units of this small-scale industry were visited. The effluent drainage and disposal systems were seen and marked for effluent/waste collection.

Sampling

For various experiments dye industry effluent was collected from different small units and a composite sample was prepared. The effluent collection was made in a 5-litre PVC container, each time as and when required. The sampling was carried out in all the three seasons *viz.*, summer (Mar-June), Rainy (July-Oct) and winter (Nov-Feb) for two conservative years *i.e.*, 2008-10.

Fungal and Bacterial Identification

Fungal isolates were identified on the basis of morphological characters and reproductive structures including spores in accordance with Manual of Soil Fungi by Gilman (1957). Identification of bacterial species was done following

Bergey's Manual of Systematic Bacteriology (Boones and Castenholz 2001) and Bergey's Manual of Determinative Bacteriology (Buchanan and Gibbons 1994).

Physiochemical Analysis

Chemical oxygen demand (COD), Total suspended solids (TSS), Total dissolved solids (TDS), Hardness, pH and conductivity of the samples was carried out before and after bioremediation treatment in accordance with APHA (1992).

Results

Biodegradability (Fungi and Bacteria)

A laboratory scale bioremediation unit was designed built and tested for the bioremoval of textile dye effluent of Bhairavgarh. Experiment was carried out to assess the efficiency of bioremediation units using fungal strains namely *Aspergillus niger* (F1), *Fusarium oxysporum* (F2), *Aspergillus* spp. (F3) and *Penicillium* spp.(F4) and Bacterial strains namely *Bacillus megaterium* (B1) and *Pseudomonas fluorescens* (B2). The specific mediums were used in experiment for better fungal and bacterial growth. The fungal and bacterial biomass accumulation in the media supplemented with effluent was investigated by determining their biomass. At the end of experiment, the COD measurements were determined to assess the degradation and removal of dye and reduction in pollution load from the textile dye industry effluent of Bhairavgarh.

Table 6.1: Shows the Initial Quality Parameters and the Raw Effluent of Dye Industry at Bhairavgarh.

☆ The COD reduction of effluent at concentration 20 per cent, 40 per cent, 60 per cent, 80 per cent and 100 per cent after treatment with F1 strain were 13.2 per cent, 12.2 per cent, 10.4 per cent, 6.64 per cent and 3.08 per cent respectively (Figure 6.1). At all concentrations a gradual fall in COD value was observed. As is evident from the results COD reduction was more at lower effluent dilutions (upto 60 per cent). Similar trend was observed for other fungal isolates at different concentrations. Among all fungal isolates, F1 showed higher COD reduction than F2, F3, and F4 and appeared to be most effective fungi for bioremediation. Efficiency of all the four fungal strains to remediate the dye industry effluent was best during winter season followed by summer and monsoon.

☆ Biomass of all the test fungi decreased as concentration of effluent increased. During the treatment period fungal biomass increased from 0.9 to 7.3 (F1, 1.2 to 6.2) (F2, 0.9 to 6.8) (F3, 1.0 to 7.2) (F4, 1.1 to 7.3) at effluent concentration 100 to 20 per cent respectively (Table 6.2).

☆ The COD reduction of effluent at concentration 20 per cent, 40 per cent, 60 per cent, 80 per cent and 100 per cent after treatment with B1 strain (bacterial) were 13.47, 11.51, 10.73, 5.91 and 4.53 per cent respectively. At all concentration gradual fall in COD value was observed (Figure 6.2). Maximum COD reduction was observed at low effluent concentrations (20-60 per cent). Similar observation was made for both the bacterial isolates at different concentrations. Among the two bacterial isolates, B1 showed a

Table 6.1: COD Reduction of Dye Industry Effluent by Different Fungal and Bacterial Species (Isolated from Rhizospheric Soil) at Different Concentration during Different Seasons

Dilution %	Initial Value	Monsoon Season					
		Fungal Isolates				Bacterial Isolates	
		Final Values					
		F1	F2	F3	F4	B1	B2
20 per cent	462	401	408	414	419.3	440	405
40 per cent	521	457	465	469	476	461	464
60 per cent	596	534	542	546	547.6	532	537
80 per cent	659	615	617	622	621	620	624
100 per cent	706	684.2	685.2	687	690.1	674	682

Dilution %	Initial Value	Winter Season					
		Fungal Isolates				Bacterial Isolates	
		Final Values					
		F1	F2	F3	F4	B1	B2
20 per cent	485	329.7	371	387	434.9	426.1	428
40 per cent	581	437	497	472	525	516	513
60 per cent	641.9	525.5	472	552	594.3	572	575
80 per cent	817.5	713.3	752.4	727	766	761	767
100 per cent	897	819.6	835	822.3	862.7	856	860

Dilution %	Initial Value	Summer Season					
		Fungal Isolates				Bacterial Isolates	
		Final Values					
		F1	F2	F3	F4	B1	B2
20 per cent	549	410	421.6	422.7	429	440	442.5
40 per cent	609.2	490	502	517.2	508.3	510	516
60 per cent	672	565.9	581.9	577	587.2	574	577
80 per cent	730	647	651.7	678	673	645	650
100 per cent	900	860	872	876	879	829	834

higher COD reduction than B2 and therefore appear to be better and more effective bacteria for bioremediation. B1 strain was found to be more effective during winter, followed by summer, and rainy season while B2 strain was more effective during summer, followed by monsoon and winter (Figure 6.2).

☆ Biomass of bacterial strains decreased, as concentration of effluent increased (Table 6.3). During treatment period bacterial dry weight for B1 increased

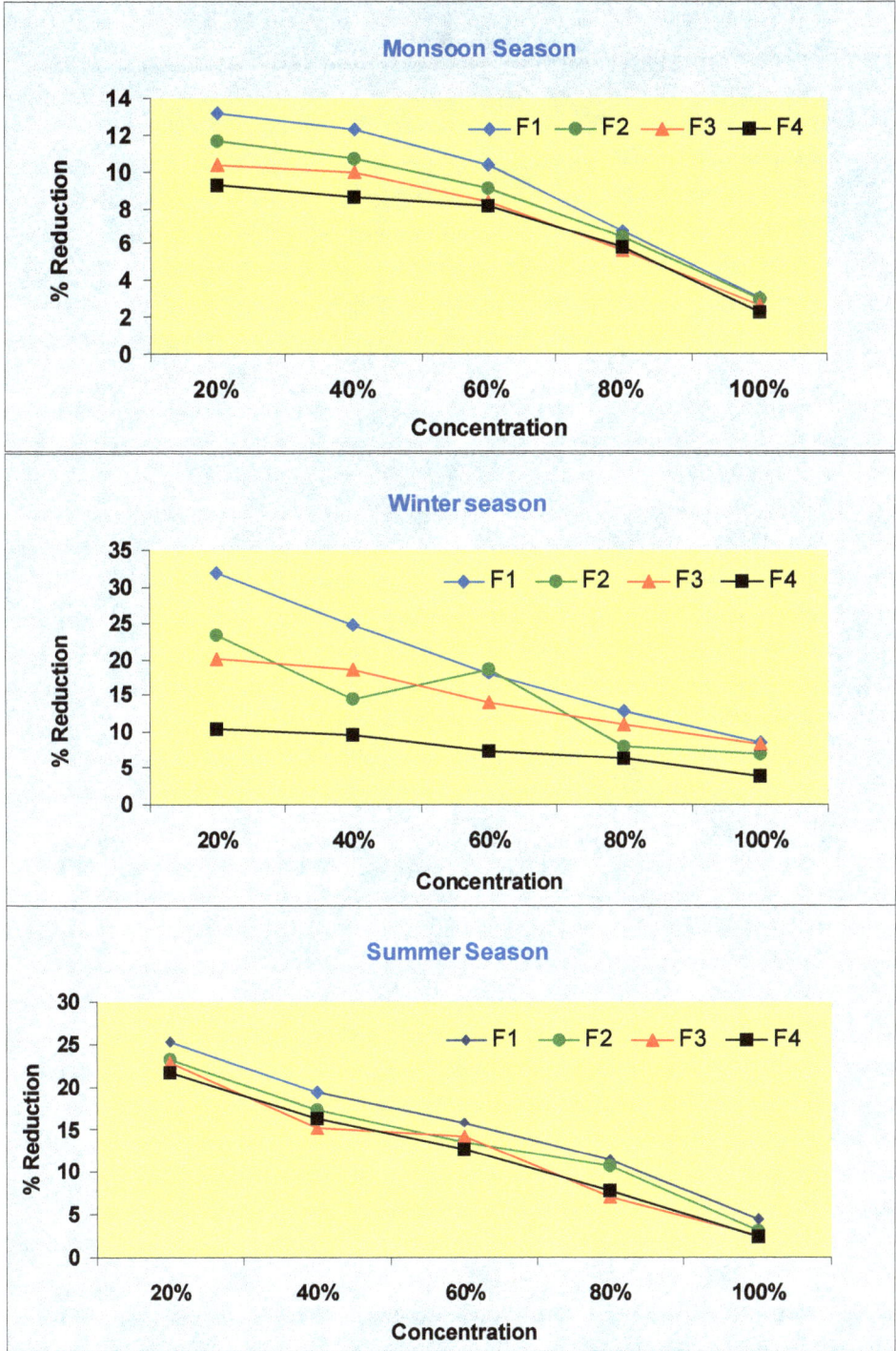

Figure 6.1: COD Reduction of Dye Industry Effluent by Different Fungal Species (Isolates from rhizosphere soil) at Different Concentration After Bioremediation Treatment during Different Seasons

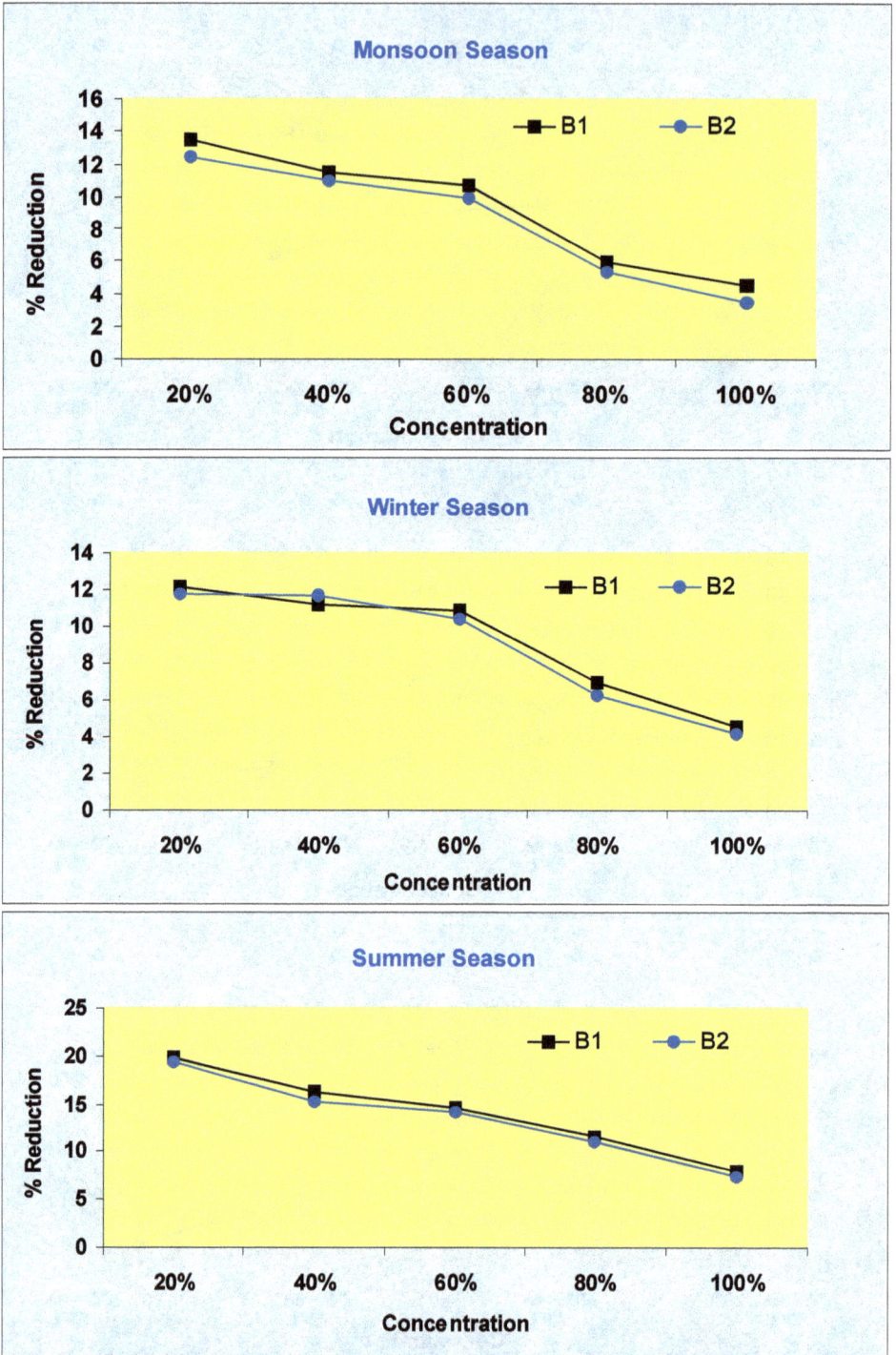

Figure 6.2: COD Reduction of Dye Industry Effluent by Different Bacterial Species (Isolates from rhizosphere soil) at Different Concentration during Different Seasons

from 0.5 to 1.0 g and for B2 0.47 to 0.9 g at concentration 100 per cent to 20 per cent respectively. These results show that B1 grows faster in effluent as compared to B2.

Table 6.2: Determination of Fungal Biomass at Different Dilution of Effluent

Sl.No.	Isolates	Effluent Concentration				
		20 per cent	40 per cent	60 per cent	80 per cent	100 per cent
1.	F1	6.2	4.6	4.5	2.5	1.2
2.	F2	6.8	4.9	4.2	2.4	0.9
3.	F3	7.2	5.2	3.9	2.1	1.0
4.	F4	7.3	6.8	4.3	2.2	1.1

Table 6.3: Dry Weight Determination of Bacterial Isolates at Different Dilution of Effluent

Sl.No.	Isolates	Effluent Concentration				
		20 per cent	40 per cent	60 per cent	80 per cent	100 per cent
1.	B1	1.0	1.2	0.9	0.7	0.5
2.	B2	0.9	0.81	0.68	0.65	0.47

Discussion

Bioremediation as a biotechnological intervention for cleaning up the residual effects of previous human activities on a site typically relies on the inherent abilities and characteristics of microbial species. In the present study, the emphasis concentrated on the contribution made by the microbes.

Bioremediation using *Aspergillus niger*, *Fusarium oxysporum*, *Aspergillus* spp., *Penicillium* spp., *Bacillus megaterium* and *Pseudomonas fluorescens* was found responsible for reduction of pollution load of the effluent as they oxidized organic and inorganic constituents present in it. Present investigation has succeeded in determining few species which showed their importance in textile effluent treatment. *A. niger* was the most prevalent fungal strain in the wastewater. *A.niger* has bioremediative potential as reported in by Sayer *et al.* (1997).

All biological approaches are expressly designed to optimise the activities of the various micro-organisms (either native to the particular soil or artificially introduced) to bring about the desired remediation. This generally means letting them to do what they would naturally do but enhancing their performance to achieve it more rapidly and/or more efficiently. Effectively it is little different from accelerated natural attenuation and typically involves management of aeration, nutrients and soil moisture, by means of their addition, manipulation or monitoring, dependent on circumstance (Biowise, 2001).

In the present study after treatment with test organisms pH of the effluent was found increased and was more of alkaline nature. This was probably due to dilution factor and may be because of nutrient salts added in it. COD is used to express the amount of oxygen needed for the chemical oxidation of the pollutants to inorganic products.

Maximum COD reduction observed 32.02 per cent, 23.5 per cent and 20.2 per cent at 20 per cent, 40 per cent and 60 per cent in winter season, while it was minimum in monsoon season at 80 per cent and100 per cent.

It may be low concentration of textile industry effluent provides, better condition for growth, multiplication of micro organism and biodegradation of organic pollutants present in effluent, while 100 per cent concentration was an unfavourable condition and not much appropriate for better use of biodegradation capability of micro organisms.

In the present study higher level of COD in initial stage was probably due to rapid utilization of oxygen in process of oxidation of organic matter and further continuous decrease in COD level was due to the low oxidation of organic matter as well as low extent of pollution load and mineralization of organic pollutant into secondary inorganic substance and products.

Decrease in COD levels is suggestive of the fact that the process of bioremediation is in progress, as earlier studies have shown that high COD levels are often indicated in wastewater containing substances that can be biologically degraded (Ademorotti *et al.*, 1992; Pathe *et al.*, 1995).

Even though the various biological mechanisms for the degradation of effluent toxins have not been studied in the present work but underline revelant processes such as biosorption, demethylation, methylation, metal-organic complexation or chelation, ligand degradation or oxidation can not be roled out. Microbes capable of utilising a variety of carbon sources and degrading a number of typical contaminants, to a greater or lesser extent, are commonly found in soils and wastewater. By enhancing and optimising conditions for them, they can be encouraged to do what they do naturally, but more swiftly and/or efficiently.

Based upon the above findings, it can be suggested that fungi and bacteria present in the environment possess a great potential for use in treatment of textile wastewater. These fungal strains can be studied for their growth in effluent and COD reduction capability and are found to be very promising in degradation of pollutants.

References

Ademorotti CMA., DO Ukponmwan and AA Omode (1992). Studies of textile effluent discharges in Nigeria. *Environ. Stud.*, 39: 291-296.

Ahmedna M, WF Marshall, AA Husseiny, RM Rao and I. Goktepe (2004). The use of nutshell carbons in drinking water filters for removal of trace metals. *Water Res.*, 38(4): 1064-1068.

APHA (1992). Standard methods for the examination of water and wastewater. American Public Health Association, Washigton DC pp-75-79, 81-89.

Boone DR, Castenholz RW (2001). Bergey's Manual of systematic Bacteriology (Edited) Springer, New York pp-49.

Biowise, UK Department of Trade and Industry (2001). *Contaminated Land Remediation: A Review of Biological Technology*, Crown copyright pp 112-114.

Buchanan RE, Gibbon NE (1994). Bergey's Manual of determinative Bacteriology. 9th Ed. Williams and Wilkins, Baltimore pp-51-52.

Gilman JD (1957). A manual of soil fungi. Iowa State University Press, Ames Iowa, USA. pp: 450.

Pathe PP, SN Kaul and T. Nandy (1995). Performance evaluation of a full scale common effluent treatment (CETP) for a cluster of small scale cotton textile units. *J. Environ. Stud.*, 48: 149-167.

Sayer JA, M. Kierans and GM Gadd (1997). Solubilization of some naturally occurring metal bearing mineral lime scale and lead phosphate by *Aspergillus niger*. *FEMS Microbiol. Lett.*, 154: 29-35.

White C, JA Sayer and GM Gadd (1997). Microbial solubilization and immobilization of toxic metals: Key biogeochemical processes for treatment of contamination. *FEMS Microbiology Review*, 20: 503-516.

2013, Environmental Technology

Editors: **D.R. Khanna, A.K. Chopra, Gagan Matta, R. Bhutiani & Vikas Singh**

Published by: **DAYA PUBLISHING HOUSE, NEW DELHI**

Pages 61–70

Chapter 7

Algal Oil Analysis by GC and GC-MS Techniques for Elucidation of its Biodiesel Potential

Varinder Kaur[1], Anju Chopra[1] and Smita Chaudhry[1]

[1]*Institute of Environmental Studies, Kurukshetra University, Kurukshetra*
[2]*Indian Oil Corporation Limited, Faridabad*

ABSTRACT

The present study was conducted to characterize lipid composition of biodiesel obtained from algae and to optimize hydrolysis procedure of algal oil by GC and GC-MS techniques to identify best algal strain for biodiesel production. Eight oil samples from different algal strains were investigated. The composition was determined at lab scale level by first hydrolyzing the lipids present in samples and then converting them into Fatty Acid Methyl Esters (FAME) and after that, fatty acids were identified by GC and GC-MS. The concept of using PTC (Phase Transfer Catalyst) helped to optimize hydrolysis method in minimum possible time which subsequently decreased total analysis time. GC and GC-MS analysis showed that most samples contained palmitic ($C16:0$) and linolenic acid ($C18:3$) as major fatty acids constituting 40-60 per cent and 20-30 per cent respectively of total fatty acids. The lack of some FAME standards restricted the determination of whole composition of some samples. The finding led to conclusion that algae vary in their fatty acid composition to a large extent and have very complex structure. The reduction in hydrolysis time from 16 hrs to only 1 hour was one of the big achievements. GC and GC-MS provided a quick and desired method for estimating lipid composition of biodiesel. Further research should concentrate on developing more efficient method for characterization of biodiesel obtained from algae.

Keywords: *Biodiesel, Fatty acid methyl esters, GC and GC-MS, Micro algae, Lipids, Phase transfer catalyst.*

Introduction

Fueling automobiles with renewable energy sources is an alternative to fossil fuels. Biofuels are important substitutes of fossil fuels (Hossai *et al.*, 2008). One of the most common biofuels is biodiesel. However, economic aspects of biodiesel production limit its development and large scale use. It is mostly made from vegetable oils but biodiesel from algae is presently making transition from a research topic to a marketed commodity. In some countries, biodiesel is less expensive than conventional fuels (Doganci *et al.*, 2005). Algae (Latin for "sea weeds", singular *alga*) are diverse group of autotrophic/heterotrophic organisms that range from single celled microalgae to macroalgae. Algae are highest yielding feedstock for biodiesel which are rich in fatty acids and can grow rapidly by doubling their biomass within 24 hrs (Harwood and Gusehine, 2009). These fatty acids can be converted into biodiesel. The oil productivity of many microalgae exceeds the productivity of best producing oil crops (Chisti, 2007). Using microalgae to generate biodiesel will not compromise the production of food, fodder and other products derived from crops. The algae derived biodiesel must comply with existing standards of ASTM (American Standards of Testing Methods) D6751 and European standards, EN14214 (Knothe, 2006). Hence, many fundamental questions relating to chemical composition of fatty acids in algae need to be answered (Hu *et al.*, 2008). Properties of biodiesel are largely determined by structure of its component fatty acid esters *e.g.* saturated fats produce biodiesel with superior oxidative stability but poor low temperature properties. Biodiesel produced from feed stocks high in PUFAs (Poly Unsaturated Fatty Acids), on the other hand, has good cold flow properties but is susceptible to oxidation. Therefore, biodiesel with these fatty acids tends to have instability problems during storage. In making biodiesel, triglycerides are reacted with methanol in "transesterification" reaction that requires 3 moles of alcohol for each mole of triglyceride to produce 1 mole of glycerol and 3 moles of methyl esters which is catalyzed by acids, alkalies and lipase enzymes. Alkali catalyzed reaction is about 4000 times faster than acid catalyzed reaction. Consequently, NaOH and KOH are commonly used catalysts. The characterization of fatty acids derived from algae is must to choose suitable lipid profile for biodiesel production (Sarpal *et al.*, 2009). GC and GC-MS (Gas Chromatography and Gas Chromatography- Mass Spectrometry) techniques have been successfully used for identifying fatty acids and some hydrocarbons present in algae (Sarpal *et al.*, 2009). The objective of study was to characterize the nature of FAME produced from algal oil by establishing a new method in minimum possible time. The study permitted to find lipid composition of target algal oil samples by GC/MS data. The analysis could be achieved in minimum possible time. In addition, the program will also be useful for choosing the best suitable strain of microalgae to produce biodiesel at large scale levels in the future.

Materials and Methods

The eight algal oil samples were prepared in IOC R and D centre only from different algal strains by their artificial harvesting. These were having code names: Algal-10A, AO-37, Algal N0-E2, AO-38, 02-CH, AO-39, HT-33 and AO-40.

Hydrolysis and FAME Production

10 ml of hydrolyzing solution was added to 0.5g of sample and was refluxed at >100°C for 1 hour. The above solution was dissolved in 5ml of dist. water and 3ml toluene. The upper toluene layer containing HCs (hydrocarbons) was separated. The lower aqueous layer containing fatty acids was added to conc. HCl and was dissolved in $CHCl_3$ to get free fatty acids. $CHCl_3$ was evaporated. After making sure by IR that the sample contained mainly acids, the acidic layer was esterified with BF_3-methanol and was left for 10 minutes on boiling water bath to form FAME.

Gas Chromatographic Analysis

0.2µL of FAME of their corresponding sample were injected and run for 60 minutes in gas chromatograph at 300°C with CP-WAX52CB column having 30m length and 0.32mm inner diameter procured from M/s Varian (USA). Perkin Elmer Clarus 500 gas chromatograph equipped with split/split less injector (1:10 split ratio) and FID (Flame Ionization Detector at 300°C) was used. Data were processed using "Total Chrom work station" software. The stationary phase used was PEG (Polt Ethylene Glycol) and carrier gas was Helium (UHP grade).

Identification of Fatty Acids

In order to calibrate gas chromatograph, FAME standards were injected into GC and analyzed, and then peaks were obtained in the form of chromatogram referring to RT (Retention Time). After that, sample was injected and peaks were obtained referring to RT again. The comparison of RT of standards and sample were done and types of FAME in sample were identified. The calculation was based upon assumption that all esters (C:8-C:24) have same response factors. The area under each FAME was calculated from chromatogram by applying following formula.

$$\text{Per cent Fatty Acid} = \frac{\text{Area of fatty acid peak} \times 100}{\text{Total area under chromatogram}}$$

GC-MS Analysis

The final analysis of FAME produced from algal oil was done by GC-MS to know complete lipid composition of samples. Out of all samples, the GC-MS of 3 samples could be done due to lack of time. Autospec Ultima High Resolution mass spectrometer procured from M/s Micro mass UK coupled with Agilent 6890N Chromatograph was used which was tuned for 1000 resolutions in the mass range of 50-300 Daltons. Per fluoro kerosene (PFK) was used as reference. The magnet scan was set at 0.5 seconds/decade and interscan delay of 0.1seconds. The ionization of samples was done using electrons of 70eV potential. The source was maintained at 225°C. The data acquisition and processing were done using OPUS3.6V software and NIST mass-spectral library was used. Capillary column DB5 (30mX0.25mm inner diameter; 0.32µm film thickness) was used. The carrier gas was Helium (99.999 per cent pure). The GC was having ramp rate of 4°C/min. Carrier gas was Helium with flow rate of 1.2ml/min. 0.2ml sample was injected at 300°C.

Results and Discussion

The characterization of fatty acids derived from algal oil is important in order to choose a suitable lipid profile for biodiesel production (Sarpal *et al.*, 2009). The "optimization of hydrolysis procedure of algal oil to make FAME for determining its biodiesel potential" was done through GC and GC-MS IR and NMR. After completion of project, some conclusions were made:

Optimization of Hydrolysis Procedure

The objective was to complete the hydrolysis of samples in minimum possible time by addition of a powerful catalyst *i.e.* Phase Transfer Catalyst. All 8 samples were found to contain free fatty acids after neutralization as confirmed by IR and were screened by GC/GC-MS. The results obtained are supported by available literature. Small amount of iso-acids identified by MS were found to occur in algal oils. Presence of these unsaturated fatty acids is not desirable in biodiesel since they are more prone to oxidation. Presence of hydrocarbons in samples indicated that prior cleaning of hydrolyzed part is required since these can interfere with GC analysis.

Selection of GC Column and Conditions

All samples after undergoing IR were esterified with BF_3 methanol and were subjected to GC and GC-MS analyses. Fused silica capillary column CP-WAX52CB (30mX0.32mmIDX0.25µm) was selected as per optimized conditions which clearly separated FAME. Composition showed the presence of C:10-C:24 fatty acids. The unsaturated fatty acids were also observed in the range of C:16-C:20. The C:16 was found to be major component in all samples (Table 7.1).

GC-MS Studies and Data Analysis

The components separated in GC were identified from their fragmented pattern in E1 mass-spectrum using NIST mass-spectral library. GC-MS analysis showed that unsaturated fatty acids constitute major portion in the algal oil. Presence of some additional peaks were also noticed in GC chromatograms which could not be identified by GC alone and were identified by GC-MS later as those of branched hydrocarbons in between C:14-C:17 fatty acids. The GC-MS also showed the presence of naphthalenes which corresponded to 27 per cent of unidentified components in GC data. Their detection shows the presence of hydrocarbons in algal oils, which impart analytical challenge to GC. 3 oil samples *viz.* Algal-10A, 02-CH and Algal No-E-2 were analyzed by GC-MS (Tables 7.2–7.4).

It may be finally concluded that microalgal biodiesel is technically feasible. The present study introduced an integrated method for hydrolysis of microalgal oil as the reduction in hydrolysis time 16 hours to only 1 hour was one of the achievements in present work. The coupling of GC with GC-MS provides a powerful analytical tool with wide applicability in algal oil analysis. The future research should concentrate in developing a highly efficient method for hydrolysis of algal oil for characterization of FAME.

Table 7.1: Relative Percentage of Fatty Acids in Each Sample by GC

Sl.No.	Component	Name	Per cent Weight							
			AO37	AO38	AO39	AO40	HT33	02CH	Algal 10A	AlgalE2
1.	C 10:0	Capric acid	11.2 (unidentified)	1.2 (unidentified)	3.0 (unidentified)	0.3 (unidentified)	0.7 (unidentified)	0.4 (unidentified)	–	1.0 (unidentified)
2.	C 11:0	Undecanoic acid							–	
3.	C 12:0	Lauric acid							–	
4.	C 12:0	Lauric acid	30.0 (unidentified)	1.7 (unidentified)	14.6 (unidentified)	4.5 (unidentified)	–			–
5.	C 14:0	Myristic acid						2.3		
6.	C 14:0	Myristic acid	6.3 (unidentified)	2.3 (unidentified)	4.7 (unidentified)	2.4 (unidentified)	0.1 (unidentified)		–	–
7.	C 14:1	Myristolic acid								
8.	C15:0	Pentadecanoic acid	–	–	–	–	–	7.6	–	–
9.	C 16:0	Palmitic acid	17.7	20.5	23.5	19.6	16.0	32.1	33.3	9.6
10.	C 16:1 cis	Palmitoleic acid	1.2	0.7	8.5	5.1	1.3	–	–	–
11.	C 16:1 trans	Palmitelaidic acid	1.2	0.8	–	0.3	0.9	–	–	–
12.	C 17:0	n-heptadecanoic acid	1.2	0.4	1.5	1.5	–	5.2	10.2	–
13.	C 18:0	Stearic acid	1.2	0.2	1.5	6.7	1.2	10.9	7.7	5.0
14.	C 18:1 cis	Oleic acid	7.2	10.5	7.8	1.0	8.2	5.1	39.1	39.9
15.	C 18:1 trans	Elaidic/Vaccenic acid	–	1.3	–	9.0	0.9	–	–	–

Contd...

Table 7.1–Contd...

Sl.No.	Component	Name	Per cent Weight							
			AO37	AO38	AO39	AO40	HT33	O2CH	Algal 10A	AlgalE2
16.	C 18:2 cis	Linoleic/Petroselenic acid	2.4	4.6	4.5	5.6	6.3	2.0	8.0	44.5
17.	C 18:3 cis	Linolenic acid	8.4	14.1	17.1	23.9	25.8	27.9	1.7	1.0
18.	C 20:0	Arachidic acid	–	0.2	–	0.7	–	–	–	–
19.	C 20:1 cis	methyl-5-eicosanoic acid	–	–	–	0.3	–	–	–	–
20.	C20:2	methyl-11,14-eicosadienoic acid	–	–	–	–	–	4.2	–	–
21.	C 22:0	Behenic acid	–	0.5	–	–	0.3	–	–	–
22.	C 24:0	Lignoceric acid	–	12.3	–	0.6	–	1.0	–	–
23.	C24:1	Nervonic acid	–	–	–	–	–	1.8	–	–

Table 7.2: Fatty Acid Composition of Algal-10A by GC-MS

Sl.No.	RT	Scan	Component Identified	Formula	M.Wt.
1.	9:18:00	2112	n-pentadecane	$C_{15}H_{32}$	212
2.	12:11:00	2398	n-hexadecane	$C_{16}H_{34}$	226
3.	26:56:00	2671	n-heptadecane	$C_{17}H_{36}$	240
4.	27:05:00	2687	Branched paraffin of C18 (t)	$C_{18}H_{38}$	254
5.	27:35:00	2736	Methyl ester of tetradecanoic acid (C14:0)	$C_{15}H_{30}O_2$	242
6.	29:32:00	2930	n-octadecane	$C_{18}H_{38}$	254
7.	29:47:00	2954	Branched paraffin of C10 (t)	$C_{19}H_{40}$	268
8.	32:02:00	3177	n-nonadecane	$C_{19}H_{40}$	268
9.	32:40:00	3240	Methyl ester of hexadecanoic acid	$C_{17}H_{34}O_2$	270
10.	34:24:00	3413	n-dodecane	$C_{20}H_{42}$	282
11.	36:02:00	3575	Methyl ester of 8,9-octadecanoic acid	$C_{19}H_{36}O_2$	296
12.	36:22:00	3622	Methyl ester of 8,11-octadecanoic acid	$C_{19}H_{34}O_2$	294
13.	36:40:00	3638	Methyl ester of 8,9-octadecanoic acid	$C_{19}H_{36}O_2$	296
14.	13:17:00	3698	Methyl ester of octadecanoic acid	$C_{19}H_{38}O_2$	298

Table 7.3: Fatty Acid Composition of Algal No-E-2 by GC-MS

Sl.No.	RT	Scan	Component Identified	Formula	M.Wt.
1.	21:18:00	2112	n-pentadecane	$C_{15}H_{32}$	212
2.	24:11:00	2398	n-hexadecane	$C_{16}H_{34}$	226
3.	26:55:00	2670	n-heptadecane	$C_{17}H_{36}$	240 (II)
4.	27:34:00	2735	Methyl ester of tetradecanoic acid (C14:0)	$C_{18}H_{38}$	254
5.	29:32:00	2929	n-octadecane	$C_{15}H_{30}O_2$	242
6.	32:06:00	3184	Methyl ester of 9-hexadecaenoic acid (C16:1)	$C_{18}H_{38}$	254
7.	32:39:00	3239	Methyl ester of hexadecanoic acid	$C_{19}H_{40}$	268
8.	34:15:00	3398	Ethyl ester of hexadecanoic acid	$C_{19}H_{40}$	268
9.	36:20:00	3621	Methyl ester of 8,11-octadecanoic acid	$C_{17}H_{34}O_2$	270 (I)
10.	36:40:00	3637	Methyl ester of8,9-octadecaenoic acid (C18:1)	$C_{20}H_{42}$	282
11.	37:16:00	3697	Methyl ester of octadecanoic acid (C18:0)	$C_{19}H_{36}O_2$	296

TABLE 7.4: Fatty Acid Composition of 02-CH by GC-MS

Sl.No.	RT	Scan	Component Identified	Formula	M.Wt.
1.	18:39:00	1849	dimethyl naphthalene	$C_{12}H_{12}$	156
2.	19:14:00	1907	dimethyl naphthalene	$C_{12}H12$	156
3.	20:00:00	1952	dimethyl naphthalene	$C_{12}H12$	156
4.	20:40:00	2050	trimethyl naphthalene	$C_{13}H_{14}$	170
5.	21:17:00	2110	trimethyl naphthalene	$C_{13}H_{14}$	170
6.	21:50:00	2166	trimethyl naphthalene	$C_{13}H_{14}$	170
7.	21:59:00	2181	Methyl ester of dodecanoic acid (C12:0)	$C_{13}H_{26}O_2$	214
8.	22:38:00	2245	Dimethyl azelate	$C_{11}H_{2004}$	216
9.	22:55:00	2272	trimethyl naphthalene	$C_{13}H_{14}$	170
10.	23:16:00	2307	trimethyl naphthalene	$C_{13}H1_4$	170
11.	23:30:00	–	Trinaphthalenes, dimethyl biphenyls, flourenes etc.	–	–
12.	26:20:00				
13.	26:47:00	2657	Methyl ester of 12-methyl tridecanoic acid	$C_{15}H_{30}O_2$	242
14.	26:54:00	2668	n-heptadecane	$C_{17}H_{30}$	234
15.	27:23:00	2736	Methyl ester of tetradecanoic acid (C14:0)	$C_{15}H_{30}O_2$	242(III)
16.	29:13:00	2898	trans methyl ester of 9-methyl teradecanoic acid	$C_{16}H_{32}O_2$	256
17.	29:28:00	2923	Trans methyl ester of12-methyl teradecanoic acid	$C_{16}H_{32}O_2$	256(III)
18.	29:57:00	2911	Octadecane+unknown	$C_{18}H_{38}$	254
19.	30:09:00	2991	Cis methyl ester of 9-methyl tetradecanoic acid/ME of pentanoic acid	$C_{16}H_{32}O_2$	256
20.	31:44:00	3148	Methyl ester of 14-methyl pentanoic acid	$C_{17}H_{34}H_2$	270(IV)
21.	32:47:00	3251	Methyl ester of hexadecanoic acid (C16:0)	$C_{17}H_{34}O_2$	270(I)
22.	33:40:00	3340	Methyl ester of methyl hexadecanoic acid (t)	$C_{18}H_{36}O_2$	284
23.	34:20:00	3351	Methyl ester of methyl hexadecanoic acid (t)	$C_{18}H_{36}O_2$	284
24.	35:00:00	3471	Methyl ester of heptadecanoic acid (C17:0) (t)	$C_{18}H_{36}O_2$	284
25.	36:39:00:	3636	Methyl ester of 8,9-octadecanoic acid (C18:1) (t)	$C_{19}H_{36}O_2$	296
26.	37:16:00	3697	Methyl ester of octadecanoic acid (C18:0)	$C_{19}H_{38}O_2$	298
27.	40:33:00	4022	Methyl ester of octadecadienoic acid (C18:2)	$C_{19}H_{34}O_2$	294

References

Antolin G, Tinaut FV, Briceno Y, Castano V, Perez C, Ramirez AI 2002. Optimisation of biodiesel production by sunflower oil transesterification. Bioresource Technol. 83: 111–114.

Asakuma Y, Maeda K, Kuramochi H, Fukui K 2007. Theoretical study of the transesterification of triglycerides to biodiesel fuel. Fuel. 86: 1201- 1207.

Banerjee A, Sharma R, Chisti Y, Banerjee UC 2002. *Botryococus braunii*: A Renewable Source of Hydrocarbons and Other Chemicals. Crit. Rev. Biotechnol. 22: 245- 279.

Basova MM 2005. Fatty acid composition in microalgae. Int. J. Algae. 7: 33- 57.

Benemann JR, Goebel RP, Weissman JC, Augenstein DC 1982. Microoagae as a source of liquid fuels. Final Technical Report. USDOE- OER.

Chisti Y 2007. Biodiesel from microalgae. Biotechnol. Adv. 25: 294- 306.

Chen W, Jhang C, Song L, Sommerfield L, Hu Q 2009. A highly throughput Nile red method for quantitative measurement of neutral lipids in microalgae. J. Microbiol. Methods. 77: 41- 47.

Chopra a, Singh D, Kalsi WR, Vatsala S, Sarpal AS, Basu B 2009. Determination of fatty acid based lubricity improver in diesel by GC. Chromatographia. 70: 1143- 46.

Chisti Y 2007. Biodiesel from microalgae. Biotechnol Adv. 25: 294–306.

Coal 21 website. (*http: //www.coal21.com.au/IGCC.php*)

Demirbas A 2007. Recent developments in biodiesel fuels. Int J Green Energy.4: 15–26.

DOE Fossil Energy website. (*http: //www.fe.doe.gov/programs/powersystems/gasification*)

Dromey RG Stefik MJ, Rindfleisch TC 1976. Extraction of Mass Spectra Free of Background and Neighboring Component Contributions from GC A.M. Anal. Chem. 9: 1368-1375.

Gao C, Xiong W, Zhang Y, Yuan W, Wu Q 2008. Rapid quantitation of lipid in microalgae by time domain NMR in alga based biodiesel production. Journal of Microbiological Methods. 75: 437- 439.

Giddings JC, 1st edition 1965. Dynamics of Chromatography. Dekker, New York

Grob RL, 3rd edition 1995. Modern Practice of Gas Chromatography. Analytical Chemistry. Villanova University. A Wiley Interscience Publication. John Wiley and Sons, Inc. ISBN 0- 471- 59700- 7.

Haiden RW 1996. Analytical Methodologies for determination of biodiesel purity-determination of total methyl esters. Technical report submitted to National Biodiesel Board wide NBB Contract.

Jame MM, 2nd edition 2005. Chromatography- Concepts and Contracts. Drew University, Medison, New Jersey. Published by John Wiley Sons, Inc.

Jeffrey D, Jeremy L. Lipid composition of Chlorarachniophytes from the genera Bigelowiella, Gymnochlora, and Lotharella 2005. Journal of Phycology. 41: 311- 21.

Kalsi WR, Tiwari AK, Singh AP, Sarpal AS 2007. Determination of biodiesel content in petroleum diesel by Gas Chromatography. Report no. 04003.

Katherine Stenerson 2005. The analysis of positional geometric fatty acid methyl ester isomers using the CP- 2560 capillary column. The Reporter. 22: 7- 14.

Kaul SN 2005. Quality criteria of biodiesel. Asian Publishers. New Delhi. India.

Key World Statistics, International Energy Agency, 2008. (*http: //www.iea.org*)

Knothe g. Dependence of biodiesel fuel properties on the structure of fatty acid alkyl esters 2005. Fuel processing technol. 86: 1059- 70.

Lin CY, Li RJ 2009. Fuel properties of biodiesel produced from the crude fish oil from the soapstock of marine fish. Fuel Process Technol. 90: 130- 136.

McNeff CV, McNeff FC, Yan B, Nowlan DT, Rasmussen M, Gyberg AE, Krohn BJ, Fedie RL, Hoye TR 2008. A continuous catalytic system for biodiesel production. Appl. Catal. 343: 39- 48.

Metting FB 1996. Biodiversity and applications of microalgae. J. Ind. Microbiol. 17: 477- 489.

Miao X, Wu Q 2006. Biodiesel production from heterotrophic microalgal oil. Bioresourc. Technol. 97: 841- 846.

Montiero MR, Ambrozin ARP, Liao LM, Ferriera AG 2008. Critical review on analytical methods for biodiesel characterization. Talanta. 77: 593-0596.

Patil V, Reitan KI, Knudsen G, Mortensen L, Kallqvist T, Olsen E, Vogt G, Gislerod HR 2005. Microalgae as Source of Polyunsaturated Fatty Acids for Aquaculture. Curr. Topics Plant Biol.6: 57-65.

Patil V, Tran KQ, Giselrod HR 2008. Towards Sustainable Production of Biofuels from Microalgae. Int J Mol Sci. 9: 1188–1195.

Sharif AB, Nasrulhaq HM, Majid S, Zuliana R 2007. Biodiesel production from waste cooking oils as environmental benefits and recycling process. A Review. Asia biofuel conference book.

Sharma YC, Singh B 2008. Advancements in development and characterization of biodiesel. A Review. Fuel. 87: 2355- 2373.

Slover, Lanza 1979. Quantitative analysis of food fatty acids by GC. JAOCS. 56: 933.

Tiwari AK, Chopra A, Vatsala A, Kalsi WR, Kagdiyal V, Upreti M, Singh AP, Mukharjee S, Sarpal AS 2009. Compositional aspects of fatty acids derived from algal oil by analytical techniques. Report No. TR- 10- 1260. R and D, Indian Oil Corporation Limited.

WTO data base. (*www.wto.org*)

2013, Environmental Technology
Editors: D.R. Khanna, A.K. Chopra, Gagan Matta, R. Bhutiani & Vikas Singh
Published by: DAYA PUBLISHING HOUSE, NEW DELHI

Pages 71–78

Chapter 8

Prediction of Groundwater Level using Artificial Neural Networks

Sumant Kumar, N.C. Ghosh and Surjeet Singh

Groundwater Hydrology Division
National Institute of Hydrology, Roorkee – 247 667

ABSTRACT

The dependence on groundwater as a reliable source to meet the requirements for irrigation, drinking and industrial uses in India has been risen rapidly during the last few decades. This has resulted in depletion of groundwater table in many areas causing concerns for the long-term sustainability of groundwater based supplies. For the effective management of groundwater, it is important to know fluctuations of groundwater level (GWL). In hydrology, the artificial neural network (ANN) models have been satisfactorily applied for prediction of nonlinear hydrologic processes such as; rainfall- runoff, stream flow, precipitation, and water quality modelling but their application to the groundwater sector has been found to be limited. In this study, an ANN model has been developed to predict the groundwater level fluctuations. The structure of the ANN is comprised of three layers: an input layer that considers rainfall and past GWL as inputs, one hidden layer, and an output layer that gives the predicted GWL. The data for this modelling task have been used from a watershed located in the Sagar distt., Madhya Pradesh, India. The performances of the developed ANN model has been evaluated using standard statistical measures *viz.* Root mean square error (RMSE), Average Arithmetic Relative Error (AARE), Correlation coefficient (R) etc. The results obtained for the study indicated that the ANN technique can successfully be used for prediction of GWL fluctuations.

Keywords: Artificial neural networks, Groundwater level, Prediction, Statistical methods, Sagar.

Introduction

In India, groundwater serves about 80 per cent of rural population, 50 per cent of urban population and about 60 per cent of agricultural area. There are more than 20 million groundwater extraction structures in place which are being used to meet requirement for domestic, industrial, and agricultural activities. The demand of groundwater is continuously in the rise because of its some fascinating features, such as; slow moving, large storage volume and long retention time, can be drawn on demand, less risk free than surface water sources, etc. For effective management of groundwater resources, one has to know the behaviour of groundwater level (GWL) fluctuations. To predict the GWL many researchers have developed a number of prediction models. Those are, physically based numerical models normally used for simulating groundwater flow profiles and predicting GWL. These numerical models are derived based on governing equations which simplify the physical processes of flow in the subsurface and usually solve with proper initial and boundary conditions using numerical methods. Both spatial and temporal GWL can be simulated using the numerical models within a given domain. However, usages of these models require a large quantity of precise data related to physical properties of the domain, and parameters to calibrate the model simulations. In practice, sufficient data for such model development are not readily available because of limitations in cost and time, resulting in model uncertainties. Another approach for predicting GWL fluctuations is based on the time-series modelling of groundwater levels and meteorological data. These models are limited to forecasting of temporal variation at a fixed location.

In recent years, the artificial neural network (ANN) technique has been found successfully using to solve various water resource problems including time-series forecasting (Zealand *et al.*, 1999; Sudheer *et al.*, 2002; Cigizoglu, 2003; Almasri and Kaluarachchi, 2005; Yoon *et al.*, 2007). The ANNs are considered to be a standard nonlinear estimator and their abilities have been verified in a variety of fields. In many hydrological problems, the ANN models have been satisfactorily applied to the prediction of nonlinear hydrologic processes such as rainfall runoff, stream flow, precipitation, and water quality modeling in the 1990s (ASCE, 2000b). Maier and Dandy (2000) reviewed 43 papers of ANN applications to water resource variables that had been published until the end of 1998. Among those papers, surface water flow and quality were the topic in 28, and rainfall forecasting in 13. Only two papers were associated with water tables forecasting using synthetic data. During the years 2000 and onwards, studies on the prediction of GWL fluctuations in the real environment have come into the lime light. Nayak *et al.* (2006) and Krishna *et al.* (2008) successfully predicted the GWL fluctuation in coastal aquifers using ANN models with meteorological information with GWL data as the input variables. In that paper, neural network models were developed, and performances evaluated were found to be efficient for groundwater level prediction. The present paper considers development and performance evaluation of an ANN model for prediction of GWL in a water scarce region.

ANN Model Development

Study Area and Data

The study area is located in the Sagar district of Madhya Pradesh, India. The Sagar city is located at the latitude of 23°50' N and longitude of 78°45' E. The catchment area of the basin is 18 km². The land use pattern of the basin is 40.9 per cent barren land, 20.9 per cent agriculture, 18.7 per cent settlement, 11.5 per cent open forest and 8.1 per cent water body. The soils of this area are of two types: the red or reddish brown lateritic soil on hill tops and the black soil at the foothills.

The data acquired from the area are consisted of rainfall and GWL. The GWL time series data were measured at a particular location in the Sagar city and the precipitation data were collected from the meteorological station located in the Sagar city. The daily rainfall and weekly GWL data were available for the period from June 1999 to July 2000. For the study, the weekly GWL data were converted into daily data for prediction of future daily GWL fluctuations.

Network Architecture

A feed-forward multilayer perceptron type ANN model architecture as given in Figure 8.1 is considered to develop the time series models. The model is comprised of three layers (Figure 8.1): (i) an input layer consists of explanatory variables, (ii) one hidden layer and (iii) an output layer consists of a single neuron representing the groundwater level to be modelled at different time, GWL(t).

The activation function employed at hidden and output neurons is considered to be sigmoid function. A computer program written in 'C' language is developed for ANN simulation. The output from all the ANN models is time varying GWL,

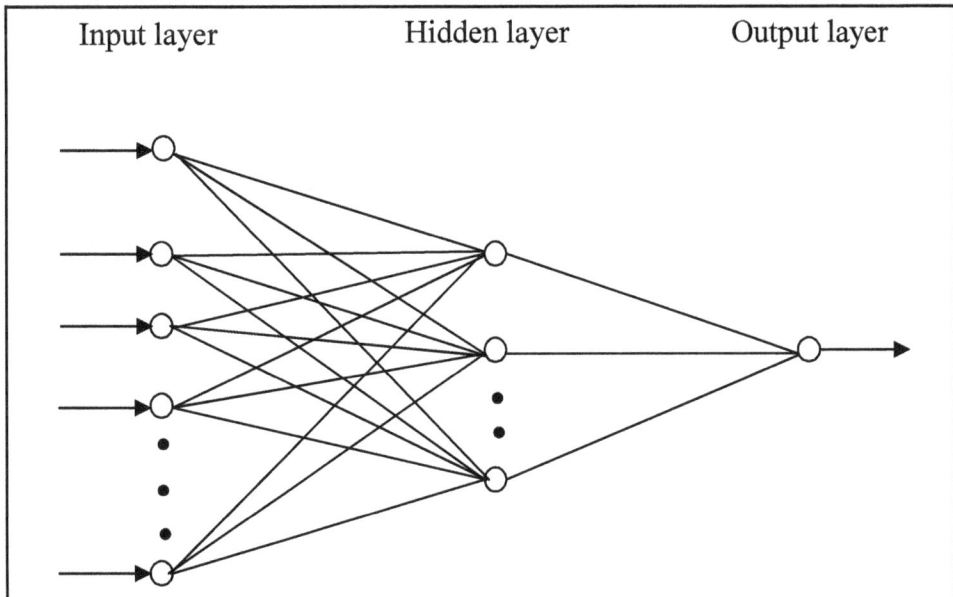

Figure 8.1: The Conceptual Structure of the ANN Model

i.e.,GWL(t). The input and output data are scaled in the range of 0.1 and 0.9 using the following equation to prevent saturation during training:

$$X_n = \frac{(b-a)(X_0 - X_{min})}{(X_{max} - X_{min})} + a$$
(1)

where,

X_n is the normalized value; a is the lower limit of the normalized value; b is the upper limit of the normalized value; X_0 is the normalized value of the X_{max} and X_{min}; in which X_{max} is the maximum and X_{min} is the minimum values of the data series. The value of a and b is taken as 0.1 and 0.9, respectively.

Inputs to ANN models are determined using auto- and cross-correlation analyses in which the values of these coefficients in excess of certain threshold values are considered as the input vector for the ANN model development. Based on the statistical analyses, the inputs selected for the models are as follows: daily rainfalls P(t), P(t-1), P(t-2), P(t-3) and P(t-4) and daily preceding groundwater level GWL(t-1), GWL(t-2), GWL(t-3), GWL(t-4) and GWL(t-5). Thus, an ANN structure of 10-N-1 is explored to simulate the time-series data set of the watershed. The optimum ANN structure is determined based on the trial and error procedure. First, a number of ANN model structures with varying number of hidden neurons (from 1 to 15) are trained. Then a plot of Co-relation Coefficient (R), Normalized Root Mean Squares (NRMSE) and Average Arithmetic Root Error (AARE) versus the number of hidden neurons (Figure 8.2) is prepared. The number of hidden neurons that provided the best performance

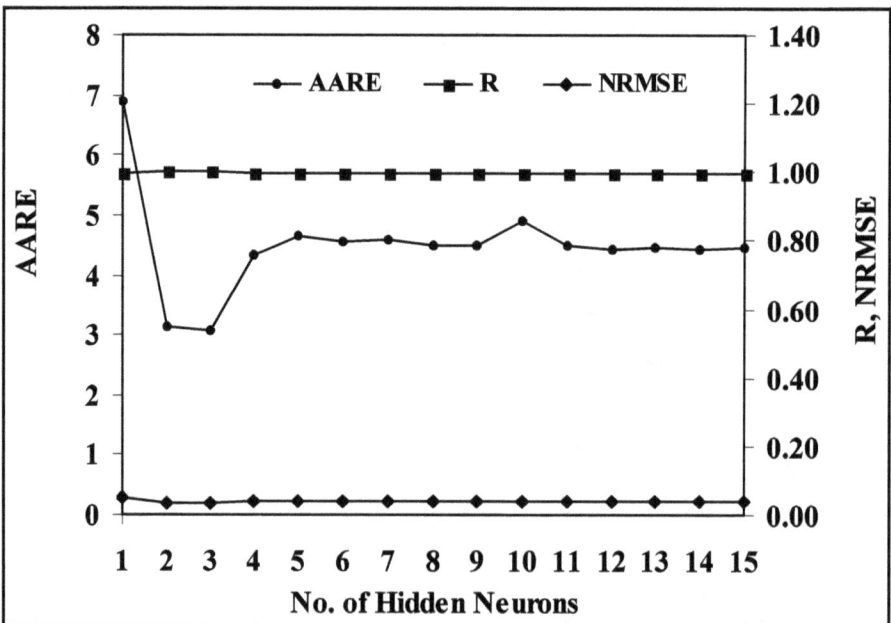

Figure 8.2: Variation of Errors Estimated by Different Statistical Methods with Different Hidden Neurons

during the training for a given data set in terms of the error statistics is selected as the structure of the ANN model. For the present case, the structure of the model is found to be 10-3-1.

Model Performance Evaluation

Five different standard statistical performance evaluation criteria are used to evaluate the performance of various models. These are: Average Absolute Relative Error (AARE), Coefficient of Correlation (R), Nash-Sutcliff Efficiency (E), Normalized Root Mean Square Error (NRMSE), and Threshold Statistics (TS). Mathematically, these statistical evaluation criteria are explained by the following equations:

$$AARE = \frac{1}{N}\sum\left|\frac{GWLp - GWLo}{GWLo}\right| \times 100 \tag{2}$$

$$R = \frac{\sum_{t=1}^{N}\left(GWLo(t) - \overline{GWLo}\right)\left(GWLp(t) - \overline{GWLp}\right)}{\sqrt{\sum_{t=1}^{N}\left(GWLo(t) - \overline{GWLo}\right)^2 \left(GWLp(t) - \overline{GWLp}\right)^2}} \tag{3}$$

$$E = \frac{E1 - E2}{E1} \tag{4}$$

$$E1 = \sum_{t=1}^{N}\left(GWLo(t) - \overline{GWLo}\right)^2 \qquad E2 = \sum_{t=1}^{N}\left(GWLp(t) - \overline{GWLo}\right)^2$$

$$RMSE = \left(\frac{1}{N}\sum_{t=1}^{N}[GWLp(t) - GWLo(t)]^2\right)^{1/2} \tag{5}$$

$$TS_x = \frac{n_x}{N}100\% \tag{6}$$

where,

N is the total number of data points, GWLp(t) and GWLo(t) are the predicted and observed output at time t, $\overline{GWLp}$ and $\overline{GWLo}$ are the average predicted and observed output, nx is the number of data points whose AARE value is less than x per cent. Threshold statistics of AARE are computed for 5, 10, 25 per cent. The lower AARE values, and the higher TSx and R values close to 1.0, would indicate a sufficient condition for a good model. The AARE, NRMSE and TS statistical measures give the effectiveness of a model in terms of its ability to accurately predict data from a calibrated model. The other statistic, such as R, quantifies the efficiency of the model in capturing the complex, dynamic and non-linear nature of the physical process being modeled.

Results

Corresponding to the model structure of 10-3-1, the results of the various statistical methods are obtained and presented in Table 8.1. From Table 8.1, it can be seen that the results are similar for E, R and AARE values except NRMSE and TS values. The comparison between the predicted and the observed GWL values for the training and testing phases showed an excellent agreement, the R values being 0.999 and 0.998, respectively. The time series plot of the observed GWL and calculated GWL for the ANN model (10-3-1 structure) for the training and testing cases are shown in Figures 8.3 and 8.4, respectively.

Table 8.1: Results of the Various Statistical Measures

Model	NRMSE	E	R	AARE	TS 5	TS 10	TS 25
During training/Calibration							
10-3-1	0.0305	0.9981	0.9990	3.06	79.25	93.78	100.00
During testing/validation							
10-3-1	0.0505	0.9976	0.9988	3.58	81.76	92.35	98.82

From the Figures 8.3 and 8.4, it is clearly evident that both the observed and computed profiles match closely. This indicates that the derived model structure is a realistic depiction of the lumped processes and can successfully be used for prediction of the GWL of the Sagar city.

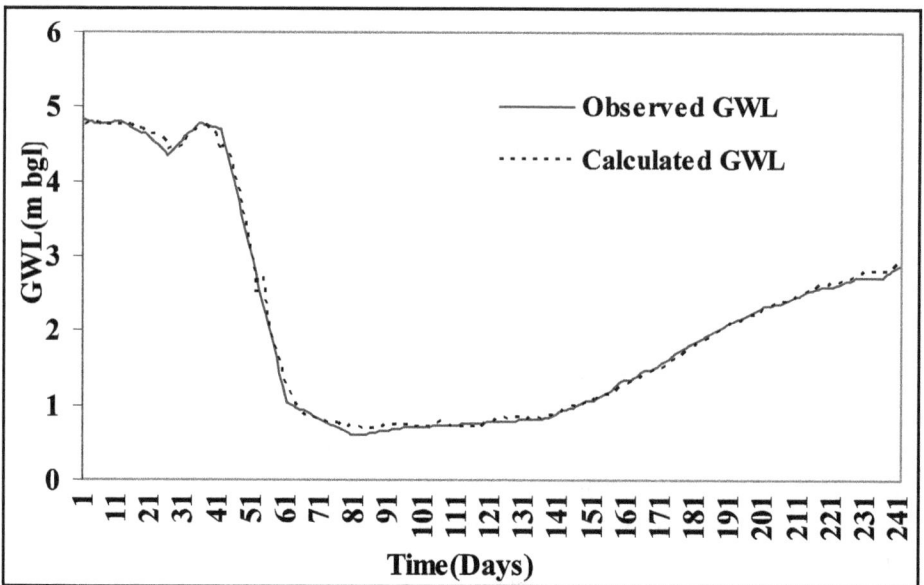

Figure 8.3: Comparison of the Observed and Calculated GWL for the Training Period

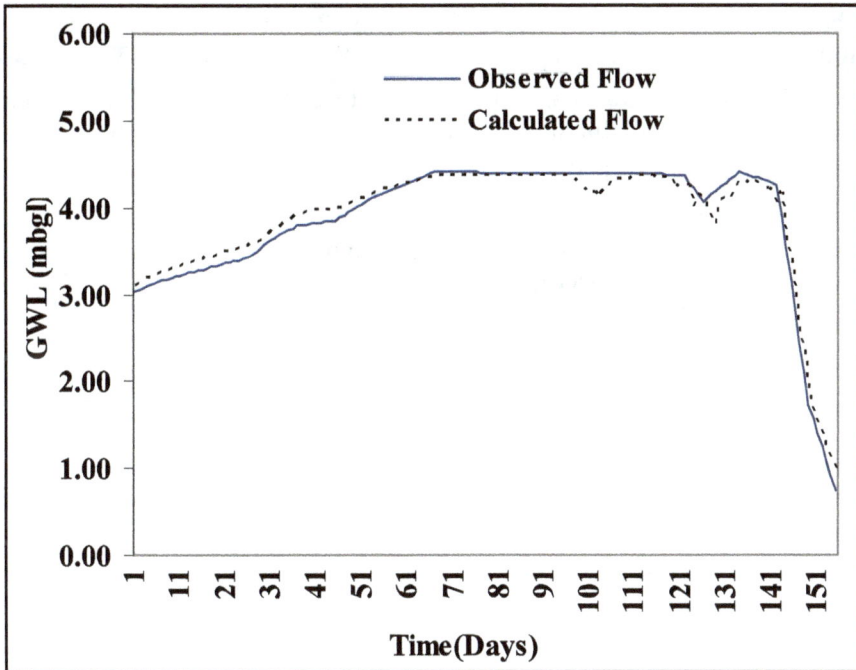

Figure 8.4: Comparison of the Observed and Calculated GWL for the Testing Period

Conclusion

A three layer ANN model has been developed based on the training and testing of the time-series data of rainfall and GWL for prediction of the GWL in the Sagar city. The model uses the rainfall and the preceding GWLas the input data to predict the proceeding GWL. The model structure has been derived using standard statistical measures, namely; Average Absolute Relative Error (AARE), Coefficient of Correlation (R), Nash-Sutcliff Efficiency (E), Normalized Root Mean Square Error (NRMSE), and Threshold Statistics (TS) as the guiding parameters. The developed ANN model can successfully be used for prediction of GWL in the Sagar city.

It is also noted that alike other hydrological problems, the ANN model can be used for prediction of GWL levels, and can be extended to resolve other complex hydrological problems adopting multi-layered ANN structure.

References

ASCE Task Committee on Application of Artificial Neural Networks in Hydrology, 2000a. Artificial neural networks in hydrology I: preliminary concepts. J. Hydrol. Eng. 5, 115–123.

Coppola, E., Szidarovszky, F., Poulton, M., Charles, E., 2003. Artificial neural network approach for predicting transient water levels in a multilayered groundwater system under variable state, pumping, and climate conditions. J. Hydrol. Eng. 8, 348–360.

Coulibaly, P., Anctil, F., Aravena, R., Bobe´e, B., 2001a. Artificial neural network modeling of water table depth fluctuations. Water Resour. Res. 37 (4), 885–896.

Coulibaly, P., Anctil, F., Aravena, R., Bobee, B., 2001. Artificial neural network modeling of water table depth fluctuations. Water Resour. Res. 37, 885–896.

Daliakopoulos, I. N., Coulibaly, P., Tsanis, I.K., 2005. Groundwater level forecasting using artificial neural networks. Journal of Hydrology 309, 229–240.

Jain, A. and Kumar, S., 2009. Dissection of trained neural network hydrologic model architectures for knowledge extraction, Wat. Resour. Res, 45, W07420, doi: 10.1029/2008WR007194.

Khalil, A., Almasri, M.N., McKee, M., Kaluarachchi, J.J., 2005. Applicability of statistical learning algorithms in groundwater quality modeling. Water Resour. Res. 41, W05010. doi: 10.1029/2004WR003608.

Krishna, B., Satyaji Rao, Y.R., Vijaya, T., 2008. Modelling groundwater levels in an urban coastal aquifer using artificial neural networks. Hydrol. Process. 22, 1180–1188.

Maier, H.R., Dandy, G.C., 2000. Neural networks for the prediction and forecasting of water resources variables: a review of modeling issues and applications. Environ. Modell. Software 15, 101–124.

Nayak, P.C., Satyaji Rao, Y.R., Sudheer, K.P., 2006. Groundwater level forecasting in a shallow aquifer using artificial neural network approach. Water Resour. Manage. 20, 77–90.

Yoon, H., Hyun, Y., Lee, K.K., 2007. Forecasting solute breakthrough curves through the unsaturated zone using artificial neural networks. J. Hydrol. 335, 68–77.

Yoon, H., Jun, S.C., Hyun, Y., Bae, G., Lee, K., 2011. A comparative study of artificial neural networks and support vector machines for predicting groundwater levels in a coastal aquifer. Journal of Hydrology 396, 128–138.

Zealand, C.M., Burn, D.H., Simonovic, S.P., 1999. Short-term streamflow forecasting using artificial neural networks. J. Hydrol. 214, 32–48.

2013, Environmental Technology

Pages 79–91

Editors: **D.R. Khanna, A.K. Chopra, Gagan Matta, R. Bhutiani & Vikas Singh**
Published by: **DAYA PUBLISHING HOUSE, NEW DELHI**

Chapter 9

Large Scale Cultivation of Brackish Water Isolates *Scenedesmus* sp. in Raceway Pond for Biodiesel Production

Lala Behari Sukla[1], Manoranjan Nayak[1], Jayashree Jena[1],
Himansu Sekhar Panda[1], Nilotpala Pradhan[1],
Prasanna Ku. Panda[1], Santosh Ku. Mishra[1],
Biswaranjan Das[1], Chandragiri Sarika[2], B.V.S K. Rao[2],
R.B.N. Prasad[2] and B.K. Mishra[1]

[1]*CSIR-Institute of Minerals and Materials Technology,*
Bhubaneswar – 751 013, Odisha
[2]*CSIR-Indian Institute of Chemical Technology,*
Hyderabad – 500 607, A.P.

ABSTRACT

High energy prices, rising energy imports and greater recognition of environmental consequences of fossil fuels that becomes unsustainable have driven a spurring demand to look for sustainable, greener fuels that are economically competitive with environmental benefits. As an emphasis switched in production of natural oil for biodiesel, microalgae becomes most potential candidate for the research due to high oil content. Several researchers have advocated for large-scale microalgal cultivation as alternate raw material source for biodiesel production. In order to fully exploit this potential in a cost effective manner, the major challenges to be addressed are to increase the growth rate and improve the lipid content of microalgal strains in large scale operation.

The study of a potential brackish water microalga *Scenedesmus* sp. has been carried out in outdoor raceway pond for biodiesel production. The designed raceway pond is a closed circular loop made up of concrete and consists of agitation system (paddle wheels), CO_2 sparging system (CO_2 diffuser) and flow mixing system (Baffles). Cultivation was done in batch mode for 18 days. Every six day interval the microalgal culture was analyzed for growth, total lipid and fatty acid composition. Effect of agitation time on biomass growth was checked.The result shows significant increase in lipid accumulation from logarithmic phase to stationary phase *i.e.* from 7 per cent to 23 per cent and the biomass yield was found to be 0.68 g/L on 18th day of cultivation time. *Scenedesmus* sp. grows fast, contains about 23 per cent lipid with high percentage of unsaturated fatty acids and yield good biomass under optimal conditions making it favorable for biodiesel production. Further it was found that agitation in the raceway pond has significant effect on biomass growth. Increase in agitation efficiency, increases the biomass yield.

Keywords: Raceway pond, Microalgae, Total lipid, Fatty acid, Biomass.

Introduction

Continued reliance on fossil fuel energy resources is unsustainable, due to depleting world reserves and their use leads to the green house gas emissions. As sustainability is a key principle in natural resource management, there are vigorous research initiatives aimed at developing alternative renewable and potentially carbon neutral solid, liquid and gaseous biofuels as alternative energy resources (Brennan and Owend, 2010). However, alternate energy resources like first generation biofuels derived from terrestrial crops such as sugarcane, sugar beet, maize and rapeseed (FAO, 2008), place an enormous strain on world food markets, contribute to water shortages and precipitate the destruction of the world's forests. Second generation biofuels derived from lignocellulosic agriculture and forest residues and from non-food crop feedstocks address some of the above problems (Brennan and Owend, 2010). Therefore, to overcome the major drawbacks associated with first and second generation biofuels, third generation biofuels specifically derived from microalgae are considered to be a technically viable alternative energy resource based on current technology and knowledge.

Microalgae are a diverse group of prokaryotic and eukaryotic photosynthetic microorganisms that grow rapidly due to their simple structure with simple growing requirements (light, sugars, CO_2, N, P, and K). They can potentially be employed for production of lipids, proteins and carbohydrates in large amounts over short periods of time. The oil (lipid) extracted from this organism can be processed for the production of biofuels in an economically effective and environmentally sustainable manner. Microalgae have been investigated for the production of a number of different biofuels including biodiesel, bio-oil, bio-syngas, and bio-hydrogen (Li *et al.*, 2008). The production of these biofuels can be coupled with flue gas CO_2 mitigation, wastewater treatment, and the production of high-value chemicals (Li *et al.*, 2008 and Mata *et al.*, 2010). They are superior to traditional oleaginous crops due to higher photosynthetic efficiency, faster growth rate, higher biomass productivities, highest CO_2 fixation and O_2 production rate. Also it can grown in variable climates, non arable land including marginal areas unsuitable for agricultural purpose, no seasonal production, thrive in non portable water, use less water, do not compete with food crop culture (Chisti, 2007).

Developments in microalgal cultivation and downstream processing (*e.g.*, harvesting, drying, and thermochemical processing) are expected to further enhance the cost effectiveness of the biofuel from microalgae strategy. Since 1940's algae culture knowledge is well developed because of commercial applications for food and nutraceuticals, therefore the large scale algae growing parameters have been well researched (Spolaore *et al.*, 2006). Algae grown under controlled conditions can produce >20 times more oil per hectare than terrestrial oilseed crops such as soy and canola (Sheehan *et al.*, 1998; Chisti, 2007; Benemann, 2008b). Under natural growth conditions phototrophic algae absorb sunlight, and assimilate carbon dioxide from the air and nutrients from the aquatic habitats. The use of natural conditions for commercial algae production has the advantage of using sunlight as a free natural resource (Janssen *et al.*, 2003). The only practicable methods of large-scale production of microalgae are raceway ponds (Chisti, 2007) and tubular photobioreactors (Molina Grima *et al.*, 1999; and Sanchez Miron *et al.*, 1999). In respect of biomass productivity, open pond systems are less efficient when compared with closed photobioreactors (Chisti, 2007). This is due to factors including evaporation losses, temperature fluctuation in the growth media, CO_2 deficiencies, inefficient mixing, and light limitation. Compared to closed photobioreactors open pond is the cheaper method of large-scale algal biomass production (Brennan and Owende, 2010) due to lower capital and operational expenditure, utilization of wasteland with marginal or no agriculture potential (Chisti, 2008), lower energy input requirement (Tan, 2009), and ease of maintenance (Ugwu, 2008).

Orissa state is located on the east coast region of India and has a large water bodies stretches along the coast of the Bay of Bengal. Orissa has a large brackish aquatic systems constitute a mixture of fresh water and seawater. A number of microalgal strains prefer brackish conditions because of the nutritional composition of the aquatic system and the warmer temperatures (Mutanda *et al.*, 2011 and Woelfel *et al.*, 2007). If this long coast could be used for microalgal biodiesel production using the locally isolated strains then it may prove to be advantageous as they are already adapted and dominant strains.

The present work deals with the mass cultivation of potential brackish water microalgal strain *Scenedesmus* sp. for oil production. Cultivation of microalgae for commercial biodiesel production in raceway pond was carried out after conducting analogous experiments in laboratory scale to investigate the feasibility of *Scenedesmus* sp. for lipid production. The designed system was operated in batch culture, microalgae oil was extracted and its fatty acid composition was analyzed for determining their suitability for biodiesel production.

Materials and Methods

Design and Construction of Raceway Pond

The raceway ponds are extensive algae farming systems that require direct sun exposed water tracts. The raceway pond was made of concrete cement with closed loop shape at both the ends where the algae, water and nutrients circulate. The details of pond size are shown in Figures 9.1(A and B). The pond consists of illuminated surface area of 100 m², depth of 0.3 m and capacity of 40,000 L(working volume of 30,000 L). The source of light was the direct sun light having maximum 180µmol m⁻²s⁻¹ light intensity at noon time. The stirring was produced by paddle wheels (2 no.), having five set blades each; both wheels mounted on a bearing assembly with detachable belt pully drive mechanism, driven by central shaft (Figure 9.2). The paddles are made up of mild steel with epoxy coating and were operated at 13-15 rpm and the flow rate of culture media in the pond was 0.285m/s. The two paddle wheels rotate in opposite direction in order to maintain the flow of culture media in one direction. Baffles were used for proper mixing. Constant stirring was done to avoid concentration gradients, to maintain adequate mixing, agitation and recirculation of the culture. Stirring system provides light homogenously to microalgal cells due to movement across water layers. Carbon dioxide was supplied for approximately 45mins with a flow rate of 1L/min to the raceway pond by means of

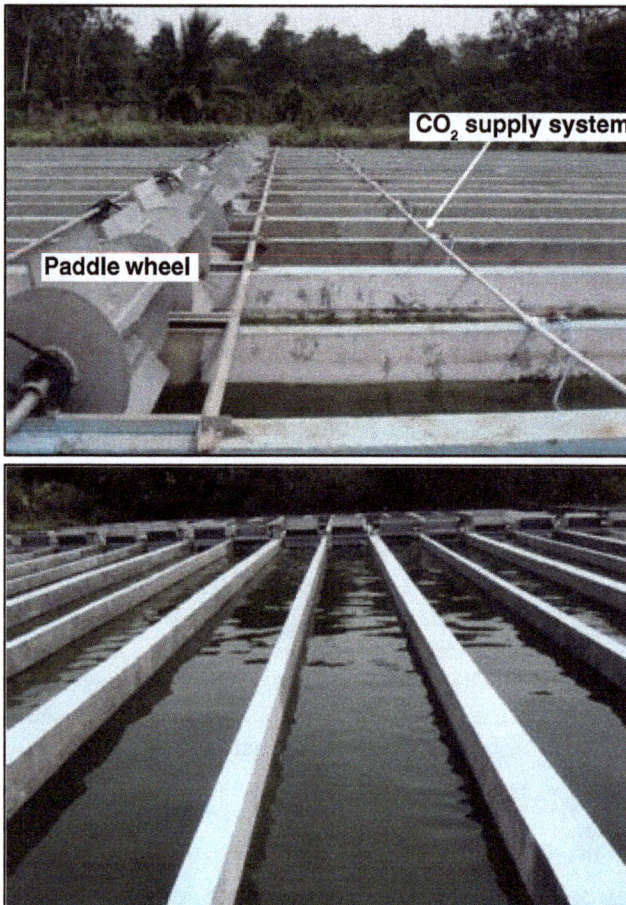

Figure 9.1: Eight Raceway Ponds with Growing Microalgal Cultures

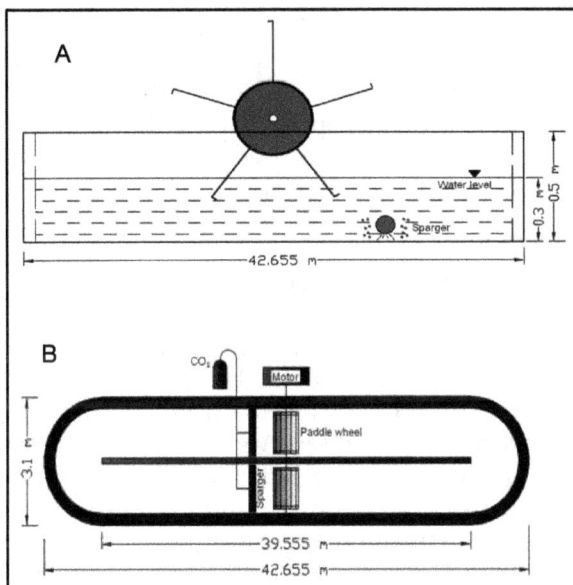

**Figure 9.2: Schematic Diagram of Raceway Pond
(A) Side View (B) Top View**

tubular pipe diffusers (Size-63mm×1000mm) having membrane made up of EPDM (Ethylene Propylene Diene Monomer) through PVC pipes from the CO_2 cylinder with a pressure regulator, until the culture media pH decreases to pHd"6.5. The raceway pond contains an inlet for water intake and an outlet for water drainage during harvesting operation at defined pH levels.

Organism and Growth Conditions

Scenedesmus sp. IMMTCC-13 from the culture collection center of IMMT was used for this experiment based on its oil content and oil type which was amenable to biodiesel production. The strain was isolated from Chilika Lake (19°06'12.61"N; 84°46'49.63"E; Altitude-15ft), Odisha, India. The samples were cultured in modified bold basal medium. The individual colonies were microscopically observed for their colonial and morphological features, incubated at 25 ± 1°C under 5000 Lux irradiance with 16:8 h light dark cycle.

Acclimatization of *Scenedesmus* sp. to Outdoor Conditions

Scenedesmus sp. established in agar plates were inoculated into 150-ml Erlenmeyer flasks and incubated for two weeks at 25 ± 1°C under 25μmol $m^{-2} s^{-1}$ light intensity with 16:8 h light dark photoperiod. The cultures were sub-cultured at an intervals of one weeks and such sub-culturing were done for five times to seven times prior to its scale up in 500, 1000 and 2000-ml Erlenmeyer flasks in appropriate low cost growth medium consisting of low cost inorganic fertilizer (modified from Palanisamy *et al.*, 1991) composed of (mgL^{-1}): Ammonium sulphate (100), Urea (15), N:P:K 10/26/26 fertilizer (15) at pH6.8. The seed cultures were exposed to open air environments in plastic jar of 10 - 20 L capacity for 2-3 weeks and then cultures were inoculated into plastic tanks of size with a culture holding capacities of 300L-1000L. The algae in the tanks were allowed to grow for more than 2 cycles for adaptation to open air condition. Cultures were observed under microscope periodically for any possible contaminants. Algal cultures in these tanks (3000L) were used as starter cultures for raceway pond (30,000L). A flow diagram of the activities is shown in Figure 9.4.

Cultivation in Raceway Pond

Modified Palanisamy media was prepared by using potable water supplied by IMMT water facility, and the pH was tentatively adjusted to 6.8-7.2. The culture was inoculated at 10 per cent (v/v) to the raceway pond of 40,000 L capacity and the volume was made up to 30,000L. The paddle wheel was set to 13-15 rpm and the flow rate of culture media in the pond was 0.285m/s from 9.30am to 5.30 pm daily. Dust and particulate matters, some of the insects were removed with the help of Strainer from raceway pond. For any possible contamination, cells were observed periodically under microscope. Light irradiance, pH, temprature and biomass yields were recorded on daily basis. Cultures were monitored for their growth for three weeks in outdoor conditions and the biomass was harvested at each sixth day by flocculation and sun dried biomass was analyzed for its lipid yield and fatty acid profile.

Dry Weight Estimation

Growth (Cell density) of cultures was monitored by measuring absorbance at 750nm (OD_{750}) using Cecil UV-vis spectrophotometer. Microalgal cells were collected by centrifugation at 5000 rpm for 15 min at 4 °C and the algal pellet were washed twice with distilled water, and dried at 65 °C for 24 h for dry weight measurement (Takagi *et al.*, 2006). In order to determine the biomass concentration as dry weight per liter a regression equation of the cell density (OD_{750}) and dry weight (Chiu *et al.*, 2008) was used. The correlations established in the Regression Equation of the Cell Density was Biomass = O.D*4012 (R^2=0.994). The microalgal cell counting was performed with an improved Neubauer hemocytometer (MARIENFELD, Germany) and a Nikon Epi-fluorescence microscope (Nikon Corporation, Tokyo, Japan).

Cell Harvest and Oil Extraction

The microalgae were harvested by centrifugation at 8,000 rpm, 4°C for 15 min on every sixth day interval till the stationary phase. The pellet was washed four times with distilled water. The total lipid of wet microalgal biomass was extracted with modified Bligh and Dyer (1959) using a mixture of chloroform and methanol (2:1v/v).To the biomass solvent mixture was added. The mixture was agitated in a vortex for 3 min and kept at room temperature for 20 min. After that distilled water was added and the mixture was again mixed in a vortex for 2 min. The layers were separated by centrifugation for 10 min. at 4,000 rpm. The lower layer was collected into a previously weighed clean vial (W_1). Another portion of chloroform was added to the upper aquous layer. The chloroform phase was collected and combined with the previous solvent phase. The bulk of the solvent was removed by rotary evaporation and residual solvent was then completely removed from the total lipid in a oven at 60°C for 30 min. The weight of the vial was again recorded (W_2).The oil content was calculated using following equation:

$$\text{Oil content (per cent dcw)} = (W_{2(g)} - W_{1(g)}/DCW_{(g)}) \times 100$$

where,

W_1 and W_2 are the weight of flask before and after extracting the oil. DCW is the dry cell weight.

Measurement of Specific Growth Rate

The specific growth rate of microalgae was calculated using the following equation:

$$\text{Specific growth rate ()} = \ln X_2 - \ln X_1 / \Delta t \text{ (day)}$$

where,

X_2 is final DCW, X_1 is initial DCW.

Analysis of Fatty Acid Composition

Procedure for the Preparation of Fatty Acid Methyl Esters: Dried biomass (0.25 g) was taken in a 50 ml round bottom flask and after addition of 15 ml of 2 per cent sulphuric acid in methanol solution refluxed for 4.0 hrs. At the end of the reaction the contents were diluted with water and the organic phase was extracted with ethyl acetate (3 × 25 ml). The ethyl acetate phase was thoroughly washed with water and dried over anhydrous sodium sulphate. Ethyl acetate was evaporated on a rotary evaporator to recover fatty acid methyl esters (0.12 g).

Gas Chromatographic Analysis of Fatty Acid Methyl Esters: The fatty acid composition analysis of algal oil methyl esters was carried out on a gas chromatograph (Agilent 6890) equipped with a flame ionization detector (FID) on a split injector system. A fused silica capillary column (DB-225, 0.25m, 30 m X 0.32 mm id) was used for the analysis. Oven temperature was programmed from 170 °C to 225°C at 1°C/min. The injector and detector temperatures were held at 250°C and 270°C respectively. Nitrogen was used as carrier gas at a flow rate of 1 ml/min. The area percentages were recorded with Agilent chemstation data processing system.

Results and discussion

Growth and pH Change of Microalgae

A pre-cultured *Scenedesmus* sp. (IMMTCC-13) was inoculated in raceway pond to reach an initial biomass concentration of 0.03 g L^{-1} as a batch culture. Growth in raceway reached stationary phase in 18 days of cultivation time (Figure 9.5A). The specific growth rate of raceway pond was high (0.18 day^{-1}) due to proper aeration and agitation.

Change in pH was observed during the growth of microalgae *Scenedesmus* sp. in raceway pond (Figure 9.3). The culture pH increases from 7.12 to 10.43 with increased in cultivation time from 1st day to 18th day. Despite of increase in culture pH,

Scenedesmus sp. was found to grow well without inhibition. In the present study increase of culture pH was controlled to some extent with the continuous aeration and agitation. Agitation helps in better mass transfer of gas facilitates better absorption of CO_2 in the system.

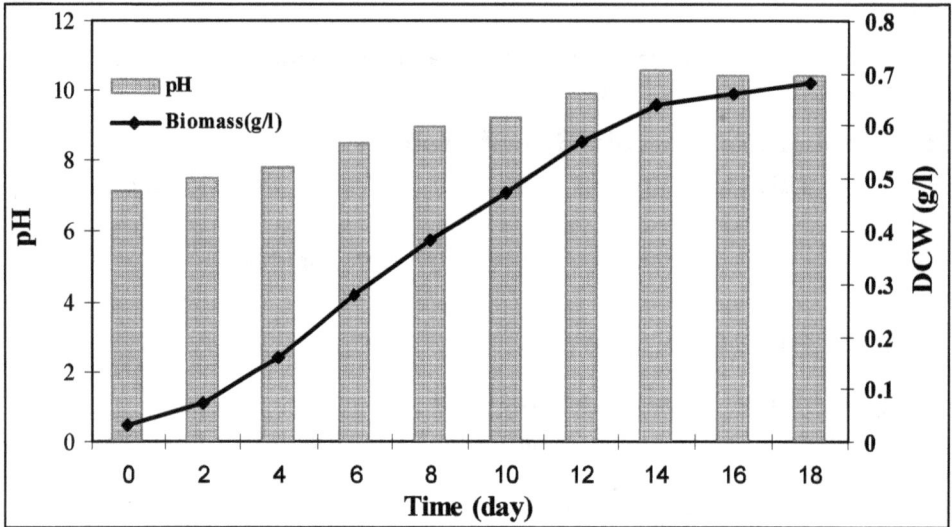

Figure 9.3: Growth and pH Change in Raceway Pond

Biomass and Lipid Productivity

From 3000 L of initial inoculums (having biomass of 0.6 g/L), 30,000 L of final microalgal culture was obtained after 18 days of cultivation with total biomass of 0.68g/L. Comparing the three harvests in different conditions, it was observed that the culture biomass productivity was improved with supply of CO_2 and continuous stirring. The culture was harvested and the microalgal biomass was recovered by two sequential unit operations: flocculation with alum $(Al_2(SO_4)_3)$, and subsequent drying of wet biomass under direct sun light. In between the microalgal cells were collected from logarithmic, early stationary phase and stationary phase of growth to measure lipid content. The result shown in Figure 9.4 indicates that the lipid accumulation in microalgal cells was associated with growth phases. Lipid accumulation increases as the culture approaches into stationary phase. Finally, microalgae oil was extracted and oil content in *Scenedesmus* sp. IMMTCC-13 was 23 per cent.

Fatty Acid Composition of Total Lipids

The fatty acid compositions of *Scenedesmus* sp. IMMTCC-13 at different culture conditions are shown in Table 9.1. Fatty acid composition of microalgae differed in three different culture stages. Microalgae cultured in 18th day *i.e.*, on stationary phase revealed high percentage of unsaturated fatty acid in comparison to 12th day and 6th day of cultivation time. The microalga contained high amount of saturated fatty acid

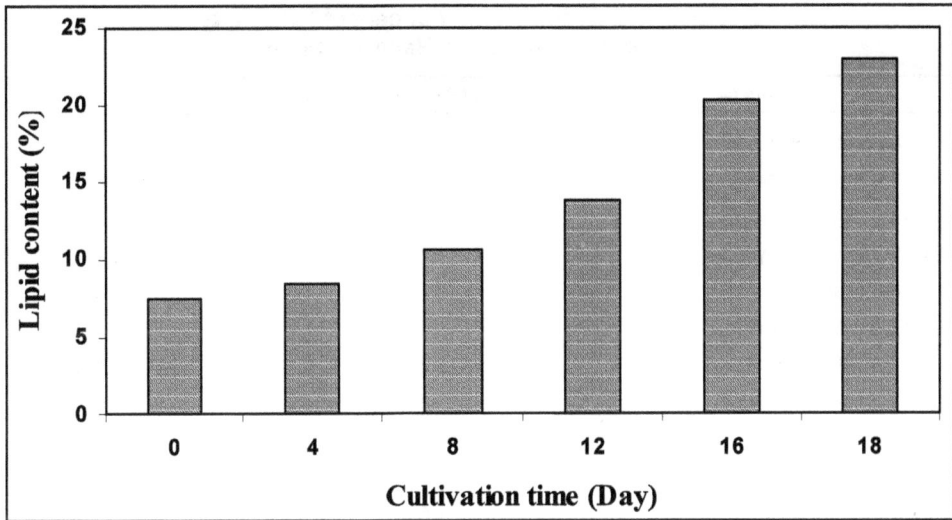

Figure 9.4: Lipid Content of *Scenedesmus* sp. in Raceway Pond

i.e. palmitic acid (C16:0) as dominant fatty acid (32.7 per cent) during first sixth day of cultivation period, followed by PUFA (34.8 per cent) and MUFA (18.3 per cent). As the cultivation time increased total PUFA content increased from 34.8 per cent to 47.7 per cent in 12th day of cultivation. At the statioinary phase of growth period *i.e.* on 18th day the PUFA content again desreases. SFA content was also decreased from 46.9 to 21.9 per cent during the cultivation period. During the experiment great variation shown in monounsaturated fatty content. The MUFA content increases from 18.3 per cent - 35.9 per cent.Among unsaturated fatty acids, special attention should be given to the linolenic (C18:3) and polyunsaturated fatty acids having four or more than four double bond content (Gouveia and Oliveira 2009). Among the unsaturated fatty acid linolenic acid was the major fatty acid, while long chain PUFA are present in small quantities.

Optimisation of Mixing Energy Requirement

With respect to biomass productivity, open pond systems are less efficient when compared with closed photobioreactors. This can be attributed to several determining factors, including, evaporation losses, temperature fluctuation in the growth media, CO_2 deficiencies, inefficient mixing, and light limitation. Generally, microalgae use CO_2 as the carbon source and light as the energy source for metabolic activity. They can also adapt to different energy conditions in different environments, So the present work was conducted to check the effect of algal culture mixing with the help of paddle wheel on biomass growth and lipid content.

As shown in Figure 9.5, as the agitation time increases, the microlalgal biomass yield also increases. Biomass yield was maximum at 24 hrs of agitation with a biomass yield of 0.83g/l. Again the result shows that there is a no significant differencein

Table 5.1: Fatty Acid Composition (Per cent of Total Fatty Acids) of *Scenedesmus* sp. in Raceway Pond

Fatty Acids	RP-6 day	RP-12 day	RP-18 day
Saturated Fatty Acids (SFA)			
C12:0 (Lauric acid)	0.9	0.7	0.3
C14:0 (Myristic acid)	1.3	0.8	1.1
C15:0 (Pentadecylic acid)	1.1	1.4	-
C16:0 (Palmitic acid)	32.7	25	15.8
C17:0 (Margaric acid)	1.4	1.5	-
C18:0 (Stearic acid)	3.4	1.4	2.3
C20:0 (Arachidic acid)	1.5	1.1	-
C21:0 (Heneicosylic acid)	-	-	1.6
C22:0 (Behenic acid)	3	0.7	0.4
C24:0 (Lignoseric acid)	1.6	0.6	0.4
Total SFAs	46.9	33.2	21.9
Monounsaturated Fatty acids (MUFA)			
C16:1 (Palmitoleic acid)	5.2	7.5	12.1
C18:1 (Oleic acid)	8.4	10.4	18
C20:1 (Eicosenoic acid)	3.4	0.5	2.4
C22:1 (Docosenoic acid)	1.3	0.7	3.4
Total MUFAs	18.3	19.1	35.9
Polyunsaturated Fatty acids (PUFA)			
C16:2 (Hexadecadieneic acid)	1.9	0.6	7.2
C16:3 (Hexadecatrienoic acid)	1.7	7.4	8.7
C18:2 (Linoleic acid)	10.0	21	10
C18:3 (Linolenic acid)	16.5	18	12.1
C20:2 (Eicosadienoic acid)	2.4	0.4	0.3
C20:3 (Eicosatrienoic acid)	-	-	0.4
C20:5 (Eicosapentaenoic acid)	-	-	1.0
C22:2 (Docosadienoic acid)	2.3	0.3	0.2
Total PUFAs	34.8	47.7	42.2
Total unsaturated fatty acid	53.1	66.8	78.1
Unidentified	-	-	0.3

biomass yield in 1-4 hr of agitation, where as rapid increase in biomass occurs at 4hrs-16 hrs of agitation.again increase in agitation time shows no significant growth,which may be due the lack of photosynthesis during night time. As during the day time equal distribution of sunlight among each microalgal cell is possible due to the proper agitation,photosynthetic rate becomes higher, with a net resulting in high biomass yield. Natural light fluctuates in either intensity or quality daily and

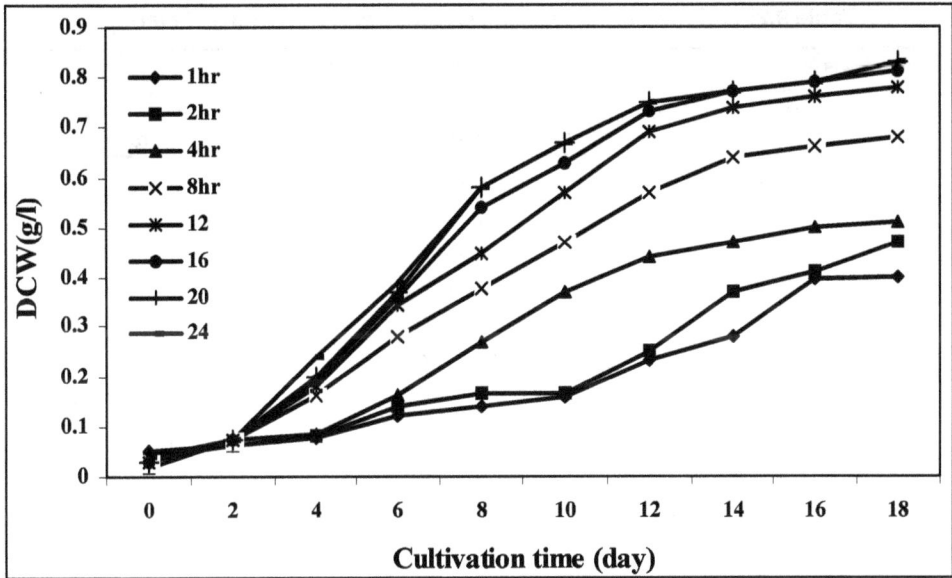

Figure 9.5: Biomass Yield at Different Mixing Duration

seasonally. Light is found to be a major limiting factor of productivity and growth when nutrition and temperature are satisfied. Light limitation due to top layer thickness may also incur reduced biomass productivity.

Again the fatty acid analysis of eight pond shows that, as the duration of agitation increases, The net saturated fatty acid content decreases with increase in unsaturated fatty acid content. Among the SFA, palmitic acid becomes the domiant fatty acid.Among the unsaturated fatty acid, MUFA content increases from 18.6 per cent - 44.2 per cent with increase in agitation time. Table 9.2 shows most important fatty acid, linolenic acid was present within the speciation range.

Conclusion

The present study deals with the culture of brackish water microalgal strain *Scenedesmus* sp.(IMMTCC-13) isolated from Orissa coast, eastern region of India in batch mode for mass cultivation, that are previously not been studied with respect to eastern region of India. The biomass yield and lipid yield during 18th day of cultivation time was found to be satisfactory. Again Fatty acids composition of the strain at different time interval at stationary phase of growth show high MUFA and PUFA content. The oil mainly comprised of palmitic acid, oleic acid as major fatty acid components, while linolenic acid was in considerable range,which show its suitability for biodiesel production in large-scale. Further process control and optimization studies are worth studying.

Further with the increase in agitation time increases the biomass yield with net increase in biomass and lipid productivity. Hence the result concludes that Odisha isolate microalga *Scenedesmus* sp. proves as one of the potential strain of eastern for

Table 9.2: Variation in Fatty Acids Content w.r.t Agitation Efficiency

Fatty Acids	Algae Grown in Different Ponds							
	Pond-11hr	Pond-22hrs	Pond-34hrs	Pond-48hrs	Pond-512hrs	Pond-616hrs	Pond-720hrs	Pond-824hrs
12:0	0.8	3.9	2.2	1.0	2.1	0.7	1.4	1.0
14:0	0.9	4.4	3.2	6.3	5.2	2.9	5.3	4.3
15:0	0.6	2.2	2.1	1.6	1.4	1.9	2.1	1.3
16:0	51.5	30.2	29.0	25.5	24.8	26.9	23.8	19.3
17:0	1.6	2.2	2.3	2.4	2.4	1.8	2.2	1.3
18:0	4.4	5.5	7.7	7.1	8.1	4.6	5.7	4.8
20:0	0.5	0.9	2.0	1.0	0.8	1.3	0.6	0.7
22:0	1.6	1.5	2.4	1.6	0.7	1.7	0.9	1.0
24:0	4.7	3.6	2.5	1.9	1.5	3.0	1.9	3.0
SFA	**66.6**	**54.4**	**53.4**	**48.4**	**47**	**44.8**	**43.9**	**36.7**
16:1	3.6	4.7	6.3	10.6	10.5	5.0	10.6	5.8
18:1	15.0	16.7	18.0	16.3	18.8	20.6	19.2	36.2
14:1	–	2.6	0.6	1.1	0.7	1.6	0.9	1.5
20:1	-	-	0.3	-	0.1	0.6	-	0.2
22:1	-	0.7	0.8	0.7	0.3	1.3	0.7	0.5
MUFA	**18.6**	**24.7**	**26**	**28.7**	**30.4**	**29.1**	**31.4**	**44.2**
16:2	2.1	3.3	2.5	5.0	0.9	4.6	3.8	2.4
18:2	5.8	8.0	9.1	7.8	11.3	8.9	9.5	8.0
20:2	2.0	3.8	0.6	1.5	0.4	1.6	1.6	1.2
22:2	-	1.8	1.2	1.1	0.4	1.7	0.8	0.9
18:3	4.9	4.0	6.4	6.2	8.4	8.4	8.2	5.9
20:5	—	–	0.8	1.3	1.2	0.9	0.8	0.7
PUFA	**14.8**	**20.9**	**20.6**	**22.9**	**22.6**	**26.1**	**24.7**	**19.1**
Total Unsaturated FA	**33.4**	**45.6**	**46.6**	**51.6**	**53**	**55.2**	**56.1**	**63.3**

commercial biodiesel production in pilot scale, due to high lipid content and high biomass yield.

Acknowledgements

Authors are thankful to the Department of Biotechnology (DBT) New Delhi, Govt. of India for their financial support. Authors are also very much thankful to Dr. R.R. Nayak, Scientist, CSIR-IICT for his valuable suggestion and comments.

References

Benemann, J.R., 2008b. Oppotunities and Challenges in Algae Biofuels Production Algae World 2008, Singapore, November 17–18, p. 15.

Brennan, L., Owende, P., 2010. Biofuels from microalgae - a review of technologies for production, processing, and extractions of biofuels and co-products. Renew. Sustain. Ener. Rev. 14, 557-577

Chisti, Y., 2007. Biodiesel from microalgae. Biotechnol. Adv. 25, 294-306.

Chisti, Y., 2008. Biodiesel from microalgae beats bioethanol. Trends Biotechnol.26, 126–131.

FAO, 2008. Biofuels -Prospects, Risks and Oppertunities. The state of food and agriculture. New York: Food and Agriculture Organization

Janssen, M., Tramper, J., Mur, L., Wijffels, R., 2003. Enclosed outdoor photobioreactors: light regime, photosynthetic efficiency, scale-up, and future prospects. Biotechnol. Bioeng. 81, 193–210.

Li Y, Horseman M, Wang B and Wu N, Lan CQ (2008) Effects of nitrogen sources on cell growth and lipid accumulation of green alga *Neochloris oleoabundans*. Appl Microbiol Biotechnol (2008).

Mata TM, Martins AA, Caetano SN (2010) Microalgae for biodiesel production and other applications: A review. Renew and sustainable Energy Reviews 14: 217-232.

Molina Grima, E., Acién Fernández, F.G., García Camacho, F., Chisti, Y., 1999. Photobioreactors: light regime, mass transfer and scaleup. J. Biotechnol. 70, 231–247.

Mutanda T., Ramesh D., Kartikeyan S., Kumari S., Anandraj A. and Bux F. (2011) Bioprospecting for hyper lipid producing microalgal strains for sustainable biofuel production. Biores Technol 102: 57-70.

Sanchez Miron, A., Contreras Gomez, A., Garcia Camacho, F., Molina Grima, E., Chisti, Y., 1999. Comparative evaluation of compact photobioreactors for large-scale monoculture of microalgae. J. Biotechnol. 70, 249-270

Sheehan, J., Dunahay, T., Benemann, J., Roessler, P., 1998. A Look Back at the U.S. Department of Energy's Aquatic Species Program – Biodiesel from Algae. National Renewable Energy Laboratory, NREL/TP-580-24190

Spolaore, P., Joannis-Cassan C., Duran, E., Isambert, A., 2006. Commercial applications of microalgae. J. Biosci. Bioeng. 101, 87-96.

Ugwu, C.U., Aoyagi, H., Uchiyama, H., 2008. Photobioreactors for mass cultivation of algae. Bioresour. Technol. 99, 4021-4028.

Woelfel J., Schumann R., Adler S., Hubener T. and Karsten U. (2007) Diatoms inhabiting a wind flat of the Baltic sea: species diversity and seasonal variations. Estuar Coast shelf Sci 75: 296-307.

2013, Environmental Technology

Pages 93–102

Editors: **D.R. Khanna, A.K. Chopra, Gagan Matta, R. Bhutiani & Vikas Singh**

Published by: **DAYA PUBLISHING HOUSE, NEW DELHI**

Chapter 10

Glimpse of Challenges and Opportunities of Energy Conservation in a Dairy Plant

J.B. Upadhyay

Department of Dairy Engineering
Sheth M.C. College of Dairy Science, AAU, Anand

ABSTRACT

About 20 per cent of world's energy is generated from coal and about 60 per cent of world's energy is generated from oil and natural gas. Because of extensive use of fossil fuel, the harmful emissions of GHG (Green House Gases) such as Carbon dioxide increases in the atmosphere which traps the extra heat reflected from the earth resulting in global warming. The energy conservation is closely related to the environmental issues in terms of its polluting effect. Energy requirement in our country increases in a very rapid rate. India's demand for commercial energy in 2020 is expected to increase by 250 per cent . Energy conservation is the quickest, cheapest and most practical method of overcoming energy shortage.

India is the highest milk producer in the world. But only about 20 per cent of milk produced is being handled by organized dairy sector. Dairy processes require mainly electrical and thermal energy. There has been an increasing consciousness regarding relationship between economic development, enhanced use of energy and adverse environmental implications due to emission of GHG. Also in the era of energy crisis, the dairy sector is compelled to optimize energy efficiency and adopt an approach of Total Energy Management (TEM). This is possible through the adoption of newer technology, process re-engineering and with the use of renewable energy sources like solar for process heating, bio-gas from anaerobic treatment of dairy effluent etc. for the reduction of conventional hydrocarbon fuel budget with simultaneous generation of a carbon credit.

Introduction:

The Government of India set up Bureau of Energy Efficiency (BEE) (Website: http://www.bee-india.nic.in) on 1st March 2002 under the provisions of the Energy Conservation Act, 2001. The mission of the Bureau of Energy Efficiency is to assist in developing policies and strategies with a thrust on self-regulation with the primary objective of reducing energy consumption. This will be achieved with active participation of all stakeholders, resulting in accelerated and sustained adoption of energy efficiency in all sectors.

About 20 per cent of world's energy is generated from coal and about 60 per cent of world's energy is generated from oil and natural gas. Because of extensive use of fossil fuel, such as coal, oil and natural gas, as primary source of energy today, the harmful emissions of GHG (Green House Gasses) such as Carbon Dioxide increases the GHG level in the atmosphere which traps the extra heat reflected from the earth resulting rise in temperature. Scientists believes that global warming will cause the average World temperature rise by one Degree Celsius by the year 2020 and four Degree Celsius by the end of 21st century. Thus energy conservation has emerged as one of the major issues in recent years. Energy requirement in our country is increasing in a very rapid rate. India's demand for commercial energy in 2020 is expected to increase by 250 per cent from today's level. Coal accounts for about 50 per cent of primary commercial energy today and is further increase its share. Energy conservation is the quickest, cheapest and most practical method of overcoming energy shortage.

Dairy industry deals with the collection of raw milk and processing it into various traditional and exotic milk products *viz.* market milk, ice cream, cheese, milk powder, khoa, paneer, etc. This processing is divided in various processing operations such as heating, cooling, churning, filtration, clarification, separation, homogenization, freezing, evaporation, drying, etc. Each of these processes requires energy input wherein the form and quantity of energy requirement is dependent on process and on the machine or equipment used. Electrical and thermal energy are mainly required for various dairy processing operations. Though dairy industry is not falling in energy intensive category, sizeable amount of energy is required for processing of milk and manufacturing of various milk products. There has been an increasing consciousness regarding relationship between economic development, enhanced use of energy and adverse environmental implications due to emission of Green House Gases (GHGs). In this regard, many national, international and intergovernmental bodies across the world have been working to formulate program to reduce the energy use in various sectors to combat the phenomena leading to global warming. It has become exceptionally critical to manage the use of energy and to adopt all possible measures to conserve it in order to boost the profitability at industrial level and to reduce emission of GHGs.

Dairy Scenario of India

If, we look at Table below, it gives an idea of average annual milk production growth rate in India. It shows that, till 1974, average growth rate was less than 2 per cent , but then after it was accelerated and to-day, India is growing at a rate of 4 per cent every year for its milk production which has crossed 100 MMTs / annum, leading to be the Top Milk Producing Country in the World. Today, Indian Dairy Industry is flourishing and is competing at International level in World market for its Dairy Products.

Sl.No.	Year	Average Annual Milk Growth Rate (per cent)
1.	1950-51 to 1960-61	1.64
2.	1950-61 to 1973-74	1.15
3.	1973-74 to 1980-81	**4.51**
4.	1980-81 to 1990-91	**5.48**
5.	1990-91 to 2000-01	**4.11**
6.	2000-01 to 2004-05	**4.05**

Dairy industry knocks-out as a major agricultural produce dominating in GDP (28 per cent) by producing around 110 million tons of milk per annum (54 per cent buffalo and 40 per cent cow, Rest 6 per cent). India contributes 14 per cent of world and 35 per cent of total Asian milk and contributes to world milk production rise @ 10 per cent which will increase up to 30-35 per cent by 2020. Indian agriculture and livestock sectors support 17.5 per cent of world population with 2.3 per cent of global land and 4.2 per cent of water. It is reported that 57 per cent of world's buffalo and 16 per cent of cattle population is in India and milk accounts for 68-70 per cent of total contribution of livestock produce. It has been estimated that about 35 per cent of milk is retained by producers and 65 per cent is traded out of which 60-70 per cent is consumed as fresh liquid milk. Organized dairy plants processes 16-17 per cent of the total milk produced in the country in about 1500 dairy plants. The productivity of Indian milch animal is 987 kg/year (world average is 2200 kg/year) while the per capita availability of milk in India is 246g/day (world average is 270g/day). Among the contribution of various states of India in milk production, Uttar Pradesh is at top position contributing 17 per cent of total milk whereas Gujarat stands at 5th position with about 7 per cent contribution. Annual milk production growth rate of India is 4.6 per cent.

Energy Scenario of India

Among the primary energy reserves world over, availability of oil and gas are estimated for just 45 and 65 years respectively, whereas the coal is likely to last about 200 years. Based on the energy consumption pattern between developed and developing countries, it is observed that 80 per cent of the world's population of the developing countries consume only 40 per cent of the world total energy consumption and the rest is consumed by developed nations. The world average energy consumption per person is equivalent to 2.2 ton of coal. American uses 32 times more commercial energy than an Indian.

India is 6th largest energy consumer, accounting for 3.4 per cent of global energy consumption. In March 2009, the installed power generation capacity of India was

148,000 MW with the per capita power consumption at 620 kWh/annum. The Indian Government has set an ambitious target to add approximately 78,000 MW of installed generation capacity by 2012. The total demand for electricity in India is expected to cross 950,000 MW by 2030. Electricity is generated 75 per cent from thermal power plants, 21 per cent from hydroelectric power plants and 4 per cent from nuclear power plants. India has also embarked in the realm of renewable energy resources. As of 2008, India's installed wind power generation capacity stood at 9,655 MW. In July 2009, India unveiled a $19 billion plan to produce 20,000 MW of solar power by 2020. Coal dominates the energy mix in India, contributing to 55 per cent of the total energy production but over the years, there has been a marked increase in the share of natural gas in primary energy production from 10 per cent in 1994 to 13 per cent in 1999.

In India, concerted efforts are underway to accomplish energy security through effective energy management accompanied by improved energy efficiency and use of cleaner forms of energy. This will not only facilitate basic need of providing access to affordable energy but also offers scope for carbon trading at international arena.

Importance of Energy Conservation

India is among the fastest growing dairy nation in the world. The demand of milk is growing steadily and expenditure on food including milk continues to be a major share of their income. The milk production in India has increased five folds in the last fifty years and we continue to be No. 1 milk producing country in the world. Currently, about 20 per cent of milk produced is being handled by organized dairy sector and most of the dairy business in organized sector is in the hands of co-operatives. As per the projections, the production as well as demand of milk by 2021-22 will be about 200 million tones and about 50 per cent of the milk will be handled by the organized dairy sector. Accordingly, there is a lot of scope for the growth of dairy industry in terms of milk processing and product making infrastructures. The processing operation in the Indian dairy sector has become very competitive.

Although, the dairy industry is not identified as energy intensive industry, to be competitive in today's global business environment, effective and efficient use of the energy as well as its conservation is pivotal to the productivity and profitability of any dairy plant. Energy Conservation is closely related to the environmental issues in terms of polluting effects caused by emission of carbon dioxide and other Green House Gases (GHG). Moreover, with the increasing fuel prices and global warming, there is an urgent need to conserve and manage energy consumption across all facets of dairy industry.

In constantly and rapidly changing global dairy business environment, the core issue of Energy Conservation in dairy sector need to be addressed as a team work among Central as well as State Government, State Departments of Animal Husbandry, Universities, Research Institutes, Co-operatives, Private Sector and NGOs with clear roles and responsibilities for goals and targets to continue in the interests of our dairy farmers. Typically, in a dairy plant, 80 per cent cost is of milk and remaining 20 per cent comprised of other variable and fixed cost. The energy cost reflects to about 4 per cent of the expenditure and the Income and Expenditure account is sensitive to energy costs as it directly affects the profitability. Hence, any attempt to efficiently manage the energy costs is an opportunity and has significant influence on the processing cost giving scope for improving the over all viability of the plant. Thus, to face the challenge of energy conservation, the need of the Indian Dairy Industry is to adopt holistic approach and take appropriate initiatives to inculcate innovative culture for improving the energy efficiency. It is reported that about 95-105 litres of milk and 10-12 litres of milk can be processed from a liter of furnace oil and one kWh electric power respectively. Thus, conservation of energy is very important for the following points of view.

1. Vital to sustain our lives
2. Economic development and profitability
3. Reducing emission of GHGs
4. Contribution in energy security
5. Possibility of carbon trading through CDM projects

Impact of Energy Conservation in Dairy Plants

Energy conservation means maximizing the processed product output from unit energy input without compromising with the quality of the finished product. It acts as an effective tool to minimize the cost of production, maximize the profit which is indispensable for the sustainability of any dairy and food processing plant in this competitive era. It is reported that average composite processing cost of milk stands at about Rs. 0.65-0.85 per litre. Estimations say that 10 to 20 per cent saving in the fuel bills can be achieved by adoption of the right energy conservation technology. Early returns of the investment, improved plant efficiency and production capacity are major benefits of energy conservation. There is also possibility of CDM projects based on energy conservation by adopting newer technology, process re-engineering and with the use of renewable energy sources.

Energy management

After Globalization in 1992, pressure had come to Indian Dairy Industry to become more competitive in Global competition with respect to Quality of its produce. Hence, Japanese philosophy called "Total Quality Management", TQM came in to picture. Through effective implementation of TQM, efforts were put to make "Zero" defect product for consumers to compete in market.

However, pressure was there on Industry to minimize/optimize its over heads to become more profitable, "TEM" concept came in to existence, *i.e.*, "Total Energy Management". Through implementation of TQM and TEM, efforts were put by Industry to manufacture their products at optimum cost and with zero defects without compromising with its quality.

Efficiency of any Industry depends on its energy being consumed for day-to-day operations. Same is the case with Dairy Industry too. Whether Thermal or Electrical Energy is used in Dairy Plant, has to be measured and monitored for its optimum usage to minimize the operating cost. Proper Energy Management will help Industry to minimize its operating cost, thereby reduced over heads and will help an Industry to be more competitive in Global competition.

In fact, Energy Management starts from designing the Dairy Plant. Correct and Realistic data for designing will help and support for selection of correct capacity in various utilities like steam, refrigeration, electricity, effluent treatment etc; Hence, due care needs to be taken while providing data for designing a Dairy Plant. Any assumption shall be based on realistic past data or nearby Dairy Plant data and based on same, proper capacity for various utilities shall be decided. Due care shall be taken while selection of equipment as now a days, energy efficient equipments are available in market. Though initially it may cost little more, but on longer run, their payback period becomes short and energy cost is reduced drastically. In addition, awareness among functionaries is also of prime importance to get better results of any Energy Management measures. In other words, savings in one energy unit, helps to reduce ten time the production units of energy. In this regard *Energy management cell* can be formed at dairy level which ensures that the industry is implementing the energy conservation act and recommendations of energy audit. Management can take adequate steps to minimize the wastage of energy. Energy Management is the strategy of adjusting and optimizing energy using systems and procedures so as to reduce energy requirement per unit of output.

Benefits

1. Framing of an Energy Management Policy.
2. Review and monitoring of energy consumption.
3. Reduce the energy consumption/set new targets.

Problems Faced in Energy management

Operational

1. No action from management level
2. Inadequate resources and improper allocation
3. Lack of organization structure-energy manager, energy management cell, etc.

Individual

1. Identification of skill gap and plugging it through training programme
2. Preparation of comprehensive proposal
3. Fear of change
4. Attitude training

Technological

1. Inadequate information base
2. Lack of instrumentation
3. Lack of information of modern equipment
4. Lack of benchmarking information and knowledge sharing

Energy Audit

Energy audit involves a systematic study undertaken on major energy consuming sections and equipments including construction of heat and mass balance with a view to identify the flow of energy, utilization efficiency of energy in each of the steps and pin-point wasteful energy used. In the Energy Conservation Act, the definition of energy audit has been expanded to mean the verification, monitoring and analysis of use of energy, including submission of technical report containing recommendations for improving energy efficiency with cost benefit analysis, and an action plan to reduce energy consumption. Implementation of recommended measures can help consumers to achieve significant reduction in their energy consumption levels.

Energy Conservation Opportunities in Dairy Plant

It has been theoretically calculated that the average specific fuel and electrical energy consumption of Indian market milk manufacturers is 1.0 GJ per ton and 200 kWh per ton (0.75 GJ per ton), respectively. These figures are higher as compared to other countries. Similarly, it is also evaluated that the specific fuel consumption and specific electricity consumption is 20.00 GJ per ton and 5,000 kWh per ton (19.00 GJ/ton) for the milk powder plant, while the specific electricity consumption in the pasteurized milk plant is between 38.0 to 43.0 kWh per ton (0.14-0.150 GJ/ton). In the milk plant, it is found that there are opportunities to reduce fuel and electricity consumption nearly by 65 per cent and 25 per cent respectively, while steam consumption could be saved by modifying the existing cleaning procedure. In the pasteurized milk plant, it is noted that improving the cleaning method can save approximate 30 per cent of electrical energy.

Steam

Selection of energy efficient boiler and its components, performance evaluation of boiler at regular interval and operational management as mentioned below can aid in reducing the losses and conservation of energy.

Monitor heat losses through

1. Flue gas
2. Radiation losses
3. Incomplete combustion of fuel
4. Blow-down
5. Excess air
6. Shoot deposits
7. Recover heat from steam condensate
8. Steam and oil leakage
9. Maintain steam pipe insulation
10. Preheat the oil and combustion air

Electricity

As a thumb rule, in a dairy plant, ~ 80 per cent of total electricity is consumed by motors. The efficiency of the motor varies with load and at the full load it is maximum.

Therefore, selection of motor for the given application is very vital as energy conservation point of view. It is suggested to use high efficiency motors in place of old motors. In addition to this, it is desirable not to use repeatedly rewound motors as rewinding leads to an efficiency loss up to 5 per cent. The electrical tariff for High Tension (H.T.) consumer of electricity is divided in to 2 categories *i.e.* Demand charges and Energy charges, so to reduce demand charges and line losses within the plant, one should improve power factor by installing capacitors. Improvement of power factor from 0.85 to 0.96 will reduce 11.5 per cent peak demand and reduce 21.6 per cent losses. Use of variable frequency drive for variable speed applications such as fans, pumps, compressors etc., avoiding use of oversize/undersize motors further help in minimizing energy usage.

Refrigeration

Vapour compression refrigeration system using ammonia as refrigerant is widely used in India for industrial refrigeration, air conditioning and cold storages. In dairy and many food processing plants, ice-bank system of refrigeration is used for chilling and processing of milk while direct expansion air chillers are employed for cold storages. Direct expansion glycol chillers are also being used for chilling of milk in many dairy plants. The refrigeration system uses electrical energy for operation of compressor and other auxiliary components of the system. It has been found that electricity consumption of refrigeration plant alone is about 50-60 per cent of total electrical consumption of the dairy depending on the nature of processing operations. The efficiency of refrigeration plant measured in Co-efficient of Performance (COP) varies from 2.5 to 4.5.

The important factors affecting the COP are as under.

1. Selection of refrigeration system and its components.
2. Design of plant components
3. Operational management of the system
    i. Evaporating temperature.
    ii. Condensing temperature.
    iii. Sub-cooling of liquid refrigerant.
    iv. Super heating of suction gas.
    v. Heat transfer at evaporator and condenser
    vi. Presence of non-condensable gases in the system
    vii. Volumetric efficiency of compressor
4. Multi-stage compression and throttling system
5. Maintenance of plant
6. Adoption of energy efficient technology
    i. Screw compressor
    ii. PHE type condensers
    iii. PHE type pre-chiller

    iv. Liquid overfeed system

    v. Fan less cooling towers

    vi. Heat recovery from discharge gas

    vii. Ice silos

Compressed Air Supply

It is the least energy efficient system of the dairy plant, where minimum output is used from given input.

1. Very energy intensive, only 5 per cent of electrical energy is converted to useful energy.
2. Air leakage is a major loss in compressed air.
3. Reduction in discharge pressure by 10 per cent saves energy consumption up to 5 per cent.
4. Decrease in inlet air temperature by 3°C decreases power consumption by 1 per cent.
5. Air output of compressors per unit of electricity input must be measured at regular intervals.
6. Use of screw compressor in place of reciprocating one will drastically reduce the electricity consumption.

Effluent Treatment System

It is an unavoidable process of the dairy industry to meet the legal requirements.

1. New technology and methods of treatment for wastewater should be adopted for generation of renewable energy.
2. Combination of anaerobic and aerobic digestion system enhances the efficiency of the ETP plant.
3. The Hydro Methane Reactor and UASB processes are anaerobic processes for the effective biodegradation of organic wastes into methane.
4. Against use of 0.5 - 0.75 kWh energy needed for removal of 1 kg of COD by aerobic process, one can generate 1.2 kWh energy from 1 kg of COD removed by anaerobic process.

Scope of Non-Conventional Sources of Energy

As the use of non-conventional sources of energy is eco-friendly, there is a scope to use solar energy in dairy plants to conserve energy. The use of solar water heating system is an established practice for heating of water required for boiler, cleaning application etc. However, due to certain economical, operational and maintenance problems, it is not exploited to its fullest extend. A solar water heating system of 50,000 liter capacity designed to heat water from 30°C to 75°C can collect 9,418,500 kJ per day which is equivalent to about 205 kg of furnace oil resulting into saving of about Rs. 25.8 Lakh every year. As a result of advancement in solar collectors and related technology, it is prudent to use solar energy in dairy now a day. Similarly, solar lighting, wind power, bio-gas etc. can also be exploited for harnessing the considerable amount of energy.

Conclusion

At National level, when the energy resources are limited, the costs are increasing day by day, and the availability of energy has continually in falling short of demand, to provide clean, abundant, reliable, and affordable energy to one and all, our national mission should be to strengthen energy security, environment quality and economic vitality that enhances energy efficiency and productivity through efficient energy conservation framework i.e effective enforcement of Energy conservation Act 2001. Improvement in the efficiency in transformation of energy from Coal, Gas, Oil, Solar and Wind Power etc., efficiency in usage and efficiency in re-conversion can be achieved through the approach of Total Energy Management (TEM).

Energy is an integral part and a major cost factor in of Dairy Industry as milk is processed either by heating or chilling and both these processes requires Energy. In the era of energy crisis, the dairy sector is compelled to optimize energy efficiency and adopt an approach Total Energy Management (TEM). There is scope for achieving immense savings through efficient handling of energy. Furthermore, our mission is also to bring clean, reliable and affordable energy technologies to the Dairy Industry.

References

Bhadania, A. G. (1998). Performance evaluation and energy conservation in Refrigeration and cold storages. A compendium, Energy conservation in dairy processing operations. SMC College of Dairy Science, Anand, Pp.45-51.

Desai, H.K and Zala, A. M. (2010). An overview on present energy scenario and scope for energy conservation in dairy industry A compendium, Energy management and carbon trading in Industry, SMC College of Dairy Science, Anand. Pp. 1–7.

Energy conservation in Dairy Industry. A Practical Guide for Dairy Operation and Technicians, Mansinh Institute of Training, Mehsana, National Dairy Development Board.

Energy management and Audit, www.bee-india.nic.in.

Rathore, N.S. (2010). Scope of renewable energy sources in dairy industries for energy conservation A compendium, Energy management and carbon trading in Industry, SMC College of Dairy Science, Anand. Pp. 23–26.

Shah, P.H.(2010). Energy management measures in dairy plant. A compendium, Energy management and carbon trading in Industry, SMC College of Dairy Science, Anand. Pp. 57-59.

Upadhyay J.B. (1998). Performance evaluation of Steam Generating System. A compendium, Energy conservation in dairy processing operations. SMC College of Dairy Science, Anand, Pp.5-25.

www.bee-india.nic.in

2013, Environmental Technology   *Pages 103–109*
Editors: D.R. Khanna, A.K. Chopra, Gagan Matta, R. Bhutiani & Vikas Singh
Published by: DAYA PUBLISHING HOUSE, NEW DELHI

Chapter 11

Role of Rainwater Harvesting Structures in Groundwater Resource Development and Management: An Example from Nagukhedi-Dewas Basaltic Terrain, Madhya Pradesh, India

Poonam Khare[1] and Pramendra Dev[2]

*[1]Department of Geology, J.V. Jain College,
Saharanpur – 247 001, U.P.
[2]School of Studies in Earth Science,
Vikram University, Ujjain – 456 010, M.P.*

ABSTRACT

The rainwater harvesting structures – major tool of artificial recharge – plays a governing role in the recharge of the groundwater resource management. The concise account of suitable rainwater harvesting structures for the augmentation of groundwater reservoir of basaltic terrain located in Nagukhedi-Dewas area, Madhya Pradesh is incorporated herein.

The systematic hydrogeological survey of Nagukhedi-Dewas area in Dewas district, Madhya Pradesh, reveals that the groundwater occurs as unconfined and confined aquifers in different hydrolgeologic units namely basaltic lava flows (Deccan Traps) and alluvium. The seasonal monitoring of static water levels in dug wells exhibits a variation range from 0.50 to 5.50 metres b.g.l. Most of the wells become dry during summer season and present trend of groundwater level depletion is resulting in acute scarcity of water supply and drought condition in the area.

Based on hydrogeologic and remote sensing data the suitable rainwater harvesting structures namely pit and trenches stop dams, nala bunds, percolation tanks, loose boulder structures, and sub-surface dykes at different sites are suggested for the augmentation of groundwater resource. It is further pointed out that the implementation of a scheme of rainwater harvesting structures supplemented by enhancing a forestation and maintaining appropriate balance between withdrawal and recharge of aquifers determination would be of assistance in minimizing the problem of acute scarcity of water supply in the area.

Keywords: *Rainwater harvesting structure, Groundwater, Development, Management, Nagukhedi-Dewas, Madhya Pradesh, India.*

Introduction

The groundwater is a dynamic and principal source of water through out the globe. It is only viable source to supplement the scarcity of surface water supply. The groundwater resources are depleting in quantity and quality degradation due to phenomenal increase in human population, industrialization, urbanization and intensive irrigation operations. The current situation of overexploitation of groundwater than its net recharge is causing drought phenomena in several parts of the country. Moreover, the excessive withdrawal of groundwater generates a number of undesirable side effects on the environmental regime resulting in land subsidence, crop failure and sea water intrusion in coastal aquifers. The rapidly developing situation of groundwater level depletion can be checked by augmentation of groundwater reservoir by implementing scheme of recharge techniques. The phenomenon of rainwater harvesting is an old conventional technique that involves the tapping of water where it falls and even at present, it is the proper strategy of the artificial recharge of groundwater system. In the light of above mentioned facts, a discussion on the problem of shortage of water supply in the Nagukhedi - Dewas area located in basaltic terrain of Dewas district, Madhya Pradesh in Indian subcontinent, using rainwater harvesting method for artificial recharge of groundwater reservoir has been incorporated herein.

Location and Environs of Study Area

The present study has been carried out in the vicinity of Nagukhedi-Dewas and adjoining area extending over 147 sq. km. of Dewas district, Madhya Pradesh in India (Figure 11.1). Physiographically, the area is characterized by the prevalence of typical topography of Malwa Plateau. The surface drainage is provided by Kshipra River and its tributaries. The climate is of tropical nature, revealing average annual rainfall as 1044 mm, temperature ranging from 3° C to 44° C, relative humidity varying from 30^ to 88 per cent and wind velocity ranging from 7.1 to 27.0 km/hr.

Data Used for Integration

The present study has been carried out by using remote sensing and other conventional techniques of georesource exploration. Thematic map preparation

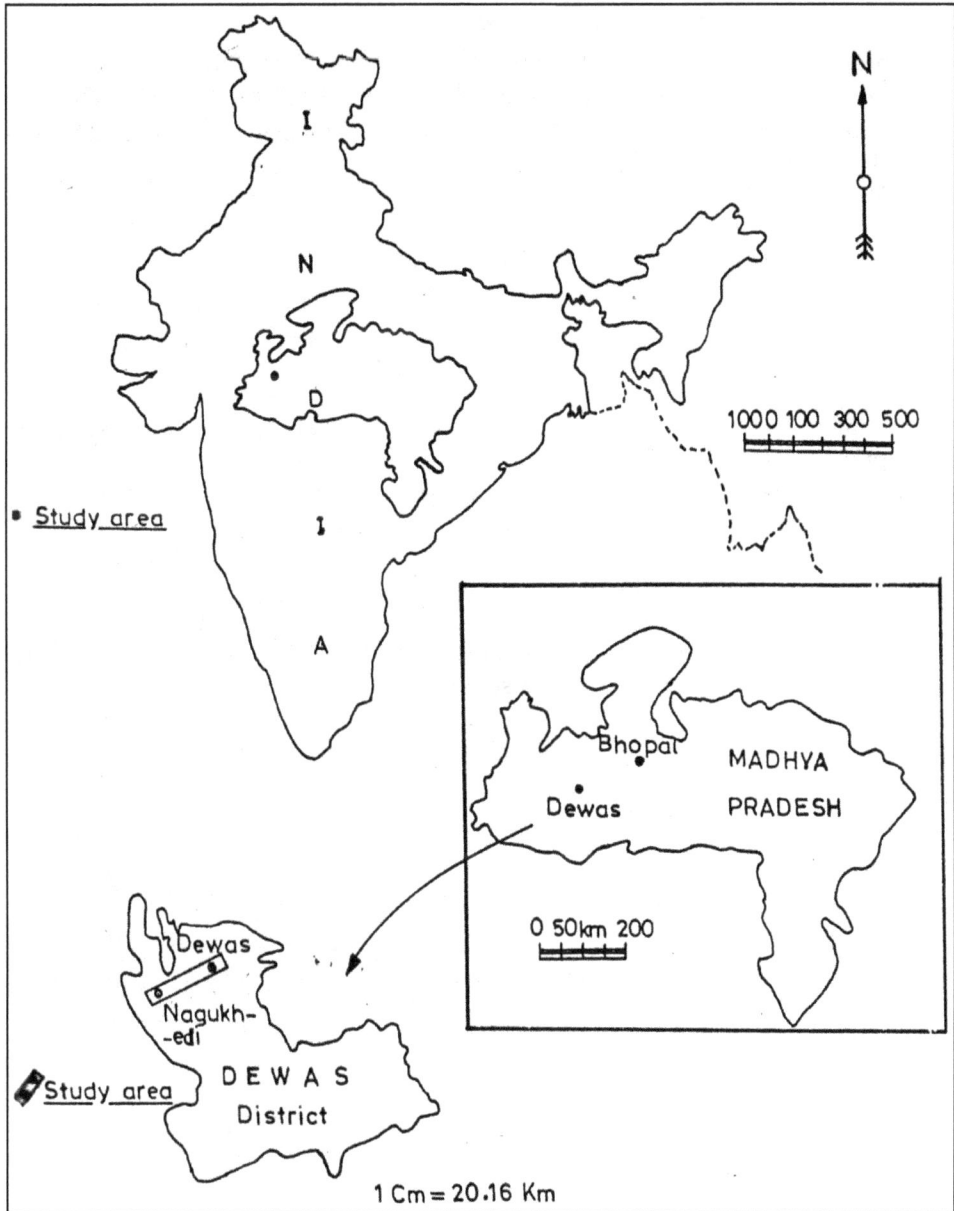

Figure 11.1: Location Map of Nagukhedi–Dewas Area, Madhya Pradesh, India

involves use of two important data sets – (1) Survey of India toposheet on 1:50000 scale and (2) Satellite data. The collateral data used for the analysis include rainfall, groundwater levels, pumping test data and water quality.

Spatial Data used for Integration

The analysis involves use of various layers of thematic information. The basic maps such as geology, geomorphology, slope, drainage, hydrological, landuse/landcover and water quality contour maps have been used. These maps were generated using SOI, Remote Sensing and GIS ARC/INFO Software. Processing of data has been done through various thematic maps, prepared by visual interpretation and were digitized using the software package ARC-INFO version 7.0 from ESRI.Hydrogeomorphological map of Nagukhedi-Dewas area has been prepared (Figure 11.2).

The above thematic maps were overlaid using the overlay module of ARC/INFO work station and the buffer zone was created for the siting of rain water harvesting structure. The detailed investigation of thematic maps generated by visual interpretation and GIS techniques, which provide valuable clues for selecting the appropriate sites for the artificial recharge. It has been possible to pin point sites for development of rain water harvesting structures to enhance the groundwater in storage in order to obtain regular water supply.

Rainwater Harvesting Structures

The parameters used for siting rainwater harvesting structures are recorded (Table 11.1).The hydrogemorphological framework of Dewas area favours the construction of rain water harvesting structures such as the pits, trenches, nallah bunds, stop dams, percplation tanks, subsurface dykes, and loose boulder structures in the study area (Table 11.2). The suggested sites for construction of rainwater

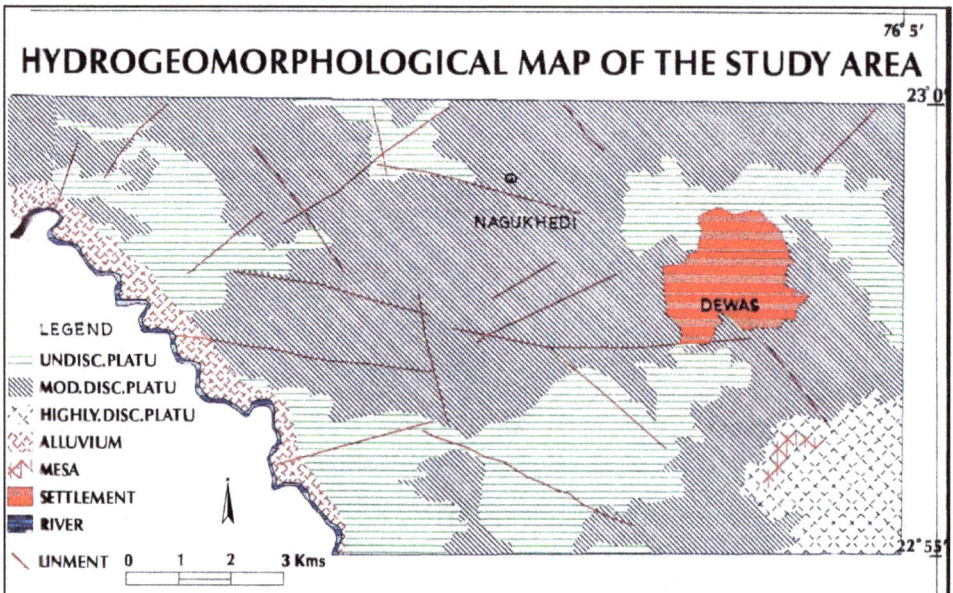

Figure 11.2: Hydrogeomorphological Map of Study Area

harvesting structures in Nagukedi - Dewas area are illustrated (Figure 11.3). The procedure of artificial recharge techniques have describe by numerous workers namely Muckel (1959), Chow (1964), Todd (1980), Raghunath (1982), Karanth (1987, 2003) and others. The salient features of suggested rainwater harvesting structures in Dewas area are described herein.

Table 11.1: Parameters Used for Siting Rainwater Harvesting Structures

Hydrogeomorphic Units (Hgm)	Land Use/ Land Cover (LS/LC)	Slope	Drainage	Suggested Rainwater Harvesting Structures
UDP, MDP, Alluvium	2.1, 2.2, 4.4	Very gentle	2nd and 3rd order stream	Stop Dams
UDP, MDPUDP	2.1, 2.24.4	Very gentle	1st and 2nd order stream	Sub-Surface Dykes
		Very gentle	2nd order stream	Nala Bunds
HDP	3.3, 4.5	Moderate and Moderately steep slopes	1st order stream	Loose Boulder Structures
HDP	3.3	Gentle to moderate	1st Order stream	Pits and trenches

UDP: Undissected Plateau; MDP: Moderately Dissected Plateau; HDP: Highly Dissected Plateau; 2.1: Cultivated Land; 2.2: Present Agriculture; 3.3: Degraded Forest; 4.4: Land with or without scrub; 4.5: Barren Sheet.

Table 11.2: Favourable Sites for Construction of Rainwater Harvesting Structures

Sl.No.	Suitable Groundwater Augmentation Sites Harvesting	Suggested Rainwater Structures
1.	Anwatpura and South of Palnagar Mendkichauki and Chandana	Sub- Surface Dyke
2.	Anwatpura and Babria	Percolation Tanks
3.	Achlukheri, Rasulpur, Chandana, Lohariya, Mareti, Panwarda and River Kshipra	Stop Dams
4.	Karnakheri	Nala Bunds
5.	South-eastern part of Nagukhedi- Dewas area	Loose Boulder Structures
6.	South-eastern part of Nagukhedi-Dewas area	Pits and trenches

Pits and Trenches

These structures can be considered at higher slopes where contours are closely spaced forming steep slopes for water and soil conservation. Monsoon run-off is collected in these trenches which retain soil moisture for longer period to facilitate the plant growth.

Nallah Bunds

These structures are constructed commonly across a nallah bund. The nallah bed must coorkntain good permeable soil cover. According to Dhokarikar (1991) the

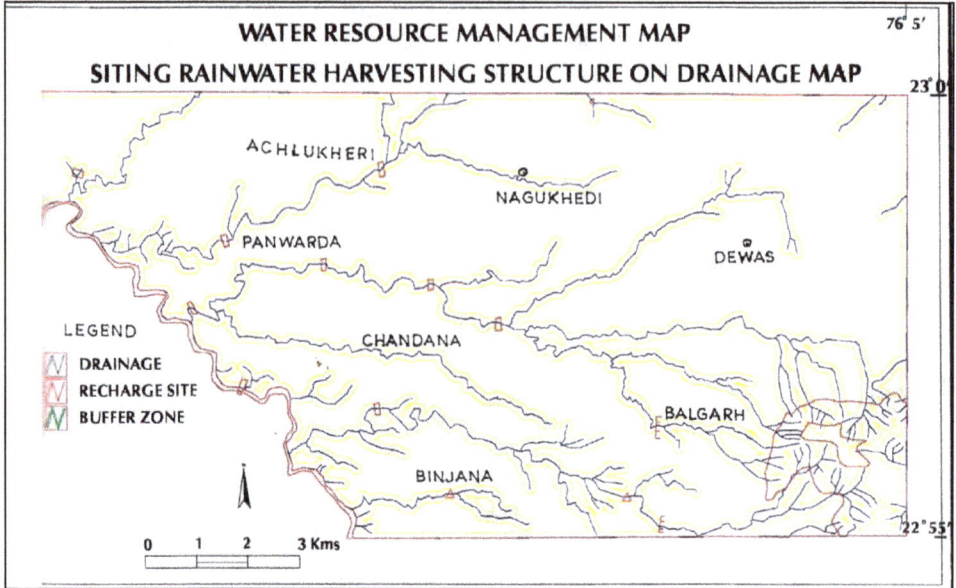

**Figure 11.3: Water Resource Management Map Siting
Rain Water Harvesting Structures**

best suitable site for nallah bund construction in basaltic terrain is that where numbers of lava flows with thin sandy cover are exposed. In the study area, the construction of small nallah bunds is recommended near the Karnakheri for arresting monsoon water flow.

Stop Dams

Pettijohn (1988) remarked that the construction of stop dam is generally used technique for augmenting recharge from existing streams of ephemeral channels In the present study area, the favourable sites for construction of stop dams for the purpose of irrigation and drinking water supply are sugeested in vicinity of the Achlukheri, Rasulpur, Chandana, Lohariya vilages and along River Kshipra. The stop dam of 2 metre height and having atleast 12 number of opening of 1.2 m and 2.0 m size to pass the silt of nallahs can be considered on the proposed sites.

Percolation Tanks

The construction of percolation is the best method for the augmentation of groundwater system to obtain a sustaiued water supply especially during the span of summer. interval. These structures are usually constructed on stream or nallahs having a sizeable catchments area. Dhokarikar (1991) recommended the design of percolation tank construction with sand or gravel filled uncased drain wells parallel to earthen dams that arrests the water received in the catchments area.The favourable sites for construction of percolation tanks have been suggested near Anwatpura and Babria with a view to divert surface run off to the aquifer. The surface water conserved after monsoon in the reservoir should percolate to aquifer adjacent to river stream.

Subsurface Dykes

The subsurface dykes are meant for arresting the groundwater flow in down stream. The subsurface dykes are impervious dykes laced in the subsurface below the stream bed level to arrest the groundwater run-off (Dhokarikar, 1991). These types of structures are recommended for the construction near Anwatpura and south of Ninjana. The subsurface dykes can be constructed upto 5- 8 m b.g.l excavation across the stream bed.

Loose Boulder Structures

The construction of loose boulder structure is generally considered in the upper reaches of high lands in an area. Such structures are favoured for construction particularly in south eastern part. part of the area.

Conclusion

Based on hydrogeologic and remote sensing data, the sustainable rainwater harvesting structures namely percolation tanks, stop dams, nala bunds, loose boulder structure, pit and trenches and subsurface dykes at different sites are suggested for augmentation of groundwater resource. It is further, recommended out that the implementation of scheme of rainwater harvesting structures supplemented by enhancing the a forestation and maintaining a balance between withdrawal and recharge of aquifers will be valuable in minimizing the problem of acute scarcity of water supply in the Dewas study area.

References

Chow, Van Te (1964). Handbook of Applied Hydrology. Mc-Graw Hill Inc., p. 13.4e-13.46.

Dhokarikar, B.G. (1991).Groundwater resource development in basltic rock terrain of Maharashtra. Water Industry Publ. Pune, 275 p.

Karanth, K. R. (1987, 2003). Groundwater assessment, develoment and management. Tata McGraw Hill Publ. Co. Ltd., New Delhi, 720 p.

Muckel, D.C. (1959). Replenishment of groundwater supplies by artificial mans.Tech. Bull. 1195, Agric Res. Service, U.S. Dept Agric, 51 p.

Pettijohn, W.A. (1088). Introduction to artificial groundwater recharge. Scientific Publ., Jodhpur, 62 p

Raghunath, H.M. (1982). Groundwater.Wiley Eastern Ltd., New Delhi, 456 p.

Todd, D.K. (1980).Groundwater hydrology. John Wiley and Sons, New York, 535 p.

2013, Environmental Technology
Editors: **D.R. Khanna, A.K. Chopra, Gagan Matta, R. Bhutiani & Vikas Singh**
Published by: **DAYA PUBLISHING HOUSE, NEW DELHI**

Pages 111–126

Chapter 12

Remote Sensing and Geographic Information System Application in Reflection of Air Pollution Environment during a Decade Span in Vicinity of Dehradun, Uttarakhand

Vartika Singh[1] and Pramendra Dev[2]

[1]*D.T.R.L., Defence Research and Development Organization,*
New Delhi – 110 054
[2]*School of Studies in Earth Science, Vikram University,*
Ujjain, M.P. – 456 010

ABSTRACT

Remote sensing satellite analysis has been employed in appraisal and monitoring of many disciplines including the air pollution aspects. Remote sensing satellite data examination and Geographic Information System technology have been employed in the identification of environmental scenario and appraisal of a decade span (1990–2000) in the vicinity of Dehradun, covering an area of 729 sq km, located in Uttarakhand, India. The main objectives of analysis involve evaluation of the air pollution, temperature and rainfall data and their effects on the Dehradun environmental system, which is presently facing pressure of several problems due to population explosion, land use expansion and forest degradation by using matrix function in ERDAS IMAGINE 8.4.

The present study has been limited to Dehradun area having a rather irregular and undulating terrain in the southern part of Doon Valley. The important feature is presence of the highland regions having cooler temperature and thick forests. Geologically, Dehradun area is dominated by rocks of the Garhwal Group. This paper deals with results of the remote sensing satellite image analysis of the air pollution environment of Dehradun area during a decade time span from 1990 to 2000 by using Remote Sensing and Geographic Information System (G.I.S.) technology, The concept and impacts of air pollution, temperature and rainfall on the environment of Dehradun area have been described herein. The concentration of SO_2 (146.8 to 293 μg/m³), NO_2 (151.6 to 407.19 μg/m³) and SPM (2904 to 9283 μg/m³) exceeds the Indian air quality standard and cause human health hazard. The consistent exposure to the air pollutants, leads to development of several diseases such as Emphysema, Pneumonia, Asthma, Cystic Fibrosis, Bronchitis, Cancer, Neurobehaviourial disorders and Heart attack. The environmental impacts of factors namely the temperature, rainfall, forest degradation and vegetation, which influence air quality are also incorporated,.

Keywords: *Satellite image, Environmental effect, Land use/Land cover change, Forest degradation, Decade, Dehradun environs, Uttarakhand, India.*

Introduction

Remote Sensing is the acquiring of data about an object without touching it (Fussell *et al.*, 1986). It is the phenomenon or technique of sensing and measuring varied objects from distance without directly coming physically in to contact with them. This technique of Remote Sensing is mainly concerned with the measurements of electromagnetic energy from the sun, which is reflected, scattered or emitted by the objects on surface of the earth (Sabins, 1987). According to Curran (1988), the term Remote Sensing concerns with the "use of electromagnetic radiation sensors to record image of the environment which can be interpreted to yield useful information's". Sabins (1999) defined remote sensing as "the science of acquiring, processing, and interpreting images, and related data, obtained from aircraft and satellites that record the interaction between matter and electromagnetic radiation". The Remote sensing has been considered as 'the science and art of obtaining information about an object, area, or phenomenon through the analysis of data acquired by a device that is not in contact with the object area, or phenomenon under investigation with the help of various sensors working on electromagnetic radiations emitted or reflected by the target' (Jensen, 2000; Lillesand and Kiefer (2002). The process of remote sensing can be classified into two stages *viz*: (a) Data Acquisition, and (b) Data Analysis.

Geographic Information System is usually known as the G.I.S. and is a computer based system used for capturing, storing, analyzing, manipulating geographical data to solve a particular problem. Gupta (2003) affirmed that Geographic Information System (G.I.S.) also called as 'Geobased Information System' is 'a higher–order computer–based system, which permits storage, manipulation, display and output of spatial information.' This technology has rapidly progressed during the period of last two decades and has become an essential tool of spatial data analysis for numerous applications including exploration, development and management of earth resources. O'Sullivan and Unwin (2003) proposed that use of term 'Geographic Information Analysis' and defined it as "concerned with investigating the patterns that arise as a result of processes that may be operating in space. Techniques and methods to enable the representation, description, measurement, comparison, and generation of spatial patterns are central to the study of geographic information analysis." The G.I.S. provides new tools for mapping the landscape and analysis functions can concurrently handle both spatial and non-spatial (attribute) data. The G.I.S.function incorporates retrieval, measurement, overlay, neighbourhood and connectivity (Gupta, 2003). For G.I.S. application the PCI Geomatica 7.0 software has been used.

Dehradun city a capital of the Uttarakhand state is placed at an altitude of 670 meter above mean sea level in the Himalayan foothills in Doon Valley having a decent climate, natural attractiveness including diverse types of forest, structures and brushwood environment. The present examination has been restricted to a rather irregular and undulating terrain in the southern part of Doon Valley in vicinity of Dehradun area within latitudes 30° 15' to 30° 30' N; longitudes 78° 00' to 78° 15' E; Survey of India, toposheet no. 53 J/3, Figure 12.1. The study area is mostly occupied by presence of the highland regions having cooler temperature and thick forests.

Physiographically, the Dehradun region is divided into two distinctive mountain and sub-mountain tracts. The mountains are atypical with very irregular steep slopes. The ridges detach drainage area of Tons River from that of Yamuna River on the west and east respectively. The sub-mountain tracts are prominent and bounded by Siwalik Hills in the south and outer scarp of Himalayan Mountain in the north. Climate of

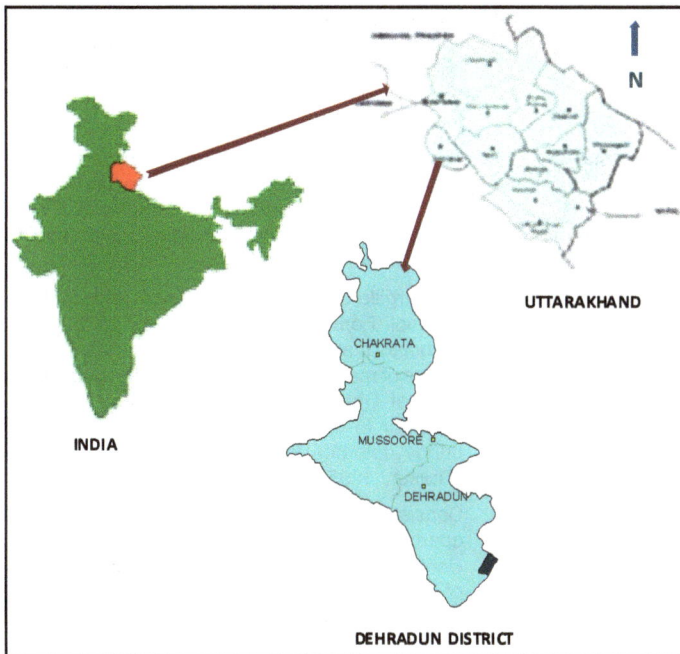

Figure 12.1: Location Map of Dehradun Area, Uttarakhand

Dehradun area is generally temperate and varies noticeably from tropical to brutal cold. In the hilly regions, summer is usually pleasing but heat is repeatedly excessive. Temperature drops below freezing point during the winter. Temperature ranges from 3°C to 22°C during winter season (December to February), and varies from 17° C to 35° C (March to June period). Dehradun experiences heavy to moderate showers during late June to mid-August. Rainfall is mainly received during June to September span and computed annual average value is 2073 mm.

Geology of Dehradun Area

The rocks of Dehradun region have been classified into Central Crystalline, Garhwal and Dutatoli Groups, which form the northern, central and southern parts of the area respectively. The main Central Thrust separates Central Crystalline from Garhwal Group, The present study area is characterized by exposures of rock formations of the Garhwal Group, which has been comprised of five formations: (1) Rudrapyrag Formation, (2) Lameri Formation, (3) Chamoli Formation, (4) Gwangarh Formation, and (5) Partoli Formation. The salient features of these formations are mentioned herein (Table 12.1).

Table 12.1: Geological Formations of Garhwal Group, Uttarakhand

Sl.No.	Formation	Characteristics of Formation
1.	Rudraprayag Formation	Argillo-arenaceous Facies.Lithounits sandwitched between the Alaknanda Fault in north and North Almora Thrust in south. Oldest stratigraphic unit (Wadia, 1919)
2.	Lameri Formation:	Conformably overlying Rudraprayag Quartzite, sequence of dolomite-limestone and slate/phyllite (Mehdi, *et al.*, 1972). Restricted to eastern closure part of the Rudraprayag Anticline and is cut-off westward near Punar dolomite exhibits stromatolite near upstream of Rudraprayag and indicates top up position of the beds.
3.	Chamoli Formation	It underlies the Lameri Formation and overlies Gwanagarh Formation. Consists of quartzite and a number of pene-contemporaneous submarine basic flows of spilitic composition and profuse intrusions of dolerites, at present metamorphosed to epidiorite.
4.	Gwangarh Formation	It is argillo-calcareous sequence and conformably overlying Chamoli Formation (Mehdi, *et al.*, 1972).Consists of marble, tremolite marble, calc silicates, intruded by granite and with lenses of magnesite. Lime-breccia and stromatolites occur south-west of Dhanpur. Normal sequence is noted southwest of Dhanpur in the old workings for copper; where local folding brings dolomite to rest over quartzite in a typical sequence (Thakur, 1995).
5.	Partoli Formation	Consists of white to greenish, fine-grained, and thickly bedded quartzite and is youngest formation. It is cut off by the Main Central Thrust to north of Kalsir. To south of Nandprayag, a bed of biotite-chlorite schist is interbedded with quartzite. It contains prophyroblasts of the sericitised. feldspars in a fine matrix of quartz, biotite and chlorite.

Rudrapyrag Formation

It belongs to an argillo-arenaceous facies. The various litho-units of Rudraprayag Formation are sandwiched between the Alaknanda Fault in the north and North Almora Thrust in the south. It appears to be the oldest stratigraphic unit of Garhwal Group, over which edifice of all younger sequence was built upon, not only in the Garhwal and Kumaun Himalaya but perhaps in other parts in lesser Himalaya (Wadia, 1919).

Lameri Formation

Conformably overlying the Rudraprayag Quartzite is a sequence of dolomite-limestone and slate/phyllite, which has been referred to as the Lameri Formation by Mehdi, *et al.* (1972). It is limited to the eastern closure part of the Rudraprayag Anticline and is cut-off westward near Punar by the Punar Fault. Due to the pinching out of lowest unit of the dolomite at about 600 m NNE of Punar, the overlying phyllites/slate directly comes over the Rudraprayag Quartzite and it is difficult to distinguish them from those associated with the quartzite. The dolomite exhibits development of stromatolite along the road section about 2.5 km upstream of Rudraprayag. The disposition of stromatolite indicates top up position of the beds.

Chamoli Formation

The one which underlies the lameri formation has been referred to as the Chamoli Formation, and the other, appearing to be the youngest, overlies the calcareous Gwanagarh Formation, as the Patroli Formation. It is characterized by quartizite and a number of penecontemporaneous submarine basic flows of spilitic composition and profuse intrusions of dolerites, now metamorphosed to epidiorite.

Gwangarh Formation

The Chamoli formation is conformably overlain by an argillo-alcareous sequence referred as Gwanagarh Formation by Mehdi, *et al.* (1972). It has attained lesser development, represented by marble, tremolite marble, calc silicates, and is intruded by granite. Lenses of magnesite are invariably associated with it. At the base of dolomite towards southwest of Dhanpur, a metre thick zone of lime-breccia contains a tendency by means of 2-6 meter thick zone of stromatolitic (*Collenia* sp.) Horizon. The lime-breccia and stromatolities both are locally phosphatic. In the area between Dhanpur and Bhainswara, the stromatolites show regional inversion due to overturned nature of the Karprayag Anticline. The western part of the formation, between Bhainswara and Toryal that the sequence assumes normal position due to the Pingalapani Syncline. Normal sequence is also observed locally southwest of Dhanpur in the vicinity of old workings for copper, where local folding brings the dolomite to rest over quartzite in a normal stratigraphic order (Thakur, 1995).

Partoli Formation

It is youngest formation of the Garhwal Group. It is mainly consists of quartzite, which is fine-grained, white to greenish and thickly bedded. Depending on the percentage of sericite or chlorie phyllite extending northwestward from east of 7156 peak to north of Kalsir where it is cut off by the Main Central Thrust. South of

Nandprayag, a bed of biotite chlorite schist is interbedded with the quartzite. It contains prophyroblasts of feldspars in a fine matrix of quartz, biotite and chlorite. The feldspars are all sericitised.

Data Used

The Survey of India Toposheet No. 53J/3, and Landsat TM false colour composite image (bands 2, 3 and 4) on the 1:50,000 scales, data of 1990 and 2000 have been used. For the preparation of vegetation map and change detation study we use PCI Geomatica 7.0, ERDAS IMAGINE 8.4 and ARC –GIS 9.1. Rainfall, temperature and air quality data of deharadun city.

Air Pollution Concept

Transport vehicles and industrial emissions are the major sources of pollutants in the Dehradun environment, have generated a problem that has been aggravated by the tremendous increase in the number of mobile sources. The air pollution has most important impacts on environmental regime. In Dehradun, the concentration of SO_2 and NO_2 has been noted within the Indian air quality standards, but the SPM concentration exceeded the Indian air quality guidelines in this area (Joshi *et al.*, 2006; Chauhan and Joshi, 2010).

Hence, there is need to evaluate the present status of air quality perfection in Dehradun area. Among the particles, those having median diameters higher than 10 m are stopped in the upper areas of respiratory system. Smaller particles with median diameters less than 10 m (PM10) can reach the lungs and aggravate respiratory problems, depending on their physico-chemical properties. The smallest ones, with diameters less than 2.5 m reach bronchial alveolus and may have long residence time inside, increasing health effects, such as asthma and respiratory allergies (Chauhan *et al.*, 2010).

Types of Air Pollution

The industries and various modes of transportation continuously stay in operation. This results into the generation of pollutants in a far greater degree than other sources. The industrial, vehicular and dwelling-related pollution are 3 important types of air pollution. Different types of air pollutions are incorporated.

Industrial Air Pollution

The different industries such as petroleum, cement, steel, thermal power plants, paper factories, atomic, pharmaceutical, sugar, other foods and pesticides play an important part towards increasing the levels of air pollution (Joshi and Chauhan, 2008).

Transportation and Air Pollution

The hydrocarbons and nitrogen oxides as air pollutants with moisture present in air create smog. A hot and humid weather condition increases the chances of smog development. This problem resulting from vehicular emissions occurs in big cities (Chauhan, 2008). The emissions of vehicles containing many poisonous gases cause health problems.

Pollution Resulting from Dwelling

The pollution resulting from dwelling is caused by aerosols and other such chemicals. High density of population is one of causes of air pollution.

Air Pollution Caused by Accidents

The causes of air pollution resulting from accidents are varied. Accidents could be anything like forest fires, blasts in industries and others. The accidents caused to vehicles that transport petroleum products also result into air pollution.

The types of air pollution can also be determined on the basis of agents which pollute the air. Different types of air pollutants are discussed below:

Carbon Monoxide

The inhalation of carbon monoxide can be life-threatening. This is because the amount of oxygen that is being delivered to the tissues and organs of body gets reduced due to carbon monoxide. Other effects of air pollution resulting from carbon monoxide are dizziness and headaches. More information on carbon monoxide poisoning should prove to be useful (Healy *et al.*, 2007).

Nitrogen Oxide

The percentage of nitrogen oxide pollution which results from vehicles is 34 per cent. Out of the 34 per cent, heavy diesel vehicles produce 42 per cent nitrogen oxide, while the light vehicles including cars account for 52 per cent. The remaining amount is contributed by other vehicles.

Hydrocarbons

Vehicles account for 29 per cent hydrocarbon production. Hydrocarbons are responsible for health problems associated with respiratory tract. Lung tissues get damaged by the inhalation of hydrocarbons.

Particle Matter

Transportation and industries are both responsible for production of particle matters. Particles (pollutants) in solid and liquid state together create a condition called haze. The air pollution facts for kids presented through above discussion should prove to be valuable (Chauhan, 2010).

Air Pollution Data of Dehradun Area

The records of air plollution data in respect of Dehradun area during the period from 1990 to 2000 have been displayed (Figure 12.2; Table 12.3). It has been observed that the concentration of SO_2 (146.8 to 293 µg/m³), NO_2 (151.6 to 407.19 µg/m³) and SPM (2904 to 9283 µg/m³) exceeds the Indian air quality standard and cause human health hazard.

Effects of Air Pollution on Human Health

The effects of air pollution on human health are fatal and life threatening. World Health Organization (W.H.O., 1999) statistics reported that over 2 million people

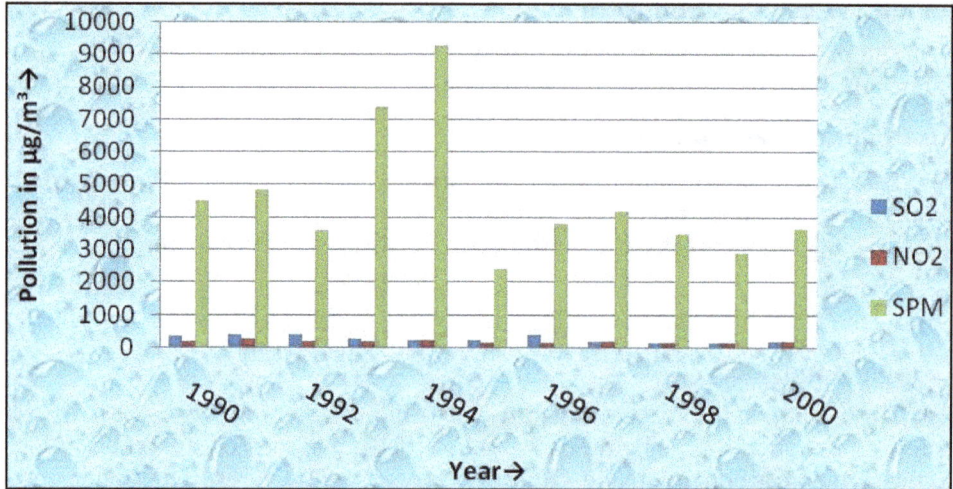

Figure 12.2: Air Pollution Data of Dehradun Area, during 1990–2000

succumb to the fatalities attributed to air pollution. Consistent exposure to the pollutants leads to development of the following:

☆ Cardiopulmonary disease

☆ Emphysema

☆ Premature mortality

☆ Pneumonia

☆ Difficulty in breathing

☆ Heart attack

Table 12.2: National Ambient Air Quality Standards (CPCB, 2001)

Sl.No.	Pollutant	Time Weighted Average	Concentration in Ambient Air		
			Industrial Areas, Residential, Rural and Other Area	Ecological Sensitive Area	Methods of Measurement
1.	Sulphur Dioxide (SO$_2$)	AnnualAverage	50µg/m^3	20 µg/m^3	- Improved West and Greek method
		24 hours	80 µg/m^3	80 µg/m^3	- Ultraviolet Fluorescence
2.	Oxides of Nitrogen as (NO$_2$)	AnnualAverage	40 µg/m^3	30 µg/m^3	- Modified Jacob and Hochheiser
		24 hours	80 µg/m^3	80 µg/m^3	- Chemiluminescence
3.	Suspended Parti-culate Matter	AnnualAverage	60 µg/m^3	60 µg/m^3	- Gravimetric
					- TOEM
		24 hours	100 µg/m^3	100 µg/m^3	- Beta attenuation

☆ Wheezing and coughing

☆ Asthma attacks

☆ Acute vascular dysfunction

☆ Cystic fibrosis

☆ Thrombus formation

☆ Bronchitis

☆ Cancer

☆ Premature death

☆ Reduced energy levels

☆ Neurobehaviour disorders

☆ Headaches and dizziness

☆ Reduced lung functioning

☆ Chronic obstructive pulmonary disease

Table 12.3: Air Pollution Data of Dehradun during 1990 to 2000

Sl.No.	Year	SO_2 $\mu g/m^3$	NO_2 $\mu g/m^3$	$SPM \mu g/m^3$
1.	1990	362	201	4527
2.	1991	398	293	4855
3.	1992	402	197	3594
4.	1993	295	185	7360
5.	1994	250	227	9281
6.	1995	248.1	179.15	2415.4
7.	1996	407.19	180.7	3812.1
8.	1997	200.3	214.5	4163
9.	1998	182.5	165.8	3475
10.	1999	151.6	146.8	2904
11.	2000	220.2	219.4	3637

Source: Central Pollution Control Board, 2001.

Temperature Impacts

Temperature is one of important atmospheric factor, which considerably affects the environmental scenario. The impact of various pollutant caps on global and hemispheric mean surface temperature changes from the percentages relative to the global-average reference case changes of 2.7°C and 0.4 meters respectively. The largest increases in temperature and sea level occur when SOx alone is capped due to the removal of reflecting (cooling) sulphate aerosols (Prinn *et al.*, 2005). The temperature affects several human responses, including thermal comfort, perceived air quality and performance at work (Seppanen, *et al.*, 2003). The temperature record of Dehradun area during the period from 1990 to 2000 indicates that the gradually increasing rate of temperature and the minimum annual temperature ranges from 15.9 to 16.36 °C and maximum annual temperature varies with range of 26.6 to 28.59 °C (Figure 12.3; Table 12.4). The higher temperatures provide favorable environment for evapo-transpiration phenomena and the unpleasant conditions for the populace. The too much cooler temperatures are also causing problems to the inhabitants of the Dehradun area.

Rainfall Impacts

Rainfall is the common term applied to atmospheric precipitation that falls to the earth's surface in the form of rains. The excess or scanty rainfall causes environmental impacts in the form of floods or droughts respectively. In India, most

Table 12.4: Temperature of Dehradun Area, during 1990 to 2000

Sl.No.	Year	Maximum Temperature in °C	Minimum Temperature in °C
1.	1990	27.55	15.9
2.	1991	28.10	15.81
3.	1992	28.08	15.79
4.	1993	28.36	15.82
5.	1994	28.59	16.15
6.	1995	28.15	16.3
7.	1996	27.87	15.9
8.	1997	26.6	15.35
9.	1998	27.77	16.36
10.	1999	28.46	16.19
11.	2000	27.31	15.95

Source: Central Pollution Contral Board, 2001.

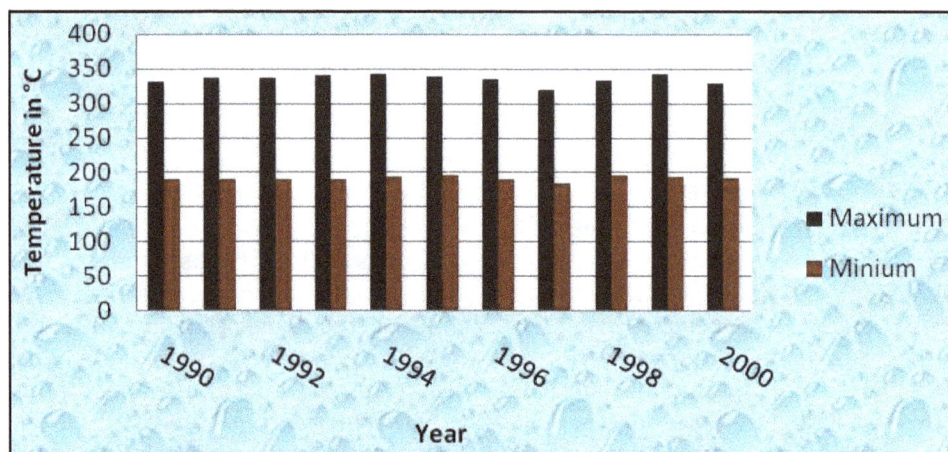

Figure 12.3: Temperature Range of Dehradun Area, during 1990–2000

of the rainfall occurs during the monsoon period from middle of June to September. In the monsoon season air quality is better than the comprasion of other days, because with the drop of rain pollutent particals are settel down. The record of rainfall data of a decade interval of Dehradun under analysis indicates a variation range from 1498.4 mm to 2881.9 mm with an average value of 2275.6 mm (Figure 12.4; Table 12.5). These values are decade period of the study area have reflected that adequate amount of rainfall was available for the recharging of groundwater system of Dehradun area.

Table 12.5: Rainfall Data of Dehradun Area, during 1990 to 2000

Year	Jan	Feb	Mar	Apr	May	Jun	Jul	Aug	Sep	Oct	Nov	Dec	Total
1990	1.2	113.3	98.2	9.9	114.4	183.8	904.4	833.3	460.7	33.6	7.2	121.9	2881.9
1991	11.6	56	50.6	55.4	21.2	265.2	304.7	455.6	236.7	0	9.1	32.3	1498.4
1992	114.1	40.5	13.5	1.8	26.3	156.1	567.8	980	187.6	7.7	6.8	0	2102.2
1993	61.5	28.9	60.6	11.2	36.3	229	466.8	735.5	406.2	0	2.4	0	2038.4
1994	57.2	56.9	1.9	78	9.4	217.7	704.2	776.6	66	0	0	2.2	1970.1
1995	33.3	73.9	39.2	14.6	0.8	62.9	489.7	630.7	310.3	2.2	0.5	9.3	1667.4
1996	40.3	106.2	38.4	13.4	10	356.8	604.2	961.1	282.3	57.7	0	0	2470.4
1997	34	21.9	65.6	111.1	130	397.4	779.6	550.3	385.8	82.1	44.8	90.8	2693.4
1998	5.4	72.4	67.5	78.6	86.3	110.4	834.2	1078	163.7	226.7	0.9	0	2724.1
1999	27.4	4.2	4.9	0	6.6	396.1	691.2	535.3	670	75.8	0	9.9	2421.4
2000	71.5	110.9	44.4	12.4	141.1	308.6	769	724.7	381.2	0.2	0	0	2564

Source: Central Pollution Contral Board, 2001.

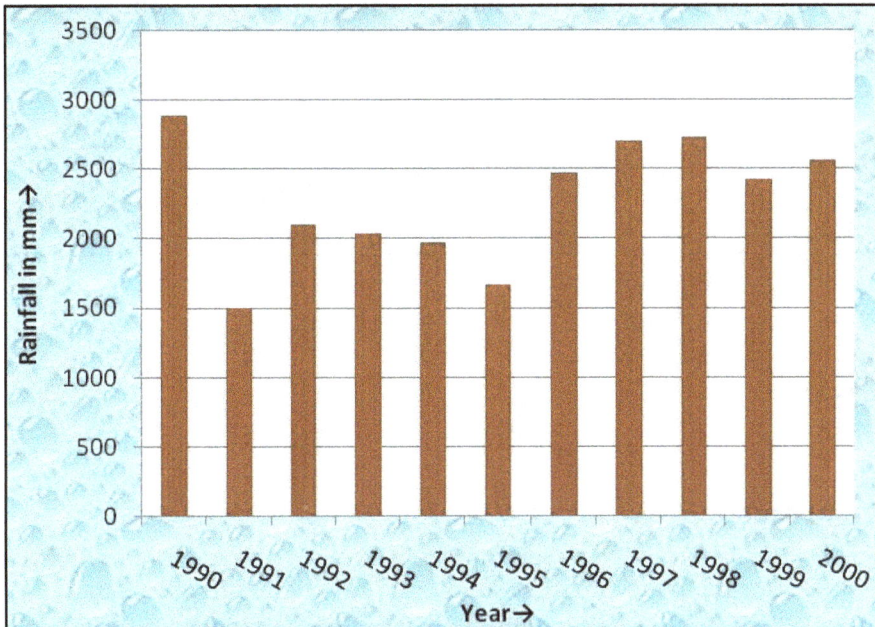

Figure 12.4: Rainfall Data of Dehradun Area, during 1990 to 2000

Forest Degradation

Urban vegetation can directly and indirectly affect local and regional air quality by altering the urban atmospheric environment. Nowak (1995) suggested that urban trees affect air quality in the following four main ways:

1. Temperature reduction and other microclimatic effects
2. Removal of air pollutants
3. Emission of volatile organic compounds and tree maintenance emissions
4. Energy effects on buildings

The air quality of an area depends oupn the character of vegetation in that area, through its influence on physical properties of the land surface properties and biogeochemical fluxes. Vegetation plays a role in capturing pollution by deposition and/or uptake a mass balance is used (Hofschreuder *et al.*, 2010). Large-scale changes in vegetation cover, for example a reduction in the current forest area, would be expected to modify the local climate. Moreover, a reduction in forest cover would also be expected to contribute to global climate change through the release of stored carbon contributing to the rise in atmospheric CO_2 (Figures 10.5 and 10.6; Table 10.6). Kutler and Strassburg (1999) measured NO and NO_2 concentrations in Aachen, Germany and found that pollution level increased with proximity to main roads and was reduced inside open green spaces where the wind speed and direction plays an important role. Lam *et al.* (2005) reported that lower values of pollutants were found inside urban parks and open spaces in comparison to the roadside stations. Vegetation has the potential to improve air quality. Trees can intercept atmospheric particles and absorb various gaseous pollutants (Bealey, 2007) and can lower air temperature through shading and evapotranspiration cooling (Givoni, 1991). The effects on air quality of very large scale planting of almost all tree species in cities would be highly positive (Hewitt, 2010).

Table 12.6: Displaying the Forest Cover Area of Dehradun Region during 1990-2000

Range	Type of Forest	Area of Forest in 1990		Area of Forest in 2000		Difference of Changes	
		(sq.kms)	%	(sq.kms)	%	(sq.kms)	%
> 80%	Dense Forest	103.89	21.85	71.78	15.78	32.11	6.07
80%–60%	Open Forest	82.98	17.45	67.19	14.77	15.79	2.68
60%–40%	Degraded Forest	76.98	16.19	84.9	18.67	-7.92	-2.8
40%–20%	Grass Land	86.97	18.29	68.51	15.06	18.46	3.23
< 20%	Agriculture Land	124.52	26.19	162.35	35.70	-37.83	-9.51

Conclusion

The examination of remote sensing satellite data in respect of Dehradun area of Uttarakhan for the period from 1990 to 2000 in Geographic Information System environment, provide significant information to depict the reflection of air pollution

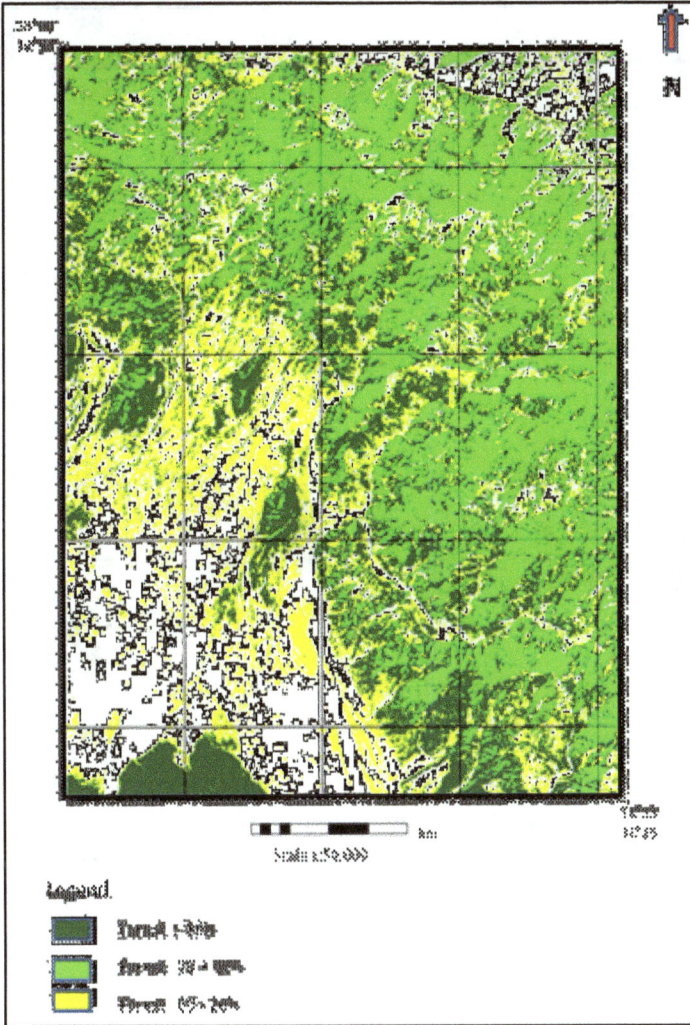

Figure 12.5: Forest Cover Area of Dehradun Region 1990

environmental scenario. The forests could be mapped from two data sets of different time periods from 1990 to 2000. A loss of 47.905 km^2 (7 per cent) was observed between 1990- 2000. An increase of 37.83 sq km area was observed in agriculture land. The spatial distribution of different forest types from 1990 to 2000 reflects that forest cover in the Dehradun area has been undergoing massive reduction with time.

The air pollution such as the concentration of SO$_2$ (146.8 to 293 µg/m^3) NO$_2$ (151.6 to 407.19 µg/m^3) and SPM (2904 to 9283 µg/m^3) exceeds the Indian air quality standard and cause human health hazards. The other environmental factors temperature, rainfall, forest degradation and vegetation also play affective role to manage air quality.

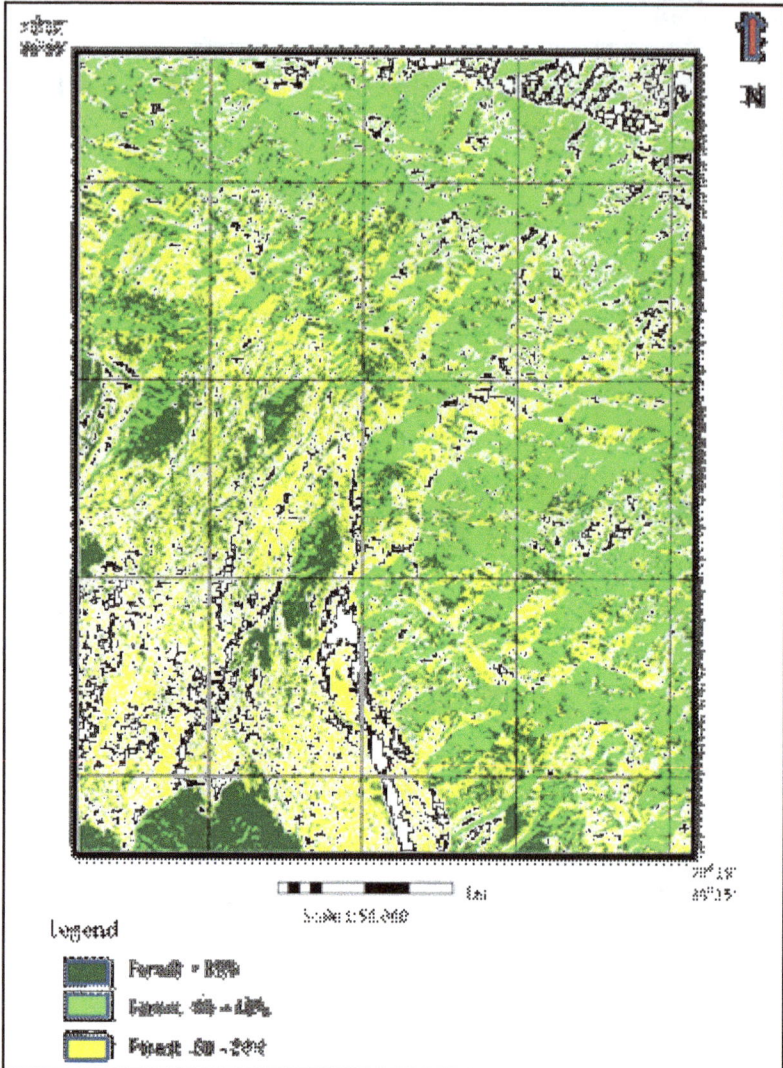

Figure 12.6: Forest Cover Area of Dehradun Region 2000

References

Bealey, W.J. McDonald, A.G. Nemitz E. Donovan, R. Dragosits, U. Duffy.T.R. and D. Fowler, (2007): Estimating the reduction of urban PM10 concentrations by trees within an environmental information system for planners, Journal of Environmental Management 85: 44–58

Central Pollution Control Board (2001): National ambient air quality statistics of India. Central Pollution Control Board, Parivesh Bhavan, Delhi, India

Chauhan, A. (2008): Effect of automobile and industrial air pollutants on some selected trees grown at the edge of road side in Haridwar. Journal of Natural and Physical Science, 22 (1-2), p. 37-47.

Chauhan, A. (2010): Photosynthetic pigment changes in some selected trees induced by automobile exhaust in Dehradun, Uttarakhand. New York Science Journal, 3 (2), p. 45-51.

Chauhan, A. and Joshi, P.C. (2010): Effect of air pollutants on wheat and mustard crops growing in the vicinity of urban and industrial areas. New York Science Journal, 3(2), p. 52-60.

Chauhan, A., Pawar, M., Kumar, R., and Joshi, P.C. (2010): Ambient air quality status in Uttarakhand (India): A case study of Haridwar and Dehradun using air quality index. Journal American Science, 6 (9), p. 565- 574.

Curran, P.J. (1988): Principal of remote sensing. English language book society, Longman, Hong Kong, 282 p.

Fussell, J., D.Rundquist and J.A.Harrington. (1986): "On Defining Remote Sensing, Photogrammetric Engineering and Remote Sensing, 52(9), pp 1507-1511.

Gupta, R. P. (2003): Remote Sensing Geology. Springer-Verlag, Berlin, Heidelberg, 655 p.

Givoni, B. (1991): Impact of planted areas on urban environment quality: A review, Atmospheric Environment 25B (3): 289-299.

Healy, D., Silvari, V., Whitaker, A., Lopez, J., Pere-Trepat, E., Heffron, E. (2007): Linking urban field measurements of ambient air particulate matter to their chemical analysis and effects on health. In Proceedings of 6 International Conference on Urban Air Quality, Limassol, Cyprus.

Hofschreuder, P., Kuypers, V., Vries, B.D., Jansen, S., Maerschalck, B.D., Erbrink, H., and Wolff, J.D. (2010): Effect of vegetation on air quality and fluxes of NO_x and PM-10 along a highway. CLIMAQS Workshop 'Local Air Quality and its Interactions with Vegetation' Antwerp, Belgium. pp 1-6.

Hewitt, N. (2010): Trees and urban air quality. CLIMAQS Workshop 'Local Air Quality and its Interactions with Vegetation' Antwerp, Belgium.

Joshi, P.C., Swami, A. and Gangwar, K.K. (2006): Air quality monitoring at two selected traffic junctions in the city of Haridwar. Him. J. Env. Zool. 20(2), p. 219-221.

Joshi, P.C. and Chauhan, A. (2008): Performance of locally grown rice plants (Oryza satiya L.) exposed to air pollutants in rapidly growing industrial area of district Haridwar, Uttarakhand, India. Life Science Journal, 2008, 5 (3). p. 57 – 61.

Jensen, J. R. (1996): Remote Sensing of the environment, An Earth Resource Perspective.240 p.

Kutler, W., Strassburger, A. (1999): Air quality measurements in Urban Green Areas – a case study, Atmospheric Environment, 33: 4101-4108.

Lam, K.C., NG, S.L., Hui, W.C., Chan, P.K. (2005): Environmental quality of urban parks and open spaces in Hong Kong, Environmental Monitoring and Assesment, 111: 55-73.

Lillisand, T.M. and Kiefer, R.W. (2000): Remote sensing and Image Interpretation. John Wiley and Sons, New York, 745 p.

Mehdi, S.H., Kumar, G. and Prakash, G. (1973): Tectonic evolution of eastern Kumaun Himalaya, A New Approach. Him. Geol., v.2, p. 481-501.

Nowak, D.J. (1995): Trees pollute? A "TREE" explains it all, in: Proc. 7th Natl. Urban For. Conf. (C. Kollin and M. Barratt, eds.), American Forests, Washington, DC, pp. 28-30.

O'Sullivan, D. and Unwin, D. (2003). Geographic Information Analysis. John Wiley and Sons, Inc., New Jersey, U.S.A., 436 p.

Prinn, R., Reilly, J., Sarofim, M., Wang, C., and Felzer, B. (2005): Effects of air pollution control on climate. MIT joint program on science and policy of global change. Report No-118, pp 1-14.

Sabins, F.F.Jr. (1987): Remote Sensing: Principles and Interpretation, W.H. Freeman and Co., New York, 429 p.

Sabins, F.F. (1999): Remote Sensing Principles and Interpretation. W. H. Freeman and Co., New York, 494 p.

Thakur, V.C. (1995): Geology of Dun Vally, Garhwal Himalaya: Neotectonics and coeval deposition with fault- propagation folds. Journal of Himalayan Geology 6920, p. 1-8.

Seppänen, O., Fisk, W.J., Faulkner, D. (2003): Cost benefit analysis of the night-time ventilative cooling. In: *Proceedings of the Healthy Buildings Conference*. Singapore 2003, Vol 3: 394-399.

Wadia, D.N. (1919): Geology of India. Mac Millan and Co., London.508 p. (Reprint, 1994).

World Health Organization (1999): Guidelines for air quality. World Health Oganization, Geneva.

2013, Environmental Technology   *Pages 127–134*
Editors: **D.R. Khanna, A.K. Chopra, Gagan Matta, R. Bhutiani & Vikas Singh**
Published by: **DAYA PUBLISHING HOUSE, NEW DELHI**

Chapter 13

Bioremediation Activity of Acclimated *Streptomyces* sp. in Solid and Liquid Waste

Ashok Kumar[1], Vishnu Dutt Joshi[2]
and Balwant Singh Bisht[1]

[1]BioReLab., Department of Zoology/Entomology,
HNBGU, Campus Badshahithaul Tehri – 249 199, U.K.
[2]Department of Zoology,
Government PG College, Kotdwara Garhwal, U.K.

ABSTRACT

In recent years, Uttarakhand has emerged as one of the most attractive industrial destinations in India. The industrial sector of the Uttarakhand state growing day by day and it is very good for the development of state and country also, but we should remember the second phage of industrialization. It is the industrial effluent, disposed by the industries. The industrial effluents have many pollutants including heavy metals. The major environmental problem is the pollution of heavy metals and they cause serious diseases in animals including human. In the present study, the *Streptomyces* strain isolated from sludge contaminated with heavy metal was trained for heavy metal remediation by exposing metal to them and used for bioremediation activity for chromium, copper and lead in the waste. The metal chromium reduction ranged from 0.386-6.42mg/l, which showed maximum reduction in chromium by *Streptomyces* sp., copper reduction ranged from 0.288-1.129mg/l and reduction in lead ranged from 0.063-0.286mg/l. In future, the gram positive filamentous Streptomyces strain can be sued for chromium ion reduction as bioremediating agent.

Keywords: Bioremediation, Streptomyces sp., Metal tolerance, Effluent, Bioremediating agent.

Introduction

In recent years, Uttarakhand has emerged as one of the most attractive industrial destinations in India. The government is encouraging private participation in all industrial activities and as a result big players such as HLL and Dabur have set up units in the state. The New Industrial Policy announced in 2003 by the state government puts in place the regulatory framework for Uttarakhand's industrialization. The New Industrial Policy indicates that private resources may be tapped while promoting integrated Industrial Estates in Uttarakhand. State Uttarakhand is a land of scenic beauty, temples, lakes, mountains, glaciers and green lush meadows. The industrial sector of the Uttarakhand state growing day by day, due to this the major concern is the industrial effluents disposed by the industries. The industrial effluents have many pollutants including heavy metals. The major environmental problem is the pollution of heavy metals and they cause serious diseases in animals including human. The main industries of Uttarakhand state are listed in Table 13.1.

Table 13.1: Industries in Uttarakhand

Sl.No.	Industry	Sl.No.	Industry
1.	Cloth Mills	11.	Dairy Products
2.	Dye and dye intermediate Industry	12.	Electroplating Industry
3.	Fermentation Industry	13.	Fertilizer Industry
4.	Fish Farming	14.	Floriculture
5.	Flour and Rice Mills	15.	Food and Fruit Industry
6.	Horticulture	16.	Paper Mills
7.	Mentha Oil Units	17.	Pharmaceuticals
8.	Pickle and Sauce Industry	18.	Poultry Farming
9.	Production of Rice and Wheat	19.	Stone Rolling Mills
10.	Sugar Mills	20.	Rubber

Heavy Metal Pollution

In early days of abundant resources and negligible development pressures, little attention was paid to environmental issue, although some environment related legislation pertaining to different sectors was authorized. Rapid economic changes have resulted in elevated level of toxic heavy metals and radionuclides entering the biosphere. The heavy metals such as lead, cadmium, copper, nickel and zinc are among the most common pollutants found in industrial effluents. Solid and/or liquid wastes containing toxic heavy metals may be generated in various industrial processes such as chemical manufacturing, electric power generating, coal and ore mining, smelling and metal refining, metal plating and others. Heavy metals pollution such as copper, cadmium, lead, mercury, arsenic and chromium has been classified as a priority pollutant by the Department of Environment. Continuous monitoring of heavy metals level in the environment is very important since it cannot be degraded and becoming public health problem when increased above acceptance level. Health problem due to heavy metals pollution include nausea, vomiting, bone complications, nervous system impairments and even death become a major problem throughout many countries when metal ions concentration in the environment exceeded the admissible limits. Due to that, various treatment technologies had been searched to reduce the concentration of heavy metals in the environment.

Material and Methods

Sample Collection

The samples (liquid and solid wastes) were collected from different sites of Uttarakhand state including Severs, Rivers and Municipality. The method of solid and liquid waste collection was followed of APHA 1998. The waste samples were carried to laboratory in well packed box sealed in ice to avoid the contamination and stored at 4°C.

Preparation of Leachate and Metal Analysis

The leachate from solid waste was prepared according to the method described by French Standard method (Ferrari *et al.*, 1999, Srivastava *et al.*, 2005 and Savitha *et al.*, 2010). The 10 per cent leachate was used for chemical analysis. The filtrate of leachate was used to analyze the metal concentrations in solid waste by ICP-MS using AR grade chemicals and high grade reference (Blank). Liquid samples were also filtered through Whatman filter paper No. 42, residue discarded and supernatant was used to determine the total metal in it by ICP-MS (Ashok *et al.*, 2010).

Microbial Strain

Streptomyces sp. strains were maintained in agar slants containing nutrient broth, which was isolated from sludge contaminated with heavy metal. They were characterized morphologically and on the basis of biochemical reactions (Holt *et al.*, 1994). They were transferred weekly to new medium in order to keep metabolic activity and checked for purity by microscopic examination.

Metal Solutions and Heavy Metal Training

The aqueous solutions of metal ions used in the present investigation were prepared by using analytical grade chemicals. Individual stock metal ion solutions of different concentration of Cu from $CuSO_4.5H_2O$, Cd from $3CdSO_4.8H_2O$, Cr from $K_2Cr_2O_7$, Ni from $NiCl.6H_2O$, Pb from Pb acetate and Zn from $ZnSO_4$ respectively was prepared. These stock solutions were used to prepare dilute solutions of these ions by dilution with double distilled water. The stock solutions were acidified to 5< pH<7 using concentrated HCl in order to prevent the formation of metal hydroxide and to return the metal ion to the dissolve state (Hammaini *et al.*, 2006). Microbial strains were allowed to grow in the supplemented media with metal concentration. The concentration of metal *viz.* Zn, Mn, Cu, Cr, Cd, Ni and Pb was increased 0 µg/ml to 350 µg/ml. Cells were inoculated in nutrient broth (100 ml/flask) and kept under agitation in a rotary shaker, at 80 rpm, for 48 hours at 35 ± 2°C.

Biosorption Experiment with Solid and Liquid Waste

Experiments of heavy metals biosorption were done in Erlenmeyer flasks containing 250 ml and 25.0 ± 1.0 mg of cells. To ensure equilibrium, cells and waste were maintained in contact for 48 hours, under constant agitation, at 30-35°C ± 2°C. In all experiments, cells were obtained from the cultivation and collected from the same flask at the same growth stage. After 48 hours, cells were separated from the medium and residual metal concentrations were monitored by ICP-MS. Experiments were done in triplicate. The optimum pH and temperature maintained for the growth of microorganisms in the batch culture (Cybulski *et al.*, 2003; Kumar *et al.*, 2010). The pH and temperature were recorded daily.

Results and Discussion

Tolerance of Microbes

The control and optimization of bioremediation processes is a complex system of many factors. These factors include: the existence of a microbial population capable of degrading the pollutants; the availability of contaminants to the microbial population; the environment factors such as temperature, pH, the presence of oxygen or other electron acceptors and nutrients. Microbes were allowed to grown in the media containing heavy metals concentration and they were adopted for the concerned metals. Understanding metal–microbe relationships has led to advances in bioremediation (Malik 2004; Bruins *et al.*, 2000). Metals are toxic to all biological systems from microbial to plant and animal, with microorganisms affected more so than other systems due to their small size and direct involvement with their environment (Patel *et al.*, 2007; Sarret *et al.*, 2005; Giller *et al.*, 1999). Metal toxicity negatively impacts all cellular processes, influencing metabolism, genetic fidelity and growth. The acclimatized culture was enriched initially with 1 mg/l of each metal and the amounts were subsequently increased to 10 mg/l each. The growth culture on the other hand had no added metals and thus encouraged maximal biomass production. The cultures were maintained at room temperature (28–30°C) at initial pH of 6.8 ± 0.2. The metal (compound form) concentration increased from 0 mg/l (control) to 300mg/l and the microorganisms showed the tolerance against the metal. The strain *Streptomyces sp* showed the tolerance for chromium and co-tolerance for lead and copper. They were used as biosorbent for bioremediation process.

Biosorption Capacities

Biosorption capacities of microorganisms for metal ions generally depend on the metal concentration, the pH of the solution, the contact time, the ionic strength and the presence of competitive ions in the solution. Significant differences were observed in the uptake capacities of gadolinium ions by the various microorganisms used and no general relationship was applicable to all microbial species. These differences could be related to the nature, the structure and the composition of the wall layers and the specific surface developed by the sorbents in suspension. Morley and Gadd (1995) concluded for fungal biomass that the different cell wall polymers have various functional groups and differing charge distributions and therefore different metal-binding capacities and affinities.

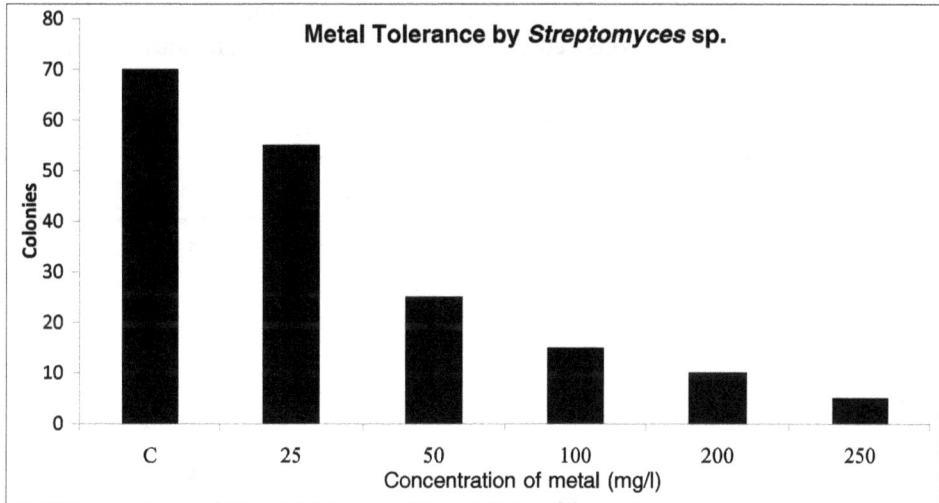

Figure 13.1: Metal Tolerance: *Streptomyces* sp. showed Tolerance Against Cr, Cu and Pb. The data presented in the figure is average of colonies having variance ±5.

Bioremediation Activity of *Streptomyces sp.*

The metal biosorption by gram positive filamentous bacteria *Streptomyces sp* at pH 7.0 and temperature 30°C is listed in Table 13.2. The copper, chromium and lead were sorbed by *Streptomyces* sp. Copper reduction ranged from 0.288-1.129mg/l, chromium reduction ranged from 0.386-6.42mg/l, which showed maximum reduction in chromium by *Streptomyces* sp. and reduction in lead ranged from 0.063-0.286 mg/l (Figure 13.2). Sharma and Loyal (2009) studied the chromium (III) removal by *Streptomyces* sp from tannery effluent and found the similar results as in case of chromium removal. Puranik PR, *et al.* (1995) found *Streptomyces pimprina* was potent biosorbent for cadmium in his study. *Streptomyces rimosus* biomass used for Cu, Zn and Cr biosorption by many investigators (Addour L, *et al.*, 1999; Chergui A, *et al.*, 2007; Ashwini C, *et al.*, 2009).

For bioremediation to be effective, microorganisms must enzymatically attack the pollutants and convert them to harmless products. As bioremediation can be effective only where environmental conditions permit microbial growth and activity, its application often involves the manipulation of environmental parameters to allow microbial growth and degradation to proceed at a faster rate.

Microorganisms are intimately involved in metal biogeochemistry with a variety of processes determining mobility and therefore, bioavailability. The balance between mobilization and immobilization varies depending on the organisms involved, their environment and physico-chemical conditions. Metal mobilization can arise from a variety of leaching mechanisms, complexation by metabolites and siderophores and methylation, where this results in volatilization. In addition, reduction of higher valency species may effect mobilization, *e.g.*, Cr(VI) to Cr(III). In the context of bioremediation, solubilization of metal contaminants provides a means of removal from solid matrices, such as soils, sediments, dumps and other solid industrial wastes (Kumar *et al.*, 2010).

Table 13.2: Biosorption with *Streptomyces* sp. (mg/l)

Sl.No.	Site	Sample	pH	Temp. (°C)	C_M (Cu)	C_{MF} (Cu)	C_M (Cr)	C_{MF} (Cr)	C_M (Pb)	C_{MF} (Pb)
1.	HR	Solid*	7.2	31	ND	ND	2.09	0.564	ND	ND
		Liquid	7.2	32	3.90	3.198	8.56	2.140	0.858	0.582
2.	DD	Solid	7.0	30	ND	ND	3.34	0.935	ND	ND
		Liquid	7.2	33	4.78	3.872	1.85	0.536	0.560	0.380
3.	MR	Solid	7.2	30	ND	ND	0.56	0.174	ND	ND
		Liquid	7.2	33	5.95	4.821	1.05	0.294	0.202	0.139
4.	RK	Solid	7.0	30	ND	ND	0.96	0.288	ND	ND
		Liquid	7.2	31	1.80	1.512	1.45	0.405	0.865	0.579
5.	KR	Solid	7.1	30	ND	ND	3.12	0.845	ND	ND
		Liquid	7.2	31	ND	ND	2.15	0.602	0.564	0.383
6.	RR	Solid	7.0	30	ND	ND	3.45	0.935	ND	ND
		Liquid	7.2	32	4.78	3.858	ND	ND	0.562	0.382

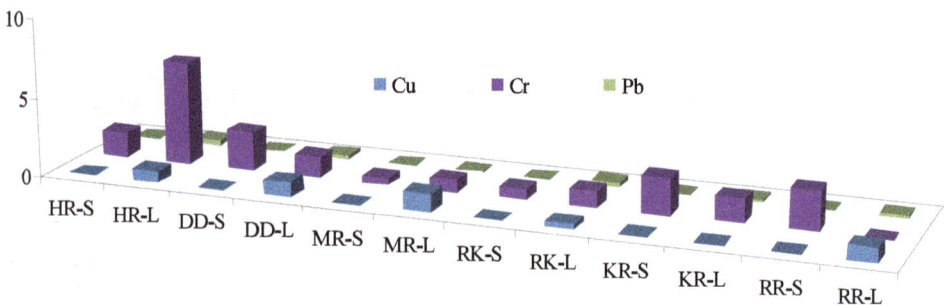

Figure 13.2: Accumulation of Cu, Cr and Pb by *Streptomyces* sp.

Conclusion

Bioremediation provides a technique for cleaning up pollution by enhancing the natural biodegradation processes. The tolerance test indicated that among experimental heavy metals maximum tolerance was shown to Cr, Pb and Cd, showing the growth of *Streptomyces* sp. up to 200-250 ug/ml. The heavy metals reduced by filamentous bacteria *Streptomyces sp* are copper, chromium and lead. The average Cu reduction by *Streptomyces* sp. was 18 per cent, Cr reduction was recorded as 72 per cent and Pb reduction was 32 per cent. The pH increases from 7.0 to 7.2 and temperature from 30°C to 33°C. The gram positive bacteria *Streptomyces* sp. was proved as strong biosorbent for chromium heavy metal. Further studies should be carried out to establish the quantity or level of bioremediation which affects the treatment efficiency.

Acknowledgements

Authors are thankful to Prof. J.P. Bhatt, HOD Biotechnology and Zoology, HNBGU (A Central University) Srinagar Garhwal (Uttarakhand) INDIA for his valuable suggestions and Dr. Gaurav Gupta, Director, Himachal Institute of Life Sciences, Panonta Sahib (HP) INDIA for providing the lab facility. The authors are also grateful to Dr. A.K Singh, Scientist C, Wadia Institute of Himalayan Geology for valuable comments and technical support.

References

Addour L, Belhocine D, Boudries N, Comeau Y, Pauss A, and Mameri N, (1999). Zinc uptake by *Streptomyces rimosus* biomass using a packed- bed column. J. Chem. Technol. Biotechnol. 74: 1089- 1095.

APHA (1998). Standard Methods for the Examination of Water and Wastewater, 18th Ed. American Public Health Association, Washington, DC: 45-60.

Ashok K., Bisht, B.S. Joshi V.D. and B. Indu (2010). Estimation of Heavy Metals and Metalloids from Wastewater of Bindal River Dehradun, Journal of Environment and Bioscience 24 (2) 195-198 ISSN 0973-6913

Ashwini C. Poopal and R. Seeta Laxman (2009). Studies on biological reduction of chromate by Streptomyces griseus, Jou. Hazar. Mat., 169(1-3): 539-545

Bruins MR, Kapil S, Oehme FW (2000). Microbial resistance to metals in the environment. Ecotoxicol Environ. Saf., 45: 198–207

Chergui A, Bakhti M.Z., Chahbob A., Haddoun S., Selatnia A and Junter G.A. (2007). Simultaneous biosorption of Cu, Zn and Cr from aqueous solution by *Streptomyces rimosus* biomass. Desalination, 206: 179-184.

Cybulski Z, Dzuirla E, Kaczorek E and Olszanowski A, (2003). The influence of emulsifiers on hydrocarbon biodegradation by Pseudomonadacea and Bacillacea strains. Spill Science and Technology Bulletin 8: 503 – 507.

Ferrari, B., Radetski, C.M., Veber, A.M. and Ferard, J.F. (1999). Ecotoxicological assessment of solid waste: a combined liquid and solid phase testing approach using a battery of bioassay and biomarkers. Env. Toxic. Chem. 18, 1195-1202.

Giller K.E., Wittwer E. and McGrath S. P. (1999). Assessing Risks of Heavy Metal Toxicity in Agricultural Soils. Human Ecol Risk Assess., 5: 683–689

Hammaini, A., F. González, A. Ballester, M. L. Blázquez and J. A. Munoz, (2006). Biosorption of heavy metals by activated sludge and their desorption characteristics. J. Environ. Manage., 84(4): 419-426.

Holt, J.G., Krieg, N.R, Senath, P.H.A. Staley, J.T. and Williams, S. T (1994). Bergey's manual of determinative bacteriology 9th Ed. Baltimore Md Williams and Wilkins Publication.

Kumar Ashok, Bisht B.S. and Joshi V.D. (2010). Biosorption of Heavy Metals by Four Acclimated Microbial species, *Bacillus* spp., *Pseudomonas* spp., *Staphylococcus* spp. and *Aspergilus niger*, J. Biol. Environ. Sci., 4(12) 97-108

Malik A (2004). Metal bioremediation through growing cells. Environ Int., 30: 261–278

Morley G.F. and Gadd G.M. (1995). Sorption of toxic metals by fungi and clay minerals. Mycol. Res. 9: 1429-1438.

Patel P.C., Goulhen F., Boothman C., Gault A.G., Charnock J.M., Kalia K. and Lloyd J.R. (2007). Arsenate Detoxification in a Pseudomanoad Hypertolerant to Arsenic. Arch Microbiol 187: 171–183.

Puranik, P.R., Chabukswar, N.S., and Paknikar, K.M. (1995). Cadmium biosorption by *Streptomyces pimprina* waste biomass, Applied Microbiology and Biotechnology, Vol. 43, pp. 1118-1121.

Sarrett G, Avoscan L, Carrière M, Collins R, Geoffroy N, Carrot F Covès J and Gouget B (2005). Chemical Forms of Selenium in the Metal-Resistant Bacterium Ralstonia metallidurans CH34 Exposed to Selinite and Selenate. Appl Environ Microbiol 71: 2331–2337.

Savitha J., Sahana N. and Praveen V.K. (2010).Metal biosorption by *Helminthosporium solani*- a simple microbiological technique to remove metal from e-waste, Curr. Sci. 98 (7), 903-904.

Srivastava Richa, Kumar Dinesh, Gupta, S.K. (2005). Bioremediation of Municipal sludge by vermitechnology and toxicity assessment by *Allium cepa*, J. Biorem. Tech., doi: 10.1016/j.biortech.2005.01.029

Sze K.F., Lu Y.J. and Wong P.K. (1996). Removal and recovery of copper ion (Cu^{2+}) from electroplating effluent by a bioreactor containing magnetic immobilized cells of *Pseudomonas putida* 5x., Global Environment. Biotech., in Proceedings of third Biennial Meetings of the International Society for Environmental Biotechnolgy, 15-20, July, U.S.A., 131-149.

2013, Environmental Technology
Editors: D.R. Khanna, A.K. Chopra, Gagan Matta, R. Bhutiani & Vikas Singh
Published by: DAYA PUBLISHING HOUSE, NEW DELHI

Pages 135–142

Chapter 14

Water Harvesting through Farm Pond and Utilization of Conserved Water for Vegetable Crops

C.R. Subudhi

CAET, OUAT, Bhubaneswar – 751 003

ABSTRACT

A trial was conducted during 2005 06 and 2006-06 at All India Coordinated Research Project for Dryland Agriculture Phulbani,Orissa,India., with an objective to obtain the water loss and economics of the lined ponds.There were three treatments T1-Lined pond with soil cement plaster (6:1) 8cm thickness,T2-Unlined pond,T-3-No pond.10 per cent of the cropped area was dug for construction of the pond in Lined and Unlined pond treatments. The size of the pond is 7m top widths, 1m-bottom width, 3m heights, and 1:1side slope. The water harvested in pond was reutilized for the pumpkin crop, which was sown only in Lined pond treatment, as there was no water available in unlined pond so the crop was not sown there. Lined pond with soil cement (6:1) plaster of 8cm thickness gave highest Tomato yield of 4.8 t/ha during *kharif* 2008-09 and radish root yield of 25.5 t/ha in *rabi* seasons of 2008-09. The water loss was 326 lit/day in lined pond and 24,000 lit/day in unlined pond. The benefit: cost ratio in lined pond was 3.04 as compared to 1.64 in unlined pond during 2008-09. The light textured well-drained upland soils in North Eastern Ghat Zone provide scope for cultivation of vegetables during rainy season. The intermittent dryspells and terminal drought affect the performance of those high value crops in most of the years. About 25 per cent of the rainfall is lost as run-off. Harvesting of this run-off water in farm pond with proper lining will conserve the run-off water and recycling of this water for life-saving irrigation will protect the crop from drought/dryspell grown in 90 per cent of land area. The ponds will be helpful for sustainability in productivity of dryland crops. Soil structure and organic matter status decide the water holding capacity of the soil. Soil physico-chemical characteristics depend on the systems of nutrient management. Keeping those points in view, the present experiment involving two water management systems (no pond and pond) has been designed.

Introduction

The light textured well-drained upland soils in North Eastern Ghat Zone provide scope for cultivation of vegetables during rainy season. The intermittent dryspells and terminal drought affect the performance of those high value crops in most of the years. About 25 per cent of the rainfall is lost as run-off. Harvesting of this run-off water in farm pond with proper lining will conserve the run-off water and recycling of this water for life-saving irrigation will protect the crop from drought/dryspell grown in 90 per cent of land area. The ponds will be helpful for sustainability in productivity of dryland crops. Soil structure and organic matter status decide the water holding capacity of the soil. Soil physico-chemical characteristics depend on the systems of nutrient management. Keeping those points in view, the present experiment involving two water management systems (no pond and pond) has been designed.

Objectives

1. To quantify the increase in land productivity and land use efficiency through on-farm water harvesting
2. To quantify the water/seepage loss in different ponds

Methodology

10 per cent of the cropped area was dug for construction of the pond in Lined and Unlined pond treatments. Size of the pond is 7m top widths, 1m-bottom widt, 3m heights, and 1:1side slope. The water harvested in pond was reutilized for the pumpkin crop, which was sown only in Lined pond treatment, as there was no water available in unlined pond so the crop was not sown there.

Rainfall (mm)	:	
a) Normal	1407mm	
b) Current year(2008)	1531.8mm	
c) Cropping season	Cauliflower-1195.1mmRadish-515.8mm	
d) Dry spells (> 10 days)	Weather was favorable to rice and groundnut but not favorable to Arhar as there was no rainfall between Sep 24, to Oct 04, and Oct 10 to Nov 15, 2008	
Crop and Variety	: Cauliflower-Hemlata	
	Radish-Pusachetki	

| Treatment details | : | T_1-Lined pond with soil cement plaster in 6:1 ratio (8 cm thickness)T_2- Unlined pondT_3- No pond (control) |
| Experimental design | : | No design |

Plot size (sq.m)	:	
a) Gross plot		30m x 15m
b) Net plot		28.2m x 12.3m
Date of sowing	:	Cauliflower-25.6.2008 and transplanted on 19.7.2008
		Radish-20.08.2008
Date of harvesting	:	Cauliflower-02.09.2008 to 26.09.2008 (14 different dates)
		Radish-03.10.08 to 28.10.08 (6 different dates)
Spacing (cm)	:	Cauliflower-45 cm X 45 cmRadish-45 cm X 5 cm
Seed rate (kg/ha)	:	Cauliflower-350gm/haRadish-6 kg/ha
Basal manuring (NPK)	:	Cauliflower-25: 40:60 kg
Top dressing		$N-P_2O_5-K_2O/ha$
		1st top dressing-50kg N/ha
		2nd top dressing-50kg N/ha
		Radish-25: 50:75 kg $N-P_2O_5-K_2O/ha$
		1st top dressing-25kg N/ha
Previous crop and fertilizer applied	:	As trial was conducted at same place so same fertilizer as mentioned above.
Layout Plan	:	

N
↑

	T_3	
Lined Pond		Unlined Pond
T_1		T_2

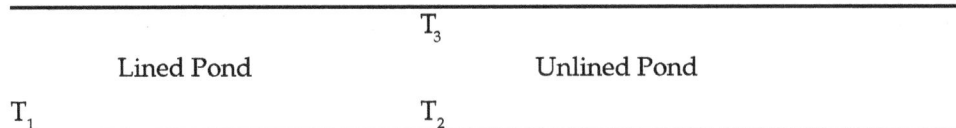

Results and Discussion

Soil texture was shown beliow,which was loamy sand from top layer to sandy loam in bottom layer.

Soil Properties		Soil Depth (cm)		
Physical		0-10	10-20	20-30
a) Texture	Class	Loamy sand	Sandy loam	Sandy loam
	Per cent Sand	87.8	77.8	75.8
	Per cent Silt	5.0	10.0	10.0
	Per cent Clay	7.2	12.2	14.2

Table 14.1(a): Yield and Economics during 2008-09

Treatments	Grain Yield (kg/ha)		Cost of Cultivation (Rs/ha)	Gross Income (Rs/ha)	Net Income (Rs/ha)	B:C Ratio
	Cauliflower (Kharif)	Radish (Rabi)				
T_1	4,800	25,500	81,700	2,49,000	1,67,300	3.04
T_2	3,822	-	46,685	76,440	29,755	1.64
T_3	3,021	-	32,750	60,420	27,670	1.85
Mean	3,881	25,500	53,712	1,28,620	74,908	2.18
2008-09	1195.1 mm (Cauliflower)	515.8 mm (Radish)				

* Market value of the Cauliflower Rs. 20 kg in 2007-08 Radish Rs. 6/- per kg (2008-09).

Table 14.1(b): Energy input and output in different treatments

Treatments	Kharif			Rabi		
	Energy Input (MJ/ha)	Energy Output (MJ/ha)	Energy Output : Input Ratio	Energy Input (MJ/ha)	Energy Output (MJ/ha)	Energy Output : Input Ratio
T_1	11977	18816	1.57	8365	99960	11.95
T_2	11817	14982	1.27	-		
T_3	11793	11842	1.00	-		

Table 14.2: Water Loss during 2005-06 and 08-09

Treatment	Water Loss (lit/day)				
	2005-06	2006-07	2007-08	2008-09	Mean
T_1-Lined pond with soil cement (6:1) plaster 8cm thickness	86	131	225	326	192
T_2-Unlined pond	37,000	33,000	28,000	24,000	30,500
T_3-No pond					

The highest B:C ratio (3.04) was obtained in lined treatment due to two crops was harvested (Table 14.1a). Highest energy output: input ratio was obtained in lined pond in both Kharif (1.57) and rabi (11.95) and lowest in T3 *i.e.*No pond treatment (Table 141b). Considering both kharif and rabi maximum energy output: input ratio was obtained in lined pond (5.83). The mean water loss and mean yield was presented in Tables 14.2 and 14.3 respectively. The mean yield was highest (8.95 t/ha) in lined pond and water loss was lowest (192 lit/day) in lined pond. The yield of Cauliflower was highest in T_1 (4.8t/ha)(2008-09) and lowest seepage loss (192 lit/day) (mean) (Tables 14.3 and 14.6).The lowest Cauliflower yield was obtained in no pond (T3) treatment (3.02t/ha). Unlined pond gave a yield of 3.82 t/ha which was 20 per cent lower than the lined pond. The seepage loss in unlined pond was highest (30,500 lit/day) over the last four years. The no of irrigation was 5 and one in case of lined and unlined pond respectively during 2008-09 (Table 14.5). The cost of lined pond was Rs9.967/- and that of unlined pond was Rs 2,993/- (Table 14.6). The water use efficiency was presented in (Table 14.7) which was highest in lined pond (4.016kg/ha/mm).The cost of lining per square meter was Rs 88/-. Figures 14.1 and 14.2 shows that the water loss in lined pond is in increasing trend may be due to some cracks where as the water loss in unlined pond is decreasing trend may be due to

Table 14.3: Yield during 2005-06 to 2008-09

Treatment	Yield of Kharif Produce (t/ha)				
	2005-06 Tomato	2006-07 Tomato	2007-08 Cauliflower	2008-09 Cauliflower	Mean
T_1-Lined pond with soil cement (6:1) plaster 8cm thickness	22.83	3.78	4.4	4.8	8.95
T_2-Unlined pond	21.33	3.55	3.82	3.82	8.13
T_3-No pond	19.83	3.52	3.53	3.02	7.48
Mean	21.33	3.62	3.92	3.88	8.19

Table 14.4: Yield during 2006-07 to 2008-09 (*Rabi*)

Treatment	2006-07 Pumpkin, kg/ha	2007-08 Radish, kg/ha	2008-09 Radish, kg/ha	Meankg/ha kg/ha
T_1-Lined pond with soil cement (6:1) plaster 8cm thickness	34,500	22,500	25,500	27,500

Table 14.5: No. and Quantity of Irrigation Applied during 2008-09

Treatment	No. of Irrigation Applied	Quantity of Irrigation Applied, lit
T_1-Lined pond with soil cement (6:1) plaster 8cm thickness	5	600X5=3,000
T_2-Unlined pond	1	300
T_3-No pond		

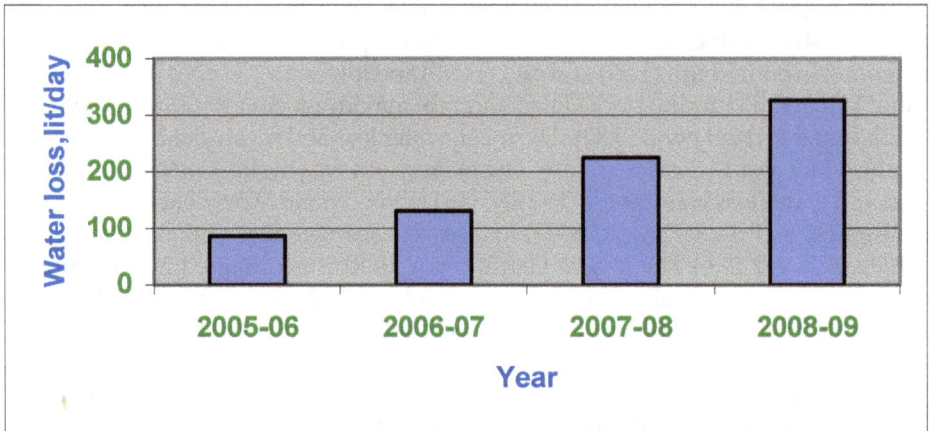

Figure 14.1: Water Loss in Lined Pond

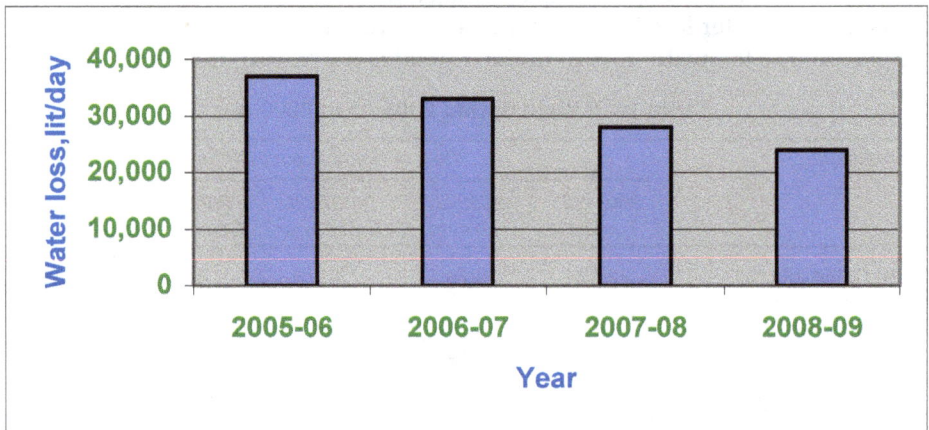

Figure 14.2: Water Loss in Unlined Pond

siltation. Figures 14.3 and 14.4 shows the radish after irrigating from lined pond and lined pond visited by farmers respectively.

Table 14.6: Seepage Loss in Different Treatments Over the Last Four Years

Treatment	Total Cost of the Pond (Rs.)	Cost of Lining (Rs./m³)	Water Loss from the Pond (lit/day)	Time taken to Dry Up the Pond (Days)	Cost of Storage (Rs/m³)	Economic Loss Due to Seepage Loss (Rs/day)
1	2	3	4	5	6=(7/4)X1000	7=2/5
T_1-Lined pond with soil cement (6:1) plaster 8cm thickness	9967	88.5	192	391	133	25.49
T_2-Unlined pond	2993	0	30,500	2.45	40	1222
T_3-No pond						

Figure 14.3: Rabi Crop Radish after giving Irrigation from Pond

Figure 14.4: Lined Pond

Table 14.7: Water Use Efficiency in 2008-09

Treatments	Grain Yield (kg/ha/mm)	
	Cauliflower (Kharif)	Radish (Rabi)
T₁	4.016	49.437
T₂	3.198	-
T₃	2.528	-
Mean	3.247	49.437
2008-09	1195.1 mm (Cauliflower)	515.8 mm (Radish)

Table 14.8 shows that growth parameters of cauliflower and radish were highest in lined pond.

Table 14.8: Growth Parameters in Different Treatments during 2008-09

Treatment	Cauliflower			Radish		
	Plant Height, cm	Spread, cm	No. of Leaves	Plant Height, cm	Spread, cm	No. of leaves
T₁-Lined pond with soil cement (6:1) plaster 8cm thickness	45	61.6	17.6	25.8	45.4	26.8
T₂-Unlined pond	38.5	53.8	16.4	-	-	-
T₃-No pond	32.2	46.8	14.2	-	-	-

References

Panda R.K., Bhattacharya R.K. 1983. Lining of small irrigation channels J. of Irrigation and Power 83: 385-391.

Subudhi, C.R. 2008. Study of lining materials for supplemental irrigation International symposium in "Agro Meteorology and food security" during 18-21Feb, 08 at CRIDA, Hyderabad.

2013, Environmental Technology
Pages 143–146
Editors: **D.R. Khanna, A.K. Chopra, Gagan Matta, R. Bhutiani & Vikas Singh**
Published by: **DAYA PUBLISHING HOUSE, NEW DELHI**

Chapter 15

Nanotechnology and Conservation: Exploring Nanomaterials for Generation of Energy

Seema K. Ubale

Department of Physics,
Dharampeth M.P. Deo Memorial Science College, Nagpur

ABSTRACT

Nanotechnology is the technology dealing with materials at the nano level, which is one billionth part of a meter. Nanotechnology based devices are built at molecular level. Nano techniques have drastically reduced the size of appliances.

Conservation means protection and preservation. It is mainly related with the environment. Conservation of environment includes conservation of natural resources such as air, water, soil and also to control and curb global warming, green house effect and to retain biodiversity.

Different electrical gadgets are available in the market right from kitchen to sky and these have become essential part of our life. But with the advancement of the technology energy requirement has also increased rapidly.

The energy is derived from two types of resources, 1. Renewable resources and 2. Non renewable resources. Non renewable resources are petrol, diesel, coal or fossil fuel. The major drawback with the non renewable energy sources is the harm they are causing to the environment due to harmful emmissions and secondly they will expire soon. Solar energy which is renewable source of energy can be a good alternative, since it is clean and abduntly available source of energy.

Solar energy is converted into electrical energy with the help of solar panels and stored in the batteries. The solar panels and the storage batteries are big in size and are very costly. Nanotechnologists are working for reducing the size of panels and storage batteries. Further with reduction in cost the technology will become user friendly thereby leading to clean environment.

Keywords: Solar energy, Nanotechnology, Clean environment, Economically viable.

Nanotechnology is the technology which deals at the nano level. 1 nano meter is equal to 10^{-9} meter *i.e.* it is equal to one billionth part of a meter. The nano devices are built at molecular level. This has drastically reduced the size of the appliances. The palm size computers, memory devices such as compact disc (cd), digital video disc (dvd), pen drives with very large storage memory are some of the examples.

This is one side of the technology and advancement, the other dark side of this issue is the ever increasing demand of the energy. This other side has driven us to the rapid fall in the environment, deforestation, green house effect, increase in average temperature of the earth and melting of ice from the polar regions.

The advancement of the technology for enhancing the quality of life and at the same time maintaining the balance of the ecosystem is the major task. Scientists are working on it continuously. For our day to day life we need different form of energy, such as electrical energy, chemical energy, mechanical energy, light energy etc. Out of these various forms electrical energy is the most important one. With the help of this energy all other forms of energy can be made available.

The electrical energy is derived from two types of energy sources, non renewable and renewable type of energy sources. The non renewable resources are fossil fuel such as coal, petrol, and diesel etc. Solar energy, wind energy, tidal energy are examples of renewable energy sources. The source of energy does not extinguish here, hence it is called as renewable energy source.

Energy from the Sun is called as solar energy. Solar energy can be tapped only in daytime. For tapping of the solar energy mirrors, glass tube panels or solar cell panels are used.

In the simple arrangement of solar cooker, mirrors or highly polished metal sheets are used to focus the sunrays on the cooking area of the cooker. The cooking vessel is painted black from outside so as to absorb more and more solar energy which can be used for cooking purpose.

The glass tube solar panels are used to heat water in solar water geysers. The solar cell panels are used to convert solar energy into electrical energy and the electrical energy is stored in the batteries so that it can be used as per requirement (Figure 15.1).

Solar energy is tapped using both solar thermal and Photovoltaic cell. In Solar thermal the solar energy is used to produce steam which is subsequently used to drive a turbo-generator to produce electricity on large scale. The photovoltaic solar cell panels are packaged interconnected assembly of solar cells or photovoltaic cells. This is an active technique of harnessing the solar energy. A single solar panel can

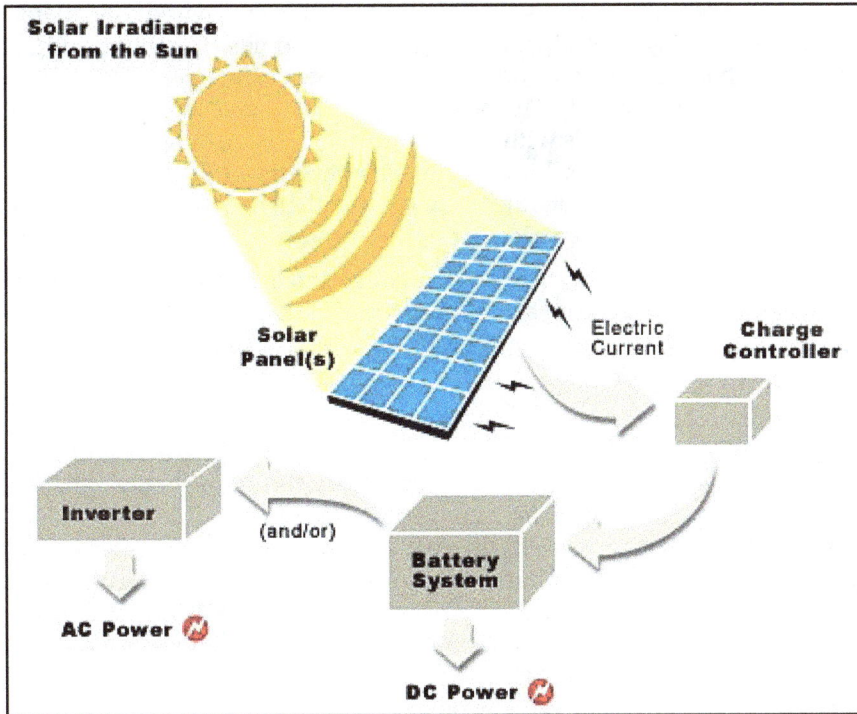

Figure 15.1: Conversion of Solar Energy to Electrical Energy

produce only a limited amount of power, thus the installations contain several panels.

Electrical Power Generation in India

In India about 65.34 per cent of the electricity consumed is generated by thermal power plants, 21.53 per cent by hydroelectric power plants, 2.70 per cent by nuclear power plants and 10.42 per cent by renewable energy sources(1). The amount of solar energy produced in India is less than 1 per cent of the total energy demand. In India we get lot of Sun shine through out the year, excluding the rainy season. There is very large scope for setting the solar power plants in our country. However the major reason for the poor percentage is the cost of its installation. The solar thermal power plant has cost 4 times as much as the coal based steam thermal power plant **(2)**. The other reason is the size and rigidity of the solar panels.

Materials presently used for photovoltaic solar cells include monocrystalline silicon, polycrystalline silicon, amorphous silicon, cadmium telluride and copper-indium selenide or sulphide. The cost of material and the preparation process are quite high. Nanotechnology in general and nano materials in particular can be of great help to overcome the problem of large size of the panels and also the rigidity of the panels.

In USA recently one company Solar ply has developed new type of solar panels. These are cheaper, more flexible and printable. The company uses copper indium gallium diselenide for the panels (3). The absorption coefficient of this material is higher than any other semiconductor used for solar modules. Nanosolar Co. based in San Jose, CA has developed and commercialized a low-cost printable solar cell manufacturing process (4). The process involves a semiconductor ink that will enable it to produce solar cells with a basic printing process, rather than using slow and expensive high-vacuum based thin-film deposition processes. The ink is deposited on a flexible substrate (the "paper"), and then nanocomponents in the ink align themselves properly via molecular self-assembly.

Thus it gives us a great hope to work on this nano material solar ink which will lead to clean form of energy and also help to conserve the eco system.

From last year, the Union government through the ministry of new and renewable energy (MNRE) has launched the JNNSM(the Jawaharlal Nehru National Solar Mission) to promote national efforts on harnessing solar energy in a big way. Towards this end, the CSIR launched a mega project titled – TAP-SUN, to focus and direct the efforts of its highly skilled scientific work force towards finding better, efficient and economic solutions to solar energy conversion, storage and conservation. NCL scientists in Pune are also working in goal-oriented research programmes in the energy sector. The primary emphasis is on non-silicon initiatives. They are working on a technology that could include organic and hybrid solar cells which can deliver fairly high efficiency at an economically viable cost. The NCL is also working on dye-sensitised solar cells, a relatively new class of low-cost solar cell, that belongs to the group of thin film solar cells, besides organic and hybrid cells.

References

As on 31-08-2011 Source: CEA Central Electricity Authority, Ministry of Power.

Ecoworld.com. 2007-05-15. Retrieved 2010-11-27.

Repins, I., Contreras, Miguel A., Egaas, Brian, Dehart, Clay, Scharf, John, Perkins, Craig L., To, Bobby, Noufi, Rommel (2008).

Vidal, John (December 29, 2007). "Solar energy 'revolution' brings green power closer". London: The Guardian. Retrieved 2007-12-31.

2013, Environmental Technology   *Pages 147–153*
Editors: **D.R. Khanna, A.K. Chopra, Gagan Matta, R. Bhutiani & Vikas Singh**
Published by: **DAYA PUBLISHING HOUSE, NEW DELHI**

Chapter 16

The Waste Hierarchy: Reduce, Reuse, Recycle–A Solution to Better Environment

Sadhana Bhoyar

Department of Resource Management,
Sevadal Mahila Mahavidyalaya, Nagpur – 440 009, M.S.

ABSTRACT

Household waste management is an issue not to be ignored. In the wake of rapidly depleting natural resources and the health hazards caused by the huge amount of waste produced, recovery and reuse have become inevitable in the present scenario. The synthetic pathways of nature are overloaded. Another important point is that the nature recycling period are very long compared with the human life span and the society is interested in recycling on a time scale which is comparable with an individual's lifetime. It is clear that new method of water reclamation and reuse must be developed. Kitchen waste or food waste or garbage is any form of biodegradable waste that was originally intended for consumption. It typically consists of vegetable scraps, fruits, food waste, etc. and other discards from kitchen. Every day we through a kitchen waste into the trash can. These kitchen wastes are full of nutrients. So we can recycle our kitchen waste into our yard or vegetable garden. The experiment on growing vegetable in kitchen waste manure, vermicompost and normal soil conducted in September to December 2010. Kitchen waste manure was prepared by collecting daily kitchen waste like egg shells, food scrap, vegetable peels etc. The collected food items were emptied in designated area and mixed with soil. After two months the dark brown soil *i.e.* manure was ready to use. Tomato plants were sown in three different mediums. Kitchen waste manure, vermicompost and normal soil. Growth and yield of tomato was highest in kitchen waste manure and scored highest marks in palatability test. Nutrient contents of tomato was comparatively equal to other medium. So this study will be helpful in preventing pollution of environment and also to earn some money by utilizing household kitchen waste. It was found that tomato grown in kitchen waste manure contains 26.1 mg Vitamin C/100gm followed by 28.4, 25.4 in vermicompost and normal soil.

Keywords: *Kitchen waste manure, Soil analysis, Nutrient contents and Vermicompost.*

Introduction

Household waste management is an issue not to be ignored. In the wake of rapidly depleting natural resources and the health hazards caused by the huge amount of waste produced, recovery and reuse have become inevitable in the present scenario. Kitchen waste or food waste or garbage is any form of biodegradable waste that was originally intended for consumption. It typically consists of vegetable scraps, fruits peels, food waste and other discards from kitchen. Everyday we through kitchen waste into trash can. These kitchen wastes are full of nutrient so we can recycle our kitchen waste into our yard. The disposal of kitchen waste is a big problem especially in country like India. There is a lack of knowledge on proper waste management and hence we see more often the wastes are being thrown outside or laying unattended. Though in many cities the urban waste is managed by City Corporation, it is still ineffective, unless each small family takes the responsibility for their own waste produced. According to analysis the total municipal solid waste collection of three cities Kanpur, Kolkata and Nagpur are 450, 1100 and 600 tons/day respectively and the vegetables wastes are 1.2 to 3.4, 2.2 and 0.6 to –13.1 respectively. The solution for this problem is composting and recycling kitchen waste into rich manure and growing vegetables in it. It is also the best solution for keeping environment clean and getting fresh vegetables from the point of health. Organic agriculture is an ecological production management system that promotes and enhances biodiversity, biological cycle and soil biological activity. It is based on minimal use of off farm inputs and management practices that restore, maintain and enhances ecological harmony.

Food waste can be reduced at point-of-purchase and in home by adopting some simple measures; planning when shopping for food is important, spontaneous purchases are shown as often the most wasteful, proper knowledge of food storage and food preparation reduces food becoming inedible and thrown away.

Materials and Methods

It was in an underdeveloped country - India that modern composting got its big start. Sir Albert Howard, a Government Agronomist developed the so called Indore Method of composting. His method calls for three parts garden clipping to one part, kitchen waste arranged in layers and mixed periodically. Howard published his ideas on organic gardening in the year 1940 in a book named 'Agricultural Treatment' and the first agriculate advocate of Howard's Method in U.S. was J. I. Rodale (1898-1971) founder of organic gardening magazine. These two men made composting popular with gardeners who prefer not to use synthetic fertilizers.

Figure 16.1: Old Cooler Tank Used for Preparation of Kitchen Waste Manure

The composting is a natural recycling process of the plant remains into dark earthly material which enriches the soil. The process of converting kitchen waste into organic manure is very simple, mostly done by nature with the help of microorganism.

The preparation of manure from kitchen waste was done by following steps:

1. Got a old cooler tank measuring 3x5 ft. having holes on all directions to have a good oxygen flow.
2. Kitchen waste nearly ½ - 1 kg was collected at the end of the day.
3. Kitchen waste was added to tank having one thin layer of soil and one layer of kitchen waste likewise.
4. Layering was repeated till the tank was full. This procedure took one month.
5. The tank was covered with sacs.
6. To keep it moist water was sprinkled on the top of the tank alternate days as it was prepared in the month of May.
7. After 2 months the manure was ready to use.

Three mediums of soil were selected for growing vegetables *i.e.* normal soil, readymade vermicompost and homemade kitchen waste manure. Two earthen pots for each soil were taken for sowing. Tomato the fruit class vegetable was selected for experiment. The sample plant of tomato grown in cocopit for 2 weeks were selected. The experiment was started in October-2010. Water was given alternate days. At the end of December fully grown tomato were seen in all three mediums of soil. No chemical fertilizers, sprays were used during the growth of tomato.

Biodegradable pollutants can be effectively treated with microorganism and decomposed to simple harmless constituents. The biological treatment is rather and inexpensive treatment. It not only eliminates the pollution but may also provide economically useful product. Little human efforts are required in process of disposal of biodegradable wastes.

Results and Discussion

The data on yield of tomato presented in Table 16.1 showed that the number of tomato grown in normal soil was 41 followed by 57 and 67 in vermicompost and kitchen waste manure. The result also showed that waste hierarchy helps to produce food of high nutritional quality in sufficient quantity.

Table 16.1: Yield of Tomato

Type of Soil	No. of Tomato	Weight of Tomato
Normal Soil	41	3 kg 600 gm
Vermicompost	57	5 kg 500 gm
Kitchen Waste Manure	67	6 kg 500 gm

The data presented in Table 16.2 showed that the height of tomato plant in normal soil was 15 inches followed by 20 and 24 in vermicompost and kitchen waste manure. The height of plant grown in kitchen waste manure was maximum compared with other medium because kitchen waste manure mixed with the soil improves the soil quality. Compost hold moisture. This is good for plants and of course it is also good for the environment because it produces soil into which rain easily soaks.

Table 16.2: Height of Fully Grown Plant

Type of Soil	Height of Fully Grown Plant(In Inches)
Normal Soil	15
Vermicompost	20
Kitchen Waste Manure	24

Table 16.3: Nutrient Contents of Tomato

Type of Soil	Nutrient Contents	
	Vitamin C (mg/100gm)	Moisture (per cent)
Normal Soil	25.4	92.9
Vermicompost	28.4	94.0
Kitchen Waste Manure	26.1	96.6

The data presented in Table 16.3 showed that tomato grown in kitchen waste manure contains 26.1mg/100gm Vitamin C. It is quite good compared to readymade vermicompost. The moisture content in kitchen waste manure is highest *i.e.* 96.6 per cent followed by 94.0 per cent and 92.9 per cent in vermicompost and normal soil.

Figure 16.2: Yield of Tomato Plant

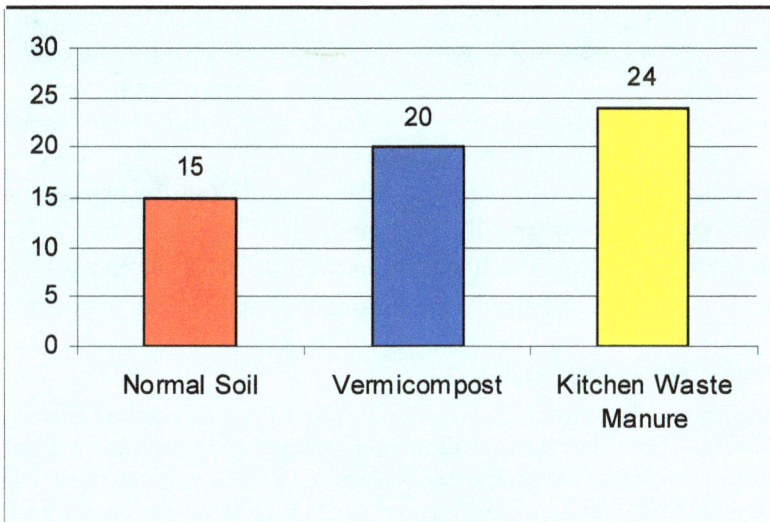

Figure 16.3: Height of Fully Grown Plant (in inches)

The moisture content of tomato grown in kitchen waste manure is highest compared to vermicompost and normal soil.

Table 16.4: Analysis of Soil Testing

Type of Soil	Characteristics of Soil			
	Nitrogen	Potash	pH	Moisture
Normal Soil	420	0.60	7.2	2.35
Vermicompost	630	0.90	7.3	4.0
Kitchen Waste Manure	987	1.41	7.0	3.30

The analysis of soil was done by the Department of Agriculture, Government of Maharashtra, Nagpur. The standard pH value of soil is 6.5 – 7.0. The data on analysis of soil in Table 16.4 shows that the normal soil is 7.2 followed by 7.3 and 7.0 in vermicompost and kitchen waste manure. Nitrogen is essential part of soil fertility. The content of Nitrogen is normal soil is 420 followed by 630 and 987 in vermicompost and kitchen waste manure. The percentage of nitrogen in kitchen waste manure is highest compared with other two mediums. Moisture holding capacity of normal soil was 2.35 followed by 4.0 and 3.30 in vermicompost and kitchen waste manure. The moisture holding capacity is not highest in kitchen waste manure but it is comparatively good.

Conclusion

☆ The problem of disposal of household kitchen waste can be solved by utilizing it for preparing manure.

☆ It will not only enhance the quality of soil but also helps in increasing the productivity of tomato.

☆ Tomatos grown in kitchen waste manure are healthy and nutritious as it contains highest percentage of moisture and Vitamin C.

☆ It helps to work as much as possible within a close system with regard to organic matter and nutrient elements.

☆ It is easy and fun way to get the whole family to take part in an environmentally friendly solution.

☆ It will be helpful to earn some money by utilizing kitchen waste as manure.

☆ This will also be helpful in preventing environmental pollution.

Acknowledgement

The author is thankful to Hon'ble Dr. Pravin Charde, Principal, Sevadal Mahila Mahavidyalaya, Nagpur for his valuable guidance and inspiration. The author is also thankful Dr. (Mrs.) Sunita Borkar, Associate Professor, Department of Resource Management, LAD College, Nagpur and Dr. S.K.U. Charjan, Assistant Professor, Punjabrao Krishi Vidyapeeth, Nagpur for their valuable guidance.

References

Abbasi, S.A. and Ramaswamy, E.V. (2001). Solid Waste Management with Earthworms, Discovery Publishing House, New Delhi, p.2, 5, 9-10.

Anjaneyulu, Y. Introduction to Environmental Science, B.S. Publications, Hyderabad, p.131.

Asthana, D.K. and Asthana M. Environment: Problems and Solutions, S. Chand and Company, Delhi, p.135.

Chatwal and Sharma, H. (2004). A Textbook of Environmental Studies, Himalaya Publishing House, Mumbai, p.216.

Websites:

info@lawn&gardenhotline.org

www.indahtrade.com

http: //www.whatwherenhow.com

2013, Environmental Technology
Editors: **D.R. Khanna, A.K. Chopra, Gagan Matta, R. Bhutiani & Vikas Singh**
Published by: **DAYA PUBLISHING HOUSE, NEW DELHI**

Pages *155–160*

Chapter 17

Application of Traditional Building Techniques for Creation of Disaster Proof Civil Structures

G.C. Mishra and Tushar Adlakha

Department of Civil Engineering,
Lingaya's University, Faridabad

ABSTRACT

In ancient times various types of sustainable structures was created by people for their shelter and other various purposes. Most of the structures of ancient period are yet strong, despite facing several natural calamities like earthquake, landslides, flooding, intense rainfall, storms etc. After studying details about the type of building material constituents used in those structure it has been found that they innovated constituents of building material for binding different unit of stones/bricks by using some extracts of plant products in combination with calcium carbonate in specific ratios which has given building structures resistant properties to various environmental degradation forces. Construction workers in ancient China developed structure using sticky rice mortar about 1,500 years ago by mixing sticky rice soup with the standard mortar ingredient. That ingredient is slaked lime, limestone that has been calcined, or heated to a high temperature, and then exposed to water. Sticky rice mortar probably was the world's first composite mortar, made with both organic and inorganic materials. The mortar was stronger and more resistant to water than pure lime mortar. Builders used the material to construct important buildings like tombs, pagodas, and city walls, some of which still exist today. Some of the structures were strong enough to shrug off the effects of modern bulldozers and powerful earthquakes. Amylopectin an important ingredients found in rice when used in the mortar acts as an inhibitor: The growth of the calcium carbonate crystal was controlled, and a compact microstructure was produced, which should be the cause of the good performance of this kind of organic-inorganic mortar."

Apart from these ingredients there are several other traditional techniques and building ingredients used by ancient people has potential to be investigated and if found suitable then can be applied in creation of sustainable building and other required civil structures. The objective of this research is to analyze different traditional building constituents used in ancient building structures and their application in appropriate ratio with modern building ingredients for creation of disaster proof building structures.

Introduction

Evaluation of new cons. techniques together with increasing understanding of seismic forces and the building response has certainly contributed positively to decrease seismic vulnerability. Even after immense progress in earthquake engineering overwhelming large proportion of the building stock still has no standards of earthquake safety and these would give way in case of earthquake loading. The common people, at present, do not adhere to correct design and construction methodology while using modern seismic design method or even the use of indigenous practices. Several case study of Uttarakhand state clearly brings forth deterioration of time tested indigenous construction practices and proliferation of non-scientific and improper use of concrete. The case study highlighted three interrelated aspects. First, these bring forth key features of local traditional knowledge and capacity of rural communities for mitigation, preparedness and recovery from earthquakes. The traditional knowledge is embodied in physical planning and building, skillfully using local resources, mutual supports systems and informal livelihood mechanisms. Second, these provide an in-depth understanding of the transformation process and their impact on traditional knowledge and capacity and resulting up-to enhanced earthquake vulnerability. Third, these show the implication of post-earthquake rehabilitation on disaster vulnerability in the long run. Severe damage and poor performance of RC buildings in India during earthquakes are a matter of serious concern. There is a need to develop suitable screening methods for seismic safety of existing building so that prioritization may be under taken on the deficient buildings. Systematic studies on calibration of damage data of the 2001 Bhuj earthquake in India has been carried out for the first time in India. Ahmedabad is located ~250 km from the epicenter and is placed in Zone III of Indian seismic zoning. Areas placed in zone III are expected to experience shaking intensity VII on MSK Scale, which is the same as the shaking intensity experience by the city during the earthquake. The present work considers only RC frame Buildings that are prevalent in urban India. Similar method needs to be developed for masonry buildings as well. Bhuj 2001, Andaman –Sumatra 2004 and Muzaffarabad 2005 have revealed that human and property loses are mainly due to collapse of very large vulnerable building stock in the region. Post Disaster studies, especially Bhuj earthquake, have given engineering community – both architect and structural engineers number of important lesson to be adequately addressed so as to mitigate the effect of such hazard in future. Poor quality of construction materials was also observed to be an important cause of failure of RC structures especially in Ahmedabad area, which is about 300km away from the source of the earthquake. Indian Standard Codes for Earthquake Resistant

Design and Construction prescribes, besides other things use of proper quality of building materials in a proper way. It is important to follow the provisions of the codes in a sincere way in construction of buildings for mitigating the effect of such hazards in future.

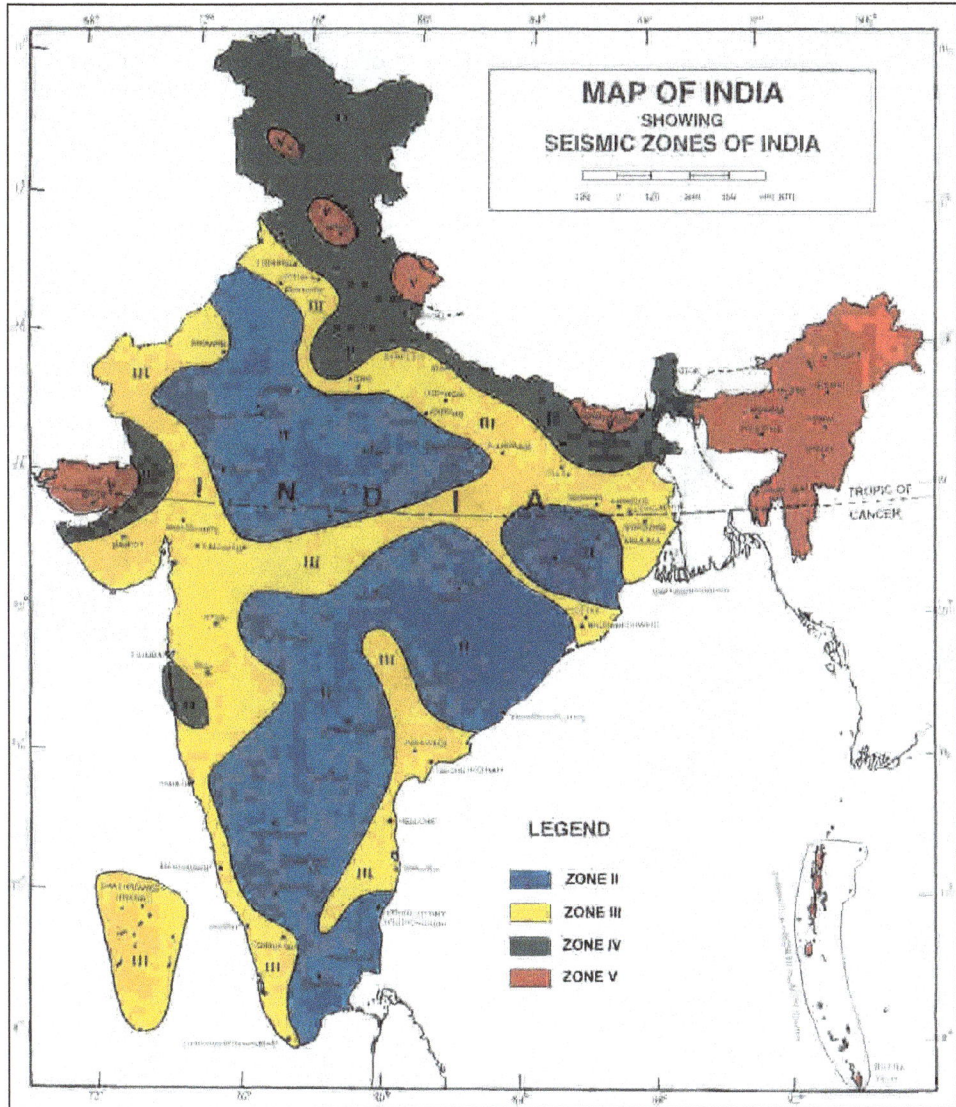

Figure 17.1: Earthquake Risk

The map of India printed here shows the types of Earthquake that can possibly occur and the risks involved. The person can locate his area on the map and become aware of the possible risk of future earthquake. There are four different zones: Numbered II, III, IV & V. Zone II has the lowest risk and Zone V has the highest risk.

Design Philosophy

It covers reason underlying our choice of loads and forces for our analytical technique and design procedure, our preferences for particular structural configuration and material and our aim for economic optimization. We typically accept higher risks of damage under seismic design forces than under any other comparable extreme loads. All the modern building codes specify any intensity of design earthquake corresponding return period of 100 to 500 years for ordinary structures like buildings. The corresponding design forces are generally too high to be resisted within the elastic range of material response and its common design structure within a fraction as low as 25 per cent of that corresponding to elastic response and to expect the structure to survive the earthquake by last inelastic deformation and energy dissipation corresponding to material distress.

Material Used

ISTOCKPHOTO

The next time you get a bowl of rice, ask yourself: would I rather eat this or put this toward my next construction project? As scientists and construction workers slowly dissect ancient construction projects, they're finding that sticky rice – sometimes used in Chinese cuisine, but a staple of many South East Asian countries' diets – was used to make super tough mortar.

Result and Discussion

In a study recently reported by scientist Bingjian Zhang in the American chemical society journal, "Accounts of Chemical Research," Zhang and fellow chemists say that they have found that a chemical in sticky rice help makes mortar strong enough to withstand earthquakes. The paper reports that when a sticky rice "soup" was mixed into mortar – the material used to bind and fill gaps between bricks as well as other construction materials – and used to restore ancient buildings, it outperformed other available substances.

This is the cheapest method to construct the building and make the building more stronger. This method is very cheaper as compare to modern methods because of costly equipments and materials used in creation of modern modern building structure makes unaffordable for common man.

The physical properties, mechanical strength, and compatibility of lime mortar were found to be significantly improved by the introduction of sticky rice. They found that sticky rice and the resulting amylopectin in the lime mortar was found to act as an inhibitor and had smaller calcium carbonate crystals than mortar without it, creating a more compact structure and causing the crystals to stick together.

Mortar with sticky rice is less permeable to water and more resistant to the stresses of changing weather than standard mortar, the authors said. This makes it more compatible with the bricks used in old buildings, and therefore the best option for conservation and restoration

It's a twist that could make a polluting substance into a way to reduce greenhouse gases. Cement, which is mostly commonly composed of calcium silicates, requires heating limestone and other ingredients to 2,640°F (1,450°C) by burning fossil fuels and is the third largest source of greenhouse gas pollution in the U.S., according to the U.S. Environmental Protection Agency. Making one ton of cement results in the emission of roughly one ton of CO_2–and in some cases much more.

Conclusion

While there is something new to learn from new earthquake, it may be said that the majority of the structure lesson should have been learned. Buildings should be constructed with the fusion of modern and ancient techniques, so that they can be more strenthen and compatible with all weather conditions. This foundation should be suitably deep to develop a good binding between building and ground, this fusion also allows a good binding between foundation, walls, floor, and roof. However, this procedure should be practiced more in labs for initial time being and do some more research and development so that we can make it possible to apply this fusion in modern age constructions i.e; multistoried skyscrapers etc. which can change the face of architecture infrastructure in the modern era for the sake of humanity.

References

Arnold, Christopher, Reitherman, Robert (1982). *Building Configuration and Seismic Design*. A Wiley-Interscience Publication. ISBN 0471861383

Bertero VV, Anderson JC &Krawinkler H. Performance of steel building structures during the Northridge earthquake. Report No. UCB/EERC-94/09. Berkeley, California: Earthquake Engineering Research Center, University of California at Berkeley. 1994.

Edited by Dr. Robert Lark (2007). *Bridge Design, Construction and Maintenance*. Thomas Telford. ISBN 0727735934.

Ekwueme, Chukwuma G., Uzarski, Joe (2003). *Seismic Design of Masonry Using the 1997 UBC*. Concrete Masonry Association of California and Nevada.

Implications of experimental on the seismic behavior of gravity load designed RC beam-column connections. Earthquake Spectra, 12(2), 185-198.

Kharrazi, M.H.K., 2005, "Rational Method for Analysis and Design of Steel Plate Walls," Ph.D. Dissertation, University of British Columbia, Vancouver, Canada,

Lindeburg, Michael R., Baradar, Majid (2001). *Seismic Design of Building Structures.* Professional Publications. ISBN 1888577525.

Omori, F. (1900). *Seismic Experiments on the Fracturing and Overturning of Columns.* Publ. Earthquake Invest. Comm. In Foreign Languages, N.4, Tokyo.

Seismology Committee (1999). *Recommended Lateral Force Requirements and Commentary.* Structural Engineers Association of California.

2013, Environmental Technology   *Pages 161–165*
Editors: **D.R. Khanna, A.K. Chopra, Gagan Matta, R. Bhutiani & Vikas Singh**
Published by: **DAYA PUBLISHING HOUSE, NEW DELHI**

Chapter 18

Clean Energy Alternatives to Fossil Fuels

Prabhash Ankit and Awasthi Shikha

University of Petroleum and Energy Studies (UPES)

ABSTRACT

The world cannot continue to rely for long on fossil fuels for its energy requirements be at the industrial levels or for domestic purposes. Fossil fuel reserves are limited across the globe. In addition, when burnt, these add to global warming, air pollution and acid rain. So switching to biogas systems are ideal for providing independent electrical power and lighting in isolated rural areas that are far away from the power grid and can also serve as a replacement for domestic fuels to some extent. These systems are non-polluting, don't deplete the natural resources and are cheaper in the long run.

This paper aims to demonstrate how and up to what extent the bio-degradable waste products from our day to day life can be utilized to serve as an alternative to conventional fossil fuels especially in the remote areas with increase in its calorific value by biogas up gradation. In short a challenge to the increasing consumption of fossil fuels in cleaner and greener way.

Keywords: Bio-degradable wastes, Global warming, Biogas upgradation.

Introduction

Increase in energy demand and the issues about current non-renewable energy sources led researchers to investigate alternative energy sources during the last two decades. Renewable energy sources draw attention all over the world because they are sustainable, improve the environmental quality and provide job opportunities in rural areas.

The Biogas production from agricultural biomass is of growing importance as it offers considerable environmental benefits and is an additional source of income for farmers. Nearly half of renewable energy sources in India stem from biomass, including wastes. Suitable substrates for the digestion in agricultural biogas plants are different energy crops, organic wastes, and animal manures. Maize (*Zea mays* L.), herbage (Poacae), clover grass (Trifolium), Sudan grass (*Sorghum sudanense*), fodder beet (*Beta vulgaris*) and others may serve as energy crops. The predominant crop for biogas production is maize. Maize is considered to have the highest yield potential of field crops grown across the globe.

Chemical Composition

Biogas is a product of the metabolism of methane bacteria and is created when the bacteria decompose a mass of organic material. The methane bacteria can only work and reproduce if the substrate is sufficiently bloated with water (at least 50 per cent). In contrast to aerobic bacteria, yeasts and fungi they cannot exist in solid phase.

Coal, natural gas and oil fired energy production plants are major contributors to CO2 emissions in the atmosphere. India has made a commitment to reduce its emissions by 10 per cent in 2020 compared to 1990 (10th IAEE European Conference "Energy, Policies and Technologies for sustainable Economies", Vienna: Austria (2009)). Mitigating the current trend of increasing CO_2 emissions relies on taking measures to reduce final energy consumption, to encourage a more rational use of primary energy sources and to exploit renewable energy sources more intensively. Specific measures have been taken at the Indian level to encourage the production of electricity from renewable sources ("green electricity"). Several countries have implemented a "green certificates" market in order to support this production.

Biogas from sewage digesters usually contains from 55 per cent to 65 per cent methane, from 35 per cent to 45 per cent carbon dioxide and less than 1 per cent nitrogen, biogas from organic waste digesters usually contains from 60 per cent to 70 per cent methane, from 30 per cent to 40 per cent carbon dioxide and less than 1 per cent nitrogen while in landfills the methane content is usually from 45 per cent to 55 per cent, carbon dioxide from 30 per cent to 40 per cent and nitrogen from 5 per cent to 15 per cent. Typically biogas also contains hydrogen sulphide and other sulphur compounds, compounds such as siloxanes and aromatic and halogenated compounds. Although amounts of trace compounds are low compared to methane, they can have environmental impacts such as stratospheric ozone depletion, the greenhouse effect and/or reduce the quality of local air.

Present Scenario

In the Indian Subcontinent

In Pakistan, India, and Nepal biogas produced from the anaerobic digestion of manure in small-scale digestion facilities is called gobar gas; it is estimated that such facilities exist in over two million households in India and in hundreds of thousands in Pakistan, particularly North Punjab, due to the thriving population of livestock. It has become popular source of fuel in many parts of Nepal. The digester is an airtight circular pit made of concrete with a pipe connection. The manure is directed to the pit, usually directly from the cattle shed. The pit is then filled with a required quantity of wastewater. The gas pipe is connected to the kitchen fireplace through control valves. The combustion of this biogas has very little odour or smoke. Owing to simplicity in implementation and use of cheap raw materials in villages, it is one of the most environmentally sound energy sources for rural needs. One type of this system is the Sintex Digester. Some designs use vermiculture to further enhance the slurry produced by the biogas plant for use as compost.

The Deenabandhu Model is a new biogas-production model popular in India. (Deenabandhu means "friend of the helpless.") The unit usually has a capacity of 2 to 3 cubic metres. It is constructed using bricks or by a ferrocement mixture. In India, the brick model costs slightly more than the ferrocement model; however, India's Ministry of New and Renewable Energy offers subsidy per model constructed.

In Developing Nations around the Globe

Depending on size and location, a typical brick made fixed dome biogas plant can be installed at the yard of a rural household with the investment between 300 to 500 US $ in Asian countries and up to 1400 US $ in the African context.

Domestic biogas technology is a proven and established technology in many parts of the world, especially Asia. Several countries in this region have embarked on large-scale programmes on domestic biogas, such as China and Japan. The Netherlands Development Organization (SNV) supports national programmes on domestic biogas that aim to establish commercial-viable domestic biogas sectors in which local companies market, install and service biogas plants for households. In Asia, SNV is working in Nepal, Vietnam, Bangladesh, Bhutan, Cambodia, Pakistan and Indonesia, and in Africa; Rwanda, Senegal, Burkina Faso, Ethiopia, Tanzania, Uganda, Kenya, Benin and Cameroon.

Biogas Upgrading

Raw biogas produced from digestion is roughly 60 per cent methane and 29 per cent CO_2 with trace elements of H_2S, and is not of high quality. The corrosive nature of H_2S alone is enough to destroy the internals of the plant. The solution is the use of a biogas upgrading or purification process whereby contaminants in the raw biogas stream are absorbed or scrubbed, leaving 98 per cent methane per unit volume of gas. There are four main methods of biogas upgrading, these include water washing, pressure swing absorption, selexol absorption, and amine gas treating.

Water Washing

The most prevalent method is water washing where high pressure gas flows into a column where the carbon dioxide and other trace elements are scrubbed by cascading water running counter-flow to the gas. This arrangement could deliver 98 per cent methane with maximum 2 per cent methane loss in the system. It takes roughly between 3-6 per cent of the total energy output in gas to run a biogas upgrading system based on water washing.

Pressure Swing Absorption

Pressure swing adsorption (PSA) is a technology used to separate some gas species from a mixture of gases under pressure according to the species' molecular characteristics and affinity for an adsorbent material. Pressure swing adsorption processes rely on the fact that under pressure, gases tend to be attracted to solid surfaces, or "adsorbed". The higher the pressure, the more gas is adsorbed; when the pressure is reduced, the gas is released, or desorbed. PSA processes can be used to separate gases in a mixture because different gases tend to be attracted to different solid surfaces more or less strongly.

Selexol Absorption

Selexol is the process of acid gas removal solvent that can separate acid gases such as hydrogen sulfide and carbon dioxide from feed gas streams. By doing so, the feed gas is made more suitable for combustion and/or further processing.

In the Selexol Process, the Selexol solvent dissolves (absorbs) the acid gases from the feed gas at relatively high pressure. The rich solvent containing the acid gases is then let down in pressure and/or steam stripped to release and recover the acid gases. The Selexol process can operate selectively to recover hydrogen sulfide and carbon dioxideas separate streams, so that the hydrogen sulphide can be sent to either a Claus unit for conversion to elemental sulphur or to a WSA Process unit for conversion to sulphuric acid while, at the same time, the carbon dioxide can be sequestered or used for enhanced oil recovery.

Selexol is a physical solvent, unlike amine based acid gas removal solvents that rely on a chemical reaction with the acid gases. Since no chemical reactions are involved, Selexol usually requires less energy than the amine based processes.

Amine Gas Treating

Amine gas treating, also known as gas sweetening and acid gas removal, refers to a group of processes that use aqueous solutions of various alkylamines (commonly referred to simply as amines) to remove hydrogen sulphide (H_2S) and carbon dioxide (CO_2) from gases

The chemistry involved in the amine treating of such gases varies somewhat with the particular amine being used. For one of the more common amines, methanolamine (MEA) denoted as RNH_2, the chemistry may be simply expressed as:

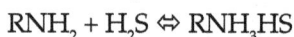

$$RNH_2 + H_2S \Leftrightarrow RNH_3HS$$

Conclusion

Biogas production offers several environmental benefits. Organic waste generated in the country has the potential to produce and displace the equivalent of over 120 million cubic meter of natural gas per year. That is approximately 3.5 per cent of the present natural gas consumption.

Biogas upgrading is a mature technology in which improvements at various level is possible, which could yield an estimated increase of its overall efficiency by 5-10 per cent.

Refrences

Evaluation of upgrading techniques for biogas, Margareta Persson, Lund University.

Indian Biogas Association.

"Asia Hits the Gas".

Food and agricultural wastes.

2013, Environmental Technology
Editors: D.R. Khanna, A.K. Chopra, Gagan Matta, R. Bhutiani & Vikas Singh
Published by: DAYA PUBLISHING HOUSE, NEW DELHI

Pages *167–173*

Chapter 19

Green Fashion: A Fashionable Way to Protect Environment

Wasia Siddiqui and Anita Rani

Department of Clothing and Textiles,
Govind Ballabh Pant University of Agriculture and Technology,
Pantnagar – 26 3145

ABSTRACT

Hurry before it starts to flurry. This is what we are counting at now days in respect of the nature and the ever increasing global warming. The green fashion or the step towards protecting the nature has turned out to be the most critical movement in the history of mankind. We all are taking measures in every field to contribute towards pushing this movement ahead for a bright green future.

Green fashion is about making clothes that take into account the environment, the health of consumers and the working conditions of people in the fashion industry. These products are made using organic raw materials, such as cotton grown without pesticides and silk made by worms fed on organic tree. These items don't involve the use of harmful chemicals and bleaches to color fabrics and are often made from recycled and reused textiles.

With the eco-fashion industry still in its infancy, the main responsibility at the moment lies with manufacturers and designers, who need to start using sustainable materials and processes. Sustainable fabrics are here to make this world green in a fashionable way. For any environment conscious citizen it can be one of the best news he would be pathetically waiting to listen to. It has got such a great impact on the users that they seem to have been caught in wear-only-green frenzy.

We should come together to hold the hands and have an environment that supports the eco friendly fashion and a fashion that also be sustained for future.

Keywords: Green fashion, Sustainable fabrics, Fashion, Clothing, Eco-friendly.

Introduction

Years of human ignorance has diminished our natural resources and aged our planet. So now making our planet green and achieving a sustainable future is foremost objective of mankind. Sustainability has 3 pillars like prediction of the environmental, economic and social consequences of changes in production structure, consumer behaviour, material and process innovations and government influence.[2]

One can forward a step towards sustainable future is by embracing green fashion. Several times a year in the world of fashion, graceful models in astounding outfits swaging down the catwalk to present the coming season's trends. Each year designers exhibit, says what's in and what's not. Chain-stores and mass retailers then adapt their ideas for the man and woman.[3]

The clothing and textiles sector is a significant part of the world's economy. Fashion is a growing industry and ranks textile and clothing as the world's second-biggest economic activity for intensity of trade. However, competition forces down costs while working conditions, more often than not in developing countries, are far from ideal. For this the environment is also paying a heavy price. Now, people are making an effort to change the way they are treating their planet. If the designers become more conscious for the environment and the impact of the material used for fashion on earth they can reduce the ill impact that help enable a sustainable future. By using eco-friendly products, being fashionable can be less harmful to our natural resources, environment as well as the planet. In today's world eco-friendly trend is not follows by the entire fashion world, but more designers are accepting the trend toward eco-fashion than ever before. If the entire fashion industry became eco-friendly, it would make a huge difference for future generations so that to make a world of green fashion. The global fashion industry generates more than a trillion dollars per year. So what we wear, we use - and how it's made and sold - can have a huge positive impact on our society, environment and our planet.[3,1]

Green Fashion

The synonyms of Green fashion are green fashion are eco fashion, ethical fashion and sustainable fashion. According to STEP (Sustainable Technology Education Project), Eco-friendly fashion is about making clothes and accessories that take into account the environment, the health of consumers and the working conditions of people in the fashion industry.[5]

Green fashion -fashion is a fairly modern concept. It was started in the 1970's. The point of the trend was to get away from conventional fashion, which was done by making their own clothes with hemp and natural dyes. The second time eco-fashion came around was in the 1990's, but this time the trend was more commercial. Clothing companies like Patagonia and J Crew were developing eco-friendly lines. The most recent trend in eco-fashion started developing in 2006. Even better known clothing companies, as well as private designers have picked up the latest trend in eco-fashion. Levi's- which is one of the world's largest clothing manufacturers, is going eco-friendly, which will expectantly set a example for companies in the fashion industry.[5]

Fashion world is showing an enormous growth in the field of eco-friendly clothing and fashion accessories. Thousands of designers are creating clothes and accessories made from natural, organic and recycled materials. Many European and Asian companies are showing a due impact by adopting different ways and creativity to achieve a sustainable future. Getting the great designers from the whole world, the fashion industry is targeting on the organic clothing rather the synthetic or fibre cloths. Celebrities, political figures and people all over the world are making eco-green statements by adapting green fashion in form of clothes, jewellery and accessories. Now days there are many boutiques and online store opens up just focusing on green fashion helping to makes it more available to the public to purchase and use. A popular celebrity wears an eco-friendly outfit causing people to open their eyes as people get more influenced by their star personality. In this way one can reduce the use of products that puts a lot pressure on the nature that pollute the environment.

The garments and products in *green fashion* will reflect at least one of six major themes:[4]

1. Material origins
2. Textile processing
3. The repurposing and recycling of materials
4. Quality of craftsmanship
5. Labour practices
6. Treatment of animals

☆ Use of organic raw materials, like cotton grown without pesticide and silk made by worms fed on organic trees. This reduces smash up to the environment, animals and people's health.

☆ Avoid harmful chemicals and bleach to colour fabrics and for their processing. This can cause long-term damage to people's health and the environment.

☆ Recycling and reusing of textiles products. Garments and accessories can be made from second-hand products and even by recycling of the products

☆ Imagination and the quality of work done by the craftsman have a greater impact. In other words it is the designer who create new things in such a creative way so that the environment can be protected

- Are made to last, so that people keep them for longer
- Fair trade and labour practices - The people who make them are paid a fair price and have decent working conditions. This means that companies
- Are paying the fair price *i.e.* Legal minimum wage
- Creating fair employment opportunities,
- Healthy working conditions

In addition, they are engaged in environment sustainable practices, making sure that product quality is maintained, honouring cultural identity as a stimulus for product development and production practices, offering business and technical expertise and opportunities for worker advancement.[3]

☆ Many people feel that green fashion includes a wish to avoid any form of animal exploitation - whereas others would disagree strongly that this is a luxury that cannot be supported in a sustainable world. Animal products used in clothing include fur, leather, silk, and wool. Earlier accessories made up of Furs, snakes and other animals were acceptable. It no longer is acceptable and now the majority of today's society frowns upon people who use those products. People are more conscious and concerned about *saving the planet* and preventing animal extinction.

Some Facts About Green Fashion

In fact many fashion experts believe that green fashion is still in its infant stage. But fashion designers have understood the importance of green fashion and have embraced it happily.

☆ T-shirt is one of the popular dresses both among youth and old. Almost 1/5 lb amount of pesticide goes in one single T-shirt, if they are made from conventionally grown cottons.

☆ Organic materials are those that are grown in 100 per cent natural way and no harmful chemicals are used which causes damage to the environment.

☆ Peace silk is one of the new and popular green fashion product. This kind of silk is also known as vegetarian silk as the silk moth is allowed to emerge out of their cocoons so that they can complete their full cycle of life. This also ensures that quality of silk remains up to the industry standard.

☆ Though green fashion is new, some brands exclusively committed to green fashion. These brands have become very popular they are Edun, Laura Miller, Loomstate and Linda Loudermilk. Some of the leading fashion brands have also decided to devote themselves in green fashion.

☆ Some of the well known designer brands that offer quality green fashion are Stella McCartney, Levis, American Apparel, Eillen Fisher etc.

☆ Some of the international celebrities who prefer green fashion are Cameron Diaz, Thandie Newton, Natalie Portman and Alicia Silverstone.

There are a variety of materials considered "environmentally-friendly" for many reasons: First reason is the re-new ability of the product. Renewable resources are items that can be replenished in a relatively short amount of time.

The second factor is the ecological footprint of the resource - how much land (usually in acres) it takes to bring one of the individuals to full growth and support it. The third thing to consider in determining the eco-friendliness of a particular product is how many chemicals it requires to grow/process it to make it ready for market.

By taking into account above mentioned content some eco- friendly fibre sources are:

Bamboo

☆ Variety of grass
☆ Biodegradable
☆ Inherent ability to breathe
☆ Fastest growing plant in the world
☆ Highly renewable
☆ Antibacterial and antifungal properties
☆ The fabric made from it is soft, luxurious, stretchy, comfortable, and strong – great for daily wear and active wear.

Corn Fibre

The starch and sugars are extracted from corn, and processed to make a fibre called

Nature works PLA. This process is currently being done by Cargill Dow Polymers, and the fabric formed is called Ingeo. The fabrics are:

☆ Comfortable
☆ Resemble cotton, silk and wool
☆ Lower cost
☆ Easier care
☆ Higher durability and superior wicking capabilities

Cotton – Organic

☆ Free from pesticides, herbicides, or insecticides during the growing cycle *i.e.* chemical free
☆ Organic cotton garments- free from chlorine bleaches and synthetic dyes

Non-organic cotton accounts for approximately 10 percent of the world's pesticides, and 25 per cent of the world's insecticides. These chemicals lead to the diseases such as cancer, birth defects, and asthma. Non-natural bleaches and dyes release further toxins.

Hemp

Hemp plants grow quickly and densely. They require only an average amount of water and are pest-resistant.

☆ Do not require herbicides, pesticides, or chemical fertilizers
☆ Can be spun into yarns with minimum processing
☆ More durable
☆ Absorbent, and insulating than cotton

Jute

☆ 100 per cent bio-degradable and thus environment-friendly

☆ Used in manufacturing different types of packaging material for agricultural and industrial products

☆ Available in abundance in India, at competitive prices

☆ Raw material for accessories and non textiles

☆ High tensile strength

☆ Resistance to heat

Milk Silk

Goats' eggs are mixed with genetic material from spiders, so that the female goats produce milk that contains silk fibres. The resulting fibre is biodegradable and durable. The downside: its genetic engineering, and kind of gross.

Recycled Polyester

☆ Polymer from recycled water bottles is melted and made new fibre.

☆ Reduces energy consumption

☆ Raves raw materials thus helpful for sustainable future

☆ Fire retardant

☆ Easy to clean

☆ Inexpensive

Soy Silk

Soy silk is made from the by-products of tofu. The liquid is extruded into fibres, and spun into yarns.

☆ High protein content of this allows easy absorption of natural dye.

☆ Fabric is soft

☆ Breathable

☆ Durable

☆ Washable

☆ Also known as vegetable cashmere

Wood Pulp

Lyocell and Modal are two fabrics produced in a solvent-spinning process from wood pulp. The resulting fabric is very smooth and supple, which is good for sensitive skin. It may also have very good wicking properties, keeping moisture away from the skin.

Green Fashion Accessories

Like garments, the green fashion also deals with accessories that are completely eco friendly. Fortunately, now eco-conscious designers have plenty of options when it comes to sustainable accessories. Accessories made from spinning yarn, using a loom and embroidery in the making of garments, jewellery or accessories made from natural items such as shells, organic cotton, hemp, jute, bamboo and recycled leather, candy wrappers and cans etc. Jewellery or accessories made locally is also an option for green fashion, which in one hand reduce the wastage of raw material and on the other hand it generate employment to the people. This means resultant clothes or accessories that are not made from industries that pollute the environment. Many online services are also available that deals with green accessories ex. the Leather-free shoes are available from Vegetarian Shoes Online Store and Green Shoes. Green accessories are the smart way to go simple, go green, and go trendy.

Conclusion

Fashion industry is an inseparable part of the ever-changing global market. Like any other sector of the market Fashion Futures is a call for a sustainable fashion industry. When manufacturers and consumers understand the importance of being eco-friendly it not only embraces green fashion from a "feel good" with regards to the environment perspective, but proves to be cost-effective too. Green fashion includes products using organic raw materials, employs ecofriendly processes, implementation of fair trade practices and proper waste management. To make our future sustainable, it is the responsibility of manufacturers as well as consumers to understand the call of nature for planet's sustainability. Save energy, money and resources, and cut your carbon footprint should be the axiom followed by everyone. Following the fashion in a green way can prove to be a milestone for protecting the environment.

Abbreviations

STEP: Sustainable Technology Education Project

PLA: Poly lactic acid

References

1. www.greenleafgoods.com/blog/2011/03/embracing-change-the-eco-friendly-fashion-revolution/

2. www.ifm.eng.cam.ac.uk/sustainability/projects/mass/uk_textiles.pdf

3. www.vboriginal.com/Me./Eco per cent 20Fashion.pdf vboriginal/About_

4. www.fitnyc.edu/7885.asp

5. www.stepin.org/casestudy.php?id=ecofashion&page=2

2013, Environmental Technology   *Pages 175–180*
Editors: D.R. Khanna, A.K. Chopra, Gagan Matta, R. Bhutiani & Vikas Singh
Published by: DAYA PUBLISHING HOUSE, NEW DELHI

Chapter 20

Eco-labels for Textiles: Benefits and Opportunities

Sonam Omar and Alka Goel

Department of Clothing and Textiles, College of Home Science,
G.B. Pant University of Agriculture and Technology,
Pantnagar – 263 145, Uttrakhand

ABSTRACT

Consumers are becoming increasingly concerned with the adverse impacts of industrial pollution on the environment and their health. Mounting pressure on industry to adopt more "eco-friendly" manufacturing processes has led to an increased demand, particularly in the textile sector, for manufacturers to have an eco-label for their products. Eco-labels certify the "eco-friendliness" of the textile product is now increasingly demanded by consumers. This certifies that their products do not contain chemicals that might be harmful to the consumer.

Introducing an eco-label allows markets to value process attributes (credence good) and to reward producers of environment-friendly attributes. The label increases the cost of production by imposing process-standards on the production of the green apparel. Conventional textile goods are at the lower bound on the quality attribute.

In addition to assisting the entry to new markets and maintaining existing ones, obtaining an eco-label can also generate financial savings through process optimization and reduced consumption of raw materials, reduce processing time, improve environmental performance and improve working conditions. Obtaining an eco-label can also help obtain ISO 14000 and ISO 9000 accreditation. Eco-labeled goods meet some process standards and convey their higher environmental quality to consumers through the label.

Keywords: Eco-textiles, Eco-labeling, Textile markets.

Introduction

Eco-labelling is becoming a differentiating factor on a worldwide scale in market for textile and apparel purchase. Consumers are becoming increasingly concerned with the adverse impacts of industrial pollution on the environment and their health. This concern is mounting a pressure on textile, fashion industry to adopt more eco-friendly, chemicals and manufacturing processes. Environmental concerns raised by production systems have been recognized since the late 1960's and attempts to move towards more sustainable and environmentally friendly approaches have been through a range of regulatory measures from green taxes to strict bans.

One approach acquiring increasing importance is that of 'environmental labelling' or 'eco-labelling'. *Environmental labelling* is a broad term and covers a range of labels and declarations of environmental performance and focus on consumption rather than the production of a given product; *e.g.* recyclable material while *Eco-labels* are a sub-group of environmental labelling and convey environmental information about a product to the consumer and communicate that the environmental impacts are reduced over the entire life cycle of a product without specifying the production practices.

In brief an eco-label:

☆ Identifies the overall environmental preferences of a product

☆ Provides information on environment related product qualities

☆ Tool for consumers to identify environmentally safe product

☆ Enables manufacturers to use eco-friendly raw material and ingredients

☆ An additional product quality which can be used as a marketing tool

☆ Can be issued by private or public body

☆ Enables to earn premium on products

Significance of Eco-Labels

Eco-labels are intended to educate and increase consumer awareness of the environmental impacts of a product and bring about environmental protection by encouraging consumers to buy products with a lower environmental impact. Consumers include individual retail consumers, as well as the procurement officers of governments and large corporations. Consumers' purchasing decisions can provide a market signal to producers about product preferences. Under effective Eco labelling regimes, producers and sellers have an incentive to compete to improve the products by changing inputs or adopting different technologies to lower the environmental burden of the product.

The eco-label also has a main role in the Integrated Product Policy (IPP) which aims to minimize the environmental degradation caused by any of the phases of a product's life cycle (tangible or intangible, such as service), *e.g.* manufacture, development, use or disposal (European Commission, 2008). All phases of a product life cycle are examined with the objective of improving their environmental performance. This approach requires all participants in this process to be engaged:

e.g. designers, industry, marketers, retailers and consumers. The US EPA (1994) defined the following five factors for measuring effectiveness of an eco-label, the first four of which serve to support the last:

1. Consumer awareness of labels
2. Consumer acceptance of labels (credibility and understanding)
3. Changes in consumer behaviour
4. Changes in manufacturer behaviour
5. Net environmental gains

Types of Eco-labels

Eco-labels, they can be generically classified into one of two categories: (1) self-declaration claims and (2) independent third-party claims. They can be both *'private'* and *'public'* (government sponsored) schemes.

Eco-labels may be voluntary or mandatory. Mandatory labelling is always third party labelling (*i.e.* an independent body is required to attest to required standards having been achieved), voluntary programmes may be established by firms or business associations as well as third party. Currently, there are no eco-labels in textiles and clothing enforced by mandatory rules. Eco-labels are normally issued either by government supported or private enterprises once it has been proved that the product of the applicant has met the criteria:

Public (Government Sponsored)	Private
Blue Angel (Germany)	Eco-tex
Eco Mark (Japan)	Oeko-Tex (textiles and clothing) (Germany)
Environmental Choice (Canada)	Green Seal (United States)
White Swan (Nordic Countries)	
EU	
Eco-Mark (India)	
Green Label (Singapore)	

Self-Declaration Claims "ozone-friendly"; "recyclable" "ozone-friendly"; "recyclable"

Independent Third-Party Verification Öko-Tex; U.S. Green Seal E.U.Eco-Label; Nordic Swan; Blue Angel

Self-declaration claims are eco-labels placed on a product by the manufacturer, retailer or marketer of such product, and may be made on a single attribute or an overall assessment of the product. Product claims could include "environmentally friendly", "ozone friendly" and "degradable". However, these claims are usually not independently verified.

Independent third-party claims are based on compliance with predetermined criteria, which are independently verified by a competent authority (usually government-sponsored). The criteria for eco-labels based on independent third-party claims are usually built on a product life-cycle approach.

The criteria for granting eco-labels are mostly based on the "cradle-to-grave" approach, *i.e.* the life-cycle analysis of the product and assessment of its impact on the environment from processing of raw materials, production, distribution, consumption and maintenance, (*i.e.* washing, ironing, dry-cleaning) and finally disposal of the product.

A 'Cradle to Cradle' certification programme assesses the sustainability of product ingredients for human and environmental health, as well as their recyclability or compostability making it easier at the design stage to create ecologically-intelligent products through choosing materials that meet key sustainability criteria for material health and material reutilization.

The *International Organisation for Standardisation* (ISO) has adopted eco-labelling as an important tool in obtaining environmental sustainability of business. It has introduced the ISO 14000 series of environmental standards, with the ISO 14020 series dealing exclusively with environmental labels and declarations. Voluntary labels are classified according to International Standards Organization (ISO). ISO is the world's largest non-governmental organization that develops and publishes International Standards and is a network of the national standards institutes of 163 countries.

There are now eco-labelling schemes both in developed and developing countries and the ISO has classified the existing environmental labels into following typologies –*Type I, and II*, specifying preferential principles and procedures for each one of them.

　☆ *Type I* environmental labels, *i.e.* "voluntary, multiple criteria-based third-party practitioner programmes that awards labels claiming overall environmental preference of a product within a particular product category based on life-cycle considerations". Well-established eco-labelling schemes falling into this category are the EU Flower (European Union Eco-Label), Green Seal (U.S.), The Nordic Swan (Nordic countries), the Blue Angel (Germany) and Environmental Choice (Canada).

　☆ *Type II* environmental labels are those that "consist of informative self-declaration claims", made without independent third-party certification by, for example, "manufacturers, importers, distributors, retailers or anyone else likely to benefit from them". The absence of third-party verification is likely to count against wide-spread acceptance of such schemes by consumers.

Many other prominent international trade and environmental organizations deal with issues related to eco-labelling, eg: the United Nations, the World Trade Organization through its International Trade Centre and Committee on Trade and Environment, the US Environmental Protection Agency, as well as the Organization for Economic Co-operation and Development.

Developing an Eco-labels

This is complex and complicated but can be generalized into four broad phases:

1. *Selection of a product category* by a labelling board through suggestions from industry, environmentalist, consumers, and other interested parties.

2. *Life-cycle analysis* to assess environmental impact of products in chosen category and examine the material and energy inputs for manufacture and use of a product and the solid, liquid, and gaseous waste generated at each stage of lifecycle, *e.g.* raw material, production, distribution, packing use and disposal.

3. *Criteria and thresholds* for the award of an eco-label set taking into consideration technical feasibility and environmental impacts in different media like air, water and soil against one another. Different eco-labels have differing methodologies, *e.g.* Oeko-tex 100 examines harmful residues on the product, while GOTS tends to look environmental as well as residual parameters.

4. *The product category and criteria* is reviewed and refined. Interested parties including industry and environmental and consumer groups are asked for their inputs, although they are often already included much earlier on in the process.

Benefits of Eco-labels

☆ *Economic Benefits*: Eco-labelling encourages decreased emissions and reduced environmental impact, the extra cost of which is passed on to the consumer in a price premium. As a result, widely acknowledged benefits of Eco-labelling for producers being certified include potential for premium market prices, access to new markets, safeguarding of existing market channels, preferred supplier status, potential to attract ethical investment in the sector as well as (co)funding of local community social and economic infrastructure.

☆ *Environmental Benefits*: Eco-labels are meant to communicate and promote environmental benefits through their positive influence on consumers' purchasing decisions. Consumer recognizes an emblem or a logo as a selection aid to make an informed decision; the manufacturer uses the product logo to communicate its good environmental practices to the consumers. This practice should be accompanied by institutional and regulatory policies that will help to differentiate justifiable environmental claims from fraudulent marketing slogans. Such a mechanism encourages industries to promote continuous improvement of their production processes thereby reducing their impact on the environment. Eco-labelling is also considered to be a useful vehicle for raising environmental awareness in some cases by highlighting the available alternatives in specific product categories.

Challenges for Eco-labelling

There are many challenges for the eco-labelling, the most serious of which are: misleading or fraudulent to uninformative claims, unfair competition and protectionism and lack of stringency or standardization in the process or mechanisms of eco-labelling. The objective of certification is to gain access to the market for environmentally sustainable products and the certification process improves the 'image of product and increases its credibility and visibility.

Interest of the consumer about eco-textiles, showed that the common consumer in the street thought themselves as environmentally aware, but they are certainly not inclined to pay extra money for eco-textiles. A small annoyance about eco-textiles is the high costs of eco-textiles. Organic products are now more expensive, because more time and care is devoted to organic farming. Furthermore, environmental labels are checked and checks cost money, but at the end of the trip an 'organic' product is reliable.

Conclusion

Eco-labels are educational tools to inform consumers about the environmental impacts of the labeled product, and thereby induce a change in purchasing behavior that mitigates the product's environmental damage.

Eco-textile is and will remain a niche market, with textile producers continue to impress with new developments such as towels made from bamboo fiber, soybean fiber and even elephant grass (mini-bamboo). Despite new products and fashions, consumers will continue to inform and ask for a eco-story or the ethical background of a textile product and an eco-label already proved a reliable and informative tool.

References

Ecolabelling – as a potential marketing tool for African Products (An overview of opportunities and challenges) by German Ministry of Environment (brochure).

Eco-labelling (2008), www.ecolabelling.org

European Commission (2008) "*What is Integrated Product Policy*" 03.06.2008 http://ec.europa.eu/environment/ipp/integratedpp.htm

Jonghe, F. D., Devaere, S., Velghe, K. 2009. Centexbel, Zwijnaarde and Howest, departement PIH, Kortrijk. Eco-Labels for Textile Products.

Naumann, E. 2001. Eco-labelling: Overview and Implications for Developing Countries, DPRU Policy Brief.

Sinha, P. and Shah, R. 2008. Creating a Global Vision for Sustainable Textiles. www.fiber2fashion.com.

2013, Environmental Technology Pages 181–187
Editors: D.R. Khanna, A.K. Chopra, Gagan Matta, R. Bhutiani & Vikas Singh
Published by: DAYA PUBLISHING HOUSE, NEW DELHI

Chapter 21

Environmental Friendly Textiles: A Road to Sustainability

Priyanka Kesarwani and Shahnaz Jahan

*Department of Clothing and Textiles, College of Home Science,
G.B. Pant University of Agriculture and Technology,
Pantnagar – 263 145, Uttrakhand*

ABSTRACT

The textile industry is considered as the most ecologically harmful industry in the world. The eco-problems in textile industry occur during some production processes and are carried forward right to the finished product. Intensive use of chemicals in the production process such as dyeing and finishing has a greater impact on the environment. Therefore controlling pollution during the production process is, as vital as making a product free from the toxic effect. Also, the utilization of wood pulp for making fibers has led to the fast depleting forests and had opened the door to the development in natural sustainable fibres like organic Cotton, Hemp Bamboo fibres etc. Petroleum-based products are harmful to the environment and in order to safeguard our environment, integrated pollution control approaches are needed. So environmentally sustainable product is an essential precondition in today's globalized world. This can preserve the long term interests of the communities by providing them luxurious, fabrics in ways that are non-toxic and sustainable.

Keywords: *Textile, Eco-friendly, Fibre, Pollution.*

Introduction

In a world where extreme use of chemical products is causing irreparable damage to the environment, use of eco-friendly textile fibres is an initiative with a vision to bring about a drastic reduction in global consumption of harmful non-biodegradable products. The world of fashion may be stylish, glamorous and exciting, but its impact on environment is worsening day by day. The pesticides that farmers use to protect plants can harm wildlife, contaminate other products and get into the food we eat. From an environmental point of view, the clothes we wear and the textiles which are being used can cause a great deal of effect to the environment. The effects are as follows.

☆ The chemicals that are used in textiles processing can affect the environment as well as the people.

☆ Old clothes that we throw away take up precious space in landfill sites, which is filling up the land rapidly.

☆ Most of the textile machineries causes' sound and air pollution.

☆ Over-usage of natural resources like plants, water, etc depletes or disturbs ecological balance.

☆ Exploitation of animals often goes hand in hand with intensive farming practices damage the environment as a whole.

Environmental Effects of Textile Fibres

Problems Caused by some of the Common Fibers

Cotton

It is one of agriculture's most water-intensive and pest sensitive crops, it is estimated to consume 11 per cent of the world's pesticides and chemical fertilizers that pollute and deplete the soil. Herbicides, and the chemical defoliants which are sometimes used to aid mechanical harvesting of the cotton, remains in the fabric after finishing, and are released from the garments,which affects both environment and human health.

Wool

The insect-resist/mothproofing treatments may cause health problems as well as producing effluent are toxic to aquatic life.

Rayon

It is a regenerated cellulosic fibre and is made from wood pulp, which seems more sustainable. However, old forests are often cleared and farmers are making way for pulpwood plantations. Often the tree planted is eucalyptus, which draws up phenomenal amounts of water, causing problems in sensitive regions, also during manufacturing of the rayon; the wood pulp is treated with chemicals such as caustic soda and sulphuric acid which causes pollution. Use of catalytic agents containing cobalt or manganese in processing also causes strong, unpleasant odor. The excessive use of rayon for clothing is resulting in to the rapid depletion of the world's forests.

Nylon and Polyester

These are made from petrochemicals and these synthetics are non-biodegradable. Nylon manufacture creates nitrous oxide, a greenhouse gas 310 times more potent than carbon dioxide. Manufacturing of polyester uses large amounts of water for cooling, along with lubricants which can become a source of contamination.

What is an Eco-friendly Fibre?

Eco-friendly fibres are those which are produced in an eco-friendly manner and processed under eco-friendly limits (defined by agencies like oekotex, ifoam etc.). The main aim of developing these fibres is to bring about a drastic reduction in global consumption of harmful non-biodegradable products.

Eco-fashion Clothes

☆ They are made by using organic raw materials, such as cotton grown without pesticides and silk made by worms fed on leaves of organic trees

☆ It does not involve the use of harmful chemicals and bleaches for the processing of the fabrics

☆ They can be made from recycled and reused textiles. Garments can be made from second-hand clothes and even recycled plastic bottles

Need to Develop an Eco-friendly Fibre

The highly competitive atmosphere and the ecological parameters are becoming more stringent, so it becomes the prime concern of the textile industry to be conscious about ecology. The main responsibility at the moment lies with clothes manufacturers and fashion designers, who need to start using sustainable materials and processes. Today companies are providing clothes made from eco-friendly fibres and the demand for these clothes is increasing day by day.

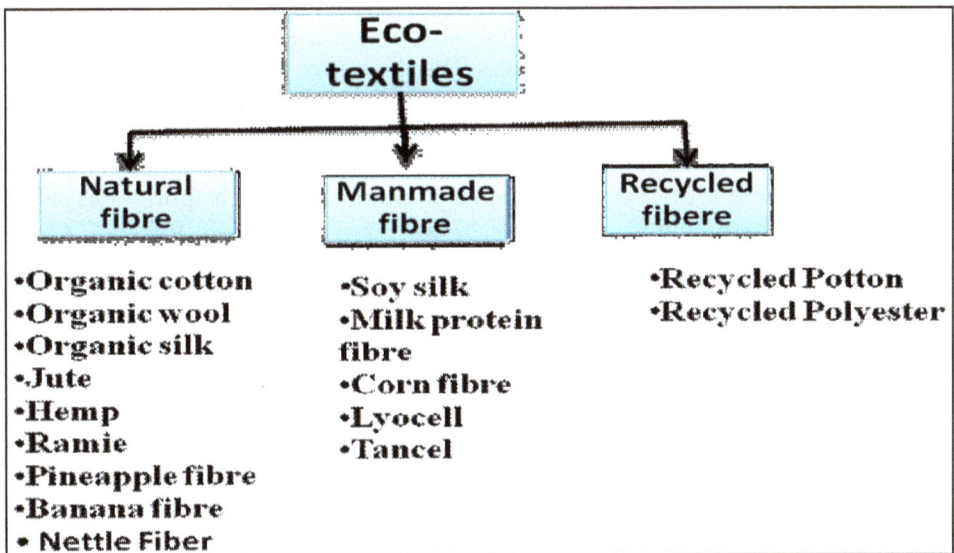

Figure 21.1: Classification of Eco-Textiles

Natural Fibres

Organic Cotton

The organic cotton is produced by organic farming in which the crops are cultivated without pesticides, chemicals and synthetic fertilizers using organic manures and bio fertilizers. As the production of Organic cotton is environmentally friendly therefore its cultivation is increasing steadily. Its cultivation helps in improving soil fertility and helps in preventing water, soil and air contamination. Equivalent/better fiber properties of organic cotton help in diversified products development- suitable for all products.

Organic Wool

Wool has been used for decades and is especially known for its durability and warmth. A sustainable alternative for the wool is Organic wool, which originates from healthy animal *e.g.* Sheep's raised on an organic farm. The animals are raised "free range" under natural, healthy conditions *i.e.* animals have been grazed on land free with pesticide and fertilizer which are harmful to humans or the livestock. The result is disease free animals with healthy immune systems. Animals are also shorn in an animal-friendly manner and the fibers produced are recyclable and biodegradable.

Organic Silk

Silk is created by the protein secretions of silkworms which solidify to create the silk fibers. Organic silk is made from the cocoons of wild and semi-wild silk moths. During the production of organic silk the pupae are not stifled or killed to obtain reeled yarn but the open ended cocoons are spun into yarn. The difference between organic and conventional silk is also based on the diet of the silkworms. With organic silk, the silkworms feed on mulberry leaves from bio-dynamically grown trees, without the use of synthetic pesticides or fertilizers. The silkworms grown in these leaves are healthier; they also produce longer silk threads of the highest quality in the world. It also means that the environment is healthier, providing the workers with higher quality lives. The organic silk has greater degree of fiber purity, resulting in better yarn luster and uniformity than normal spun silk. Since the fiber is spun, it has better fiber strength and durability.

Jute

Jute is one of the most important natural fibers after cotton. Jute is 100 per cent bio-degradable and thus, environment-friendly fibre. Fiber is extracted from the bast of the plants. It is used extensively in manufacturing different types of packaging material for agricultural and industrial products. Jute is available in abundance in India, at competitive prices. Jute is now not just a major textile fiber, but also a raw material for non-textile products, which help to protect environment. It has high tensile strength and are resistance to heat and fire. Jute fiber has a wider applications.

Hemp

It is the worlds oldest and the strongest natural fiber cultivated on earth. Organic hemp grows without fertilizer. It requires little water, and doesn't deplete soil nutrients at the same time binding and enriching the soil with its deep roots. It is easy to harvest and this plant has high potential to create eco-friendly textiles.

Banana Fiber

This fiber is extracted by hand stripping and decortications of the leaf of the banana plant. It is a 100 per cent eco-friendly fiber. It is strong, shiny, lightweight and bio-degradable. Banana fibers were used for making ropes and mats till recent past. With its many qualities getting popular, the fashion industry is also fast adopting this fiber for making various fashion clothing and home furnishing.

Pineapple Fiber

it is one of the eco- friendly fibers that are gaining popularity. It is also known as Pina fiber. It is extracted from the pineapple leaves by hand scraping, decortications or retting. Pineapple fabric is lightweight, soft, shinning, transparent and a little stiff. The fabric is used for making clothes having elegant looks.

Nettle Fiber

This fiber is obtained from the Brennessel plant which is naturally resistant to vermin and parasites. It can be grown without pesticides and herbicides. The fiber is obtained by retting of the plant stem. Fiber is separated from the stem mechanically and the fibers are used for making fabric.

Man-Made Fibres

Soy Silk

A lesser-known, eco-friendly fiber is soy silk, which is made from tofu-manufacturing waste. For the manufacturing of the Soy fibre, Soy protein is liquefied and then extruded into long, continuous fibers that are cut and processed like any other spinning fiber. Soy has high protein content, therefore the fabric is much more receptive to natural dyes – eliminating the need for synthetic dyes. The fabric is 100 per cent biodegradable.

Casein Fiber

Milk proteins have also been used to create synthetic fabric. Milk or casein fibre possesses anti-bacterial properties. These fibers are used to make eco- friendly yarn *i.e.* the milk yarn. For manufacturing the fibre the milk is dewatered, *i.e.* all the water content is taken out from it and then skimmed, with the help of bio-engineering technique, a protein spinning fluid is made. Wet spinning process converts this fluid into high-grade textile fiber. The skin friendly milk yarn is used to make glossy and luxurious fabrics similar in appearance to silk fabrics that has a pH similar to human skin, it also has antibacterial and antifungal properties and it is marketed as a biodegradable, renewable synthetic fibre.

Bamboo Fiber

The bamboo fiber is obtained from the pulp of bamboo plants. Bamboo fiber is a regenerated cellulose fiber. However, bamboo fiber is eco- friendly when it is extracted with mechanical process. When mechanically processed, the crushed bamboo is treated with biological enzymes which break it into a mushy mass after which individual fibers are combed out. Organic bamboo fabric is left unbleached by the manufacturers. Bamboo fibers make smooth, soft, antibacterial and luxurious fabric that have a very good handle

Corn Fibre

It is produced by extracting the starch and then sugars from corn, and processing them to make a fibre, which can be spun into a yarn or woven into fabric.

Recycled Fibre

Recycled Cotton

Recycled Cotton is also an eco-friendly choice in cotton clothing since recycled cotton is cotton fabric which is made from recovered cotton that would otherwise be cast off during the spinning, weaving or cutting process. The discarded cotton waste is collected, shredded into small fibers and processed again into yarns and fabrics. Truly ecofriendly – because of waste recycling process, no chemicals used during processing, it helps in generating employment and is good for the environment.

PET fibre

Polyester fibre is one of the most non-biodegradable polymers which create environmental problems. The legislation opens the door towards working over recycling of PET which is 100 per cent recyclable it can be recycled back, either to its original elements or into other products. PET fibres are used in synthetic clothing and other forms of textiles. It is light weight and can be crushed/compressed, making it easy to transport to recycling plants. Most supermarkets have collection points, making it convenient for customers to return their empty packaging. The large proportion of supermarket customer's return used PET packaging. Thus PET packaging is recycled into new products. Therefore, it could be said that PET is a sustainable material. This fiber is suitable for diversified products range such as blankets, T-shirts, sportswear, soft luggage and socks etc.

Conclusion

The global textile industry does not operate environmentally friendly and sustainably but the concern for the degrading environment conditions due to irresponsible use of chemical products have led to worldwide efforts to develop eco-friendly fibres in the ever expanding horizon of textile fibres. The demand for eco-friendly textiles is one of the driving forces in the textile industry as consumers are becoming increasingly concerned about the environment. The marketplace responds with new technology to fit the demands of a greener lifestyle.

References

Eco-friendly Fabrics Make Green Fashion Statement, Edited by Potirala. S, Bangalore University India, 2009.

Ecotextile, *The way forward for sustainable development in textiles,* Edited by Misaftab and A R Horrocks, University of Bolton, UK, 2004.

Sustainable fashion, retrived on 11oct 2011. http: //en.wikipedia.org/wiki/ Sustainable_fashion.

Organic cotton retrived on 2nd August 2011, http: //en.wikipedia.org/wiki/ Organic_cotton.

Organic wool, retrived on 10th July 2011, http: //en.wikipedia.org/wiki/ Organic_wool.

2013, Environmental Technology
Editors: D.R. Khanna, A.K. Chopra, Gagan Matta, R. Bhutiani & Vikas Singh
Published by: DAYA PUBLISHING HOUSE, NEW DELHI

Pages *189–196*

Chapter 22

Availability and Evaluation of Eco-friendly Concretes Ingredients

G.C. Mishra and Akriti Gupta

Department of Civil Engineering,
Lingaya's University, Faridabad

ABSTRACT

Concrete is a composite construction material. More than six billion tons of concrete are produced annually *i.e.*; about one ton per person on the planet. Concrete is made from cement, water, sand, water and chemical admixtures and gravel. The cement is made by heating raw materials such as limestone and clay to very high temperatures until they chemically react. This process uses massive amounts of energy (about five percent of the world's use per year) and releases about a ton of carbon dioxide per ton of cement made. It also claimed to be huge source of carbon emission in the atmosphere *i.e.*; up to 5 per cent of world's total amount of carbon emissions. So there is an urge to develop an eco friendly concrete to reduce its adverse effect on environment. Since last couple of years efforts are on to develop eco friendly concrete by using various industrial or other waste materials available in plenty and presently causing harm to the environment as an alternate ingredients. "Anything one can do to make concrete more eco-friendly will have a big impact," "simply because it's the world's most used material." Concrete is made from cement, water, sand and gravel. It is the need of the hour to make cement out of waste materials instead of new materials to reduce the carbon dioxide emitted and energy used for production. There are several waste materials available in plenty such as fly-ash, blast furnace slag, silica fumes, wood ash, de-watered sludge, rice husk ash, municipal solid waste ash, volcanic ash, cement kiln dust of the coal-burning industry etc. Fly ash is already used in concrete as a cement substitute since long time and has a composition similar to cement. In the recent past all these waste materials are in use for making cement concrete to save energy and prevent waste from piling up in landfills while making something people want.

The objective of this study is to evaluate the use of various types of substitute materials in terms of their availability, their potential environmental impact if they remain in the environment, application obstacles in cement/concrete making, and comparative strength in order to reduce its adverse effect on environment as well as for sustainable infrastructure development.

Introduction

There are several types of waste materials available which has got potential to be used as ingredient of concrete such as:

Fly Ash

It is one of the residues generated in combustion and comprises the fine particles that rise with flue gases. Ash which does not rise is termed as bottom ash. Fly ash refers to ash produced during combustion of coal. Fly ash is generally captured by electrostatic precipitator or other particle filtration equipments before the flue gases reach the chimney of coal fired plant and together with bottom ash. Depending upon the source and make up of coal being burned, the components of fly ash vary considerably but all fly ash includes substantial amount of silicon dioxide (SlO_2) and calcium oxide. Toxic constituents depend upon the specific coal bed make up. It may be Arsenic, Beryllium, Boron, Cadmium, Cobalt, Lead, and Mercury, along with Dioxins and PAH compounds. Two classes of fly ash are defined as class F fly ash and class C fly ash. The difference between both of these is the amount of Calcium, Silica, Alumina and Iron.

Class F Fly Ash

It is produced by the burning of harder, older Anthracite and Bituminous Coal. It is pozzolanic in nature and contains less than 20 per cent lime possessing pozzolanic properties, the glassy silica and alumina of class F fly ash requires cementing agent, such as Portland cement, quick lime or hydrated lime with the presence of water in order to react and produce cementitious compounds.

Class C Fly Ash

It is produced from the burning of younger lignite, sub-bituminous coal, in addition to having pozzolanic properties, also has some self cementing properties.

Total Production and Utilisation

The total production of fly ash in India is 110 million tones, out of which only 30 per cent is used amounting to 30.6 million tones. And from those 30.6 million tones, 15 per cent is used for building, 35 per cent for dykes, 30 per cent for cement, 15 per cent for land development and 5 per cent for others.

Tale 22.1

Sl.No.	Zone Name	FA Generation (million tonnes) (1997-2007)	Utilization (million tonnes)	Utilization Per cent
1.	Southern	13.5	0.8	6
2.	Western	16.5	0.8	5
3.	Central	18	2.84	15.8
4.	Eastern	10.21	2.94	28.8
5.	Northern	15.5	2.3	14.8
	TOTAL (All India)	73.71	9.96	70.4

Utilisation of Fly Ash and CO_2 Emission

Out of total utilization, about 22 per cent amounting to 7.75 million tones was used in areas of roads and embankment work last year.

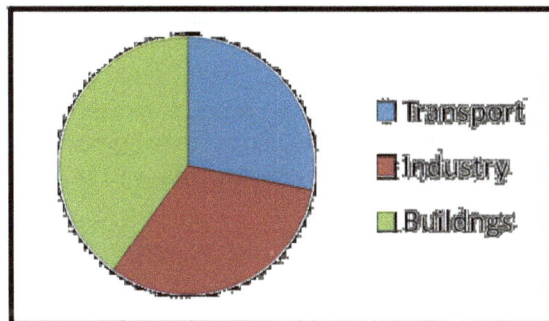

Buildings are responsible for about 40 per cent of air emission and part of those cement are due to cement used in concrete. With concrete, cement is responsible for the largest environmental burden. The stat states that one tonne of cement produces about 1 tonne of CO_2 and the world cement production was 1.3 billion tones in 2005. So we are putting a larger amount of CO_2 in environment. This is about 6 per cent of CO_2 cemission worldwide

Contaminants

Fly ash contains trace concentrations of heavy metals and other substances that are known to be detrimental to health in sufficient quantities. Potentially toxic trace elements in coal include arsenic, berrylium, cadmium, barium, chromium, copper, lead, mercury, molybdenum, nickel, radium, selenium, thorium, uranium, vanadium and zinc. Fine crystalline silica present in fly ash has been linked with lung damage in particular silicosis. Another fly ash component of some concerns is lime (CaO). This chemical reacts with water to form calcium hydroxide[Ca(OH)], giving fly ash a pH somewhere between 10 and 12, a medium to strong base which can cause lung damage if present in sufficient quantities.

Result and Discussion

Advantages of using Fly-ash

☆ Use of fly ash increases the absolute volume of cement containing materials compared to normal concrete, leading to reduction in aggregate particles interference and enhancement in workability.

☆ It also reduces bleeding by providing greater finer volume and lower water content for a given workability.

☆ Fly ash increases the time of setting of concrete.

☆ Strength of fly ash concrete is influenced by type of cement, quality of fly ash and curing temperature compared to that of non-fly ash concrete proportional for 28 days compressive strength, fly ash concrete usually have higher ultimate strength when properly cured.

☆ Considering all these advantages it is extremely essential to promote use of fly ash in concrete.

Environmental Impact of Fly Ash Usage

Utilization of fly ash will not only minimize the disposal problem but will also help in utilizing precious land in a better way. Since there is no seepage of rain water into the fly ash core, leaching of heavy metals is also prevented.

Economy in Use of Fly Ash

Use of fly ash in road works results in reduction in construction cost by about 10 to 20 per cent. Fly ash is available free of cost at the power plant and hence only transportation cost, laying and rolling cost are there in case of fly ash.

Blast Furnace Slag

It is obtained by quenching molten iron slag from a blast furnace in water or in steam, to produce a glassy, granular that is then dried and grind into fixed powder. It is used to make durable concrete structures in combination with ordinary Portland cement or other pozzolanic materials.

Production of Blast Furnace and Its Utilisation

Typically for ore feed containing 60-65% iron blast furnace slag production ranges from about 300to 540 kg per tonne of pig or crude iron produced. Lower grades ores yield much higher slag fractions, sometimes as high as one tonne of slag per tonne of pig iron produced. Steel slag output is approximately 20% by mass, of the crude steel output. In India, iron production of 718 MT, produces slag of 180-220 MT. The current blast slag production is 400kg/tonne.

Table 22.2: Plant-wise Average Generation of Slag 2006-07 and 2007-08 (In kg/tonne of hot metal)

Steel Plant	Production
Bhilai Steel Plant	408
Rourkela Steel Plant	-
Visvesvarya Steel Plant	318
Durgapur Steel Plant	365
IISCO Steel Plant	582
IDCOL Kalinga Iron Works Ltd.	380
JSW Steel Limited	348
Tata Steel Ltd., Jamshedpur	239

CO_2 Emission by Blast Furnace Slag

It was assume that Portland blast furnace slag cement accounted for all of the 16 per cent increase in the production ratio of blended cement, this would contribute to an annual reduction in CO_2 emissions of 640,000 tons. Specifically, the use of 20 per cent Portland blast furnace slag cement in the construction of a single apartment complex would result in a per-household CO_2 reduction of approximately 1, 200 kg.

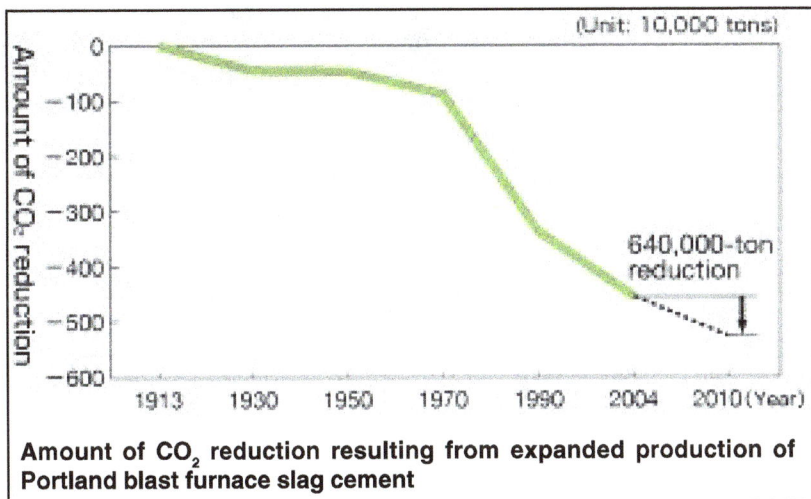

Amount of CO_2 reduction resulting from expanded production of Portland blast furnace slag cement

Advantages of Using Blast Furnace Slag in Cement/Concrete

☆ It increases the durability of concrete as well as its use usually improves workability and decreases the water demand due to increase in paste volume caused by relative density of slag.

☆ The higher strength potential of slag may allow for a reduction of total material.

☆ An increase of slag in cement from 35-65 per cent by mass, it can increase the setting time by 60 minutes.

☆ It has low heat of hydration and it has low alkali aggregate reaction.

☆ It has high resistance to chlorides and sulphates and it can substitute the use of 43 and 53 grades of ordinary Portland cement.

Environmental Impact of Blast Furnace Slag

Increased utilization of slag benefits the Portland cement producers. Producers can enhance the production capacity without generating additional greenhouse gas emission like CO_2. It offers the possibility of considerable energy recycling in the form of hot water and heated air.

Cost

The BF slag is not gaining much ground because of availability of other minerals at cheaper prices than the prices of BF slag. In 2006-2008 the prices of BF slag varies from 300-700rs.

Comparison of Concrete Having Blast Furnace Slag and One Containing Fly Ash

Five basic concrete mixes were considered. These were: conventional mix with no material substitutions, 50 per cent replacement of cement with fly ash, 50 per cent replacement of cement with blast furnace slag, 70 per cent replacement of cement with blast furnace slag and 25 per cent replacement of cement with fly ash and 25 per cent replacement with blast furnace slag. Recycled concrete aggregate was investigated in conventional and slag-modified concretes. Properties investigated included compressive and tensile strengths, elastic modulus, coefficient of permeability and durability in chloride and sulphate solutions.

Table 22.3: Chemical Composition and Physical Properties of Cement, Fly Ash and Blast Furnace

		Slag	
	Cement	Fly Ash	Slag
CaO	63.25	5.54	39.84
SiO_2	20.8	47.58	38
Al_2O_3	4.61	26.42	7.52
Fe_2O_3	2.59	12.19	0.31
MgO	4.17	0.9	10.54
Na_2O	0.16	1.5	0.32

Workability

Fly ash gave the most significant improvement in workability and required shorter vibration time.

Tensile Strength

The splitting tensile strength of the natural aggregate mixes was increased by incorporation of 50 per cent and 70 per cent slag and decreased by incorporation of 50 per cent fly ash. The 28 day mean splitting tensile strength ranged from 1.69 to 2.83 MPa for the 50 per cent fly ash and 50 per cent slag mixes with natural aggregate, respectively. The FA/BFS blend had a mean strength in between that of the mixes with either 50 per cent slag or 50 per cent fly ash with a value of 2.41 MPa. The mean tensile strengths of the RCA mixes were 0.02–0.32 MPa lower than those of the equivalent natural aggregate formulations.

Rice Husk Ash

Rice Husk Ash (RHA) is a carbon neutral green product. This super-pozzolana can be used in a big way to make special concrete mixes. There is a growing demand for fine amorphous silica in the production of special cement and concrete mix high performance concrete, high strength, low permeability concrete, for use in bridges, marine environments, nuclear power plants etc. RHA is a highly reactive pozzolanic material suitable for use in lime-pozzolana mixes and for Portland cement replacement. RHA contains a high amount of silicon dioxide, and its reactivity related to lime depends on a combination of two factors, namely the non-crystalline silica content and its specific surface.

Production and Utilisation

The world rice harvest is estimated in 500 million tons per year. Considering that 20 per cent of the grain is husk, and 20 per cent of the husk after combustion is converted into ash, a total of 20 million tons of ash can be obtained. RHA has high levels of silicon dioxide, approximately 93 per cent, and the specific gravity is $2.16 cm^2/g$. The average article size distribution was 13.34m. Thus the RHA is finer than cement and should be expected to work not only a pozzolanic role, but also a micro filler effect.

Advantages of Using Rice Husk Ash in Concrete

☆ Higher substitution amounts of RHA results in lower water absorption values, it occurs due to the RHA is finer than cement. Adding 10 per cent of RHA to the concrete, a reduction of 38.7 per cent in water absorption.

☆ The addition of RHA causes an increment in the compressive strength due to the capacity of the pozzolana, of fixing the calcium hydroxide, generated during the reactions of hydrate of cement. Replacement of RHA increased the compressive strength. For a 5 per cent of RHA, 25 per cent of increment is observed.

☆ Replacement of RHA researched; achieve results in splitting tensile strength. According to the results, may be realized that there is no interference of adding RHA in the tensile strength.

☆ A decreasing in the module is realized when the levels of RHA are increasing.

Magnesium Silicate Cement

Cement, based on magnesium silicates, not only requires much less heating, it also absorbs large amounts of CO_2 as it hardens, making it carbon negative. This cement can absorb, over its lifecycle, around 0.6 tonnes of CO_2 per tonne of cement. This compares to carbon emissions of about 0.4 tonnes per of standard cement.

Total Production

The industry supplies 2.9bn tonnes of magnesium silicate cement to support global economic development every year. Supply is expected to grow at 3-5 per cent per annum, particularly driven by development needs in India and China.

CO_2 Emissions

1. Non-carbonate feedstock –uses magnesium silicates with reserves of 10,000 billion tonnes. No CO_2 from the raw material. Limestone and its stored carbon left in the ground.

2. Lower temperature chemical process (700°C) can utilise biomass fuel.

Cost

Cost may remain nominal. It will not exceed much in comparison to ordinary Portland cement.

Advantages of Magnesium Silicate Cement

☆ Mechanical properties already enough for initial applications with large markets, *e.g.*, blocks and pavers.

☆ Adjustable pH system –allows wider choice of potential aggregates, *e.g.* lower cost waste aggregates, glass or plastic.

☆ Completely recyclable at end of life.

References

Bethke, C. M. 1996. Geochemical Reaction Modeling. Oxford University Press, Inc., NY, 397 p.

EPRI. 1987. Chemical characterization of fossil fuel combustion wastes. Electric Power Research Institute final report EPRI EA-5321.

Fishman, N. S., C. A. Rice, G. N. Breit, R. D. Johnson. 1999. Sulfur-bearing coatings on fly ash from a coal-fired power plant: composition, origin, and influence on ash alteration. Fuel, 78, 187-196

Groenewold, G. H., D. J. Hassett, R. D. Koor and O. Manz. 1985. Disposal of western fly ash in the northern Great Plains. Materials Research Society, Symposium Proceedings, Fall 1984, Symposium M, 213-226.

ASTM C150, C595, C1012, C1157 Annual Book of ASTM Standards Volume01, ASTM International, West Conshohocken, Pa., www.astm.org

Thomas, Michael, "Optimizing the Use of Fly Ash in Concrete, " Portland Cement Association, Publication IS 548, 2007, 24 pages

A forum held 8 December 1998 - Sponsored by EHDD Architecture and Pacific Energy

Center Hill, R.L., and Folliard, K.J. (2006), "The Impact of Fly Ash on Air-Entrained Concrete, " *Concrete InFocus,* Fall 2006, pp. 71-72.

Roy, W. R., R. A. Griffin, D. R. Dickerson, and R. M. Schuller. 1984. Illinois Basin coal fly ashes. 1. Chemical characterization and solubility. Environmental Science and Technology, 18, 734-739.

Spears, D. A. and S. Lee. 2004. The geochemistry of leachates from coal ash. Geological Society, London, Special Publications, 236, 619-639.

2013, Environmental Technology   *Pages 197–201*
Editors: D.R. Khanna, A.K. Chopra, Gagan Matta, R. Bhutiani & Vikas Singh
Published by: DAYA PUBLISHING HOUSE, NEW DELHI

Chapter 23

Mathematical Modeling of Lightning: An Application to Earth Resistivity Interpretation

P.P. Pathak and Jyotika

Department of Physics,
Gurukul Kangari Vishwavidyalaya, Haridwar

ABSTRACT

Lightning is a powerful natural electrostatic discharge produced during a thunderstorm. The atmospheric disturbance created by a stroke of lightning interacts appreciably with the earth surface and beneath. This paper by and large models the interaction of vertical component of lightning induced electric field with the earth surface. It has been found that the lightning radiated electric field construes information regarding the subsurface structure of earth and consequently resistivity variation of earth surface.

Introduction

The upper most surface of earth has a stratified structure comprising of different layers. The resistivity feature of these layers characterizes their respective nature. This interpretation is of utmost importance in different exploration realms ranging from geophysical to archeological. The basic principle involving all these exploration modules requires some sources for energizing the earth surface.

The current conducting nature of different materials forms the basis of this study. The lightning radiations emanating from the clouds towards the earth induces current in the earth surface. The current flowing through the ground explores different layers characterized by different resistivity tendencies (Delfino *et al.*, 2008).

The synthesis follows the simple rule that characteristics of lightning induced electric field is chiefly governed by the properties of the sub-surface materials through which they permeate while depending on various important lightning parameters. A suitable variant of lightning discharge (Very Low Frequency radiations in our case) yields relevant electric fields (Pathak, 1990). Moreover, the features arising due to different mediums available under the earth's surface are of immense importance as they act like mirrors for imaging the earth interior (Wait, 1969).

The field generated on the surface of earth has a spatial variability. It has also been analyzed that the field changes appreciably as it interacts with the surface. This change has been studied by various authors. (Divya and Rai, 1986; Pathak, 1990; Cooray, 2003) treating the earth as a surface with a finite resistivity. An empirical relation has been envisaged then for accounting for the variations in the field waveforms.

On the other side, (Kumar, 1990; Shoory *et al.*, 2010) assumed that the earth has a two layer structure beneath the surface with finite resistivity variation. The yielded results were beneficial in accounting for the field changes. In this paper a three layer earth model is assumed with the bottom layer extending up to infinity. However, for the sake of simplicity an image of lightning channel is assumed below the surface of earth, which effectively accounts for the resistivity variation.

Theory

The estimation of vertical electric field on the ground requires the support of mathematical modeling of the lightning return-stroke. The entire feature of return-stroke has been studied extensively yielding several return-stroke models (Thottappillil and Rakov, 2001; 2007). The models tend to specify current as a function of space and time along the radiating lightning channel. The vertical component of electric field is computed using the Bruce and Golde model (Bruce and Glode, 1941).This model specifies the current variation along the lightning channel as:

$$i(z',t) = i_0(t) \quad \text{for every} \quad z' \le vt \tag{1}$$

$$i(z',t) = 0 \quad \text{for every} \quad z' > vt \tag{2}$$

The geometric configuration utilized for the computation of vertical component of electric field over a layered earth model has been illustrated in figure (1). The observation point is assumed to be situated at a distance D measured horizontally from the foot of lightning channel. At this point the contribution from an elemental dipole of current i (z', t) of length dz' located along the vertical axis at z' is computed using the Maxwell's equation as:

$$\vec{E} = \frac{1}{\epsilon^c} \int \frac{1}{\mu^c} \left(\nabla \times \nabla \times \vec{A} \right) dt \tag{3}$$

With A as the source vector potential having a form:

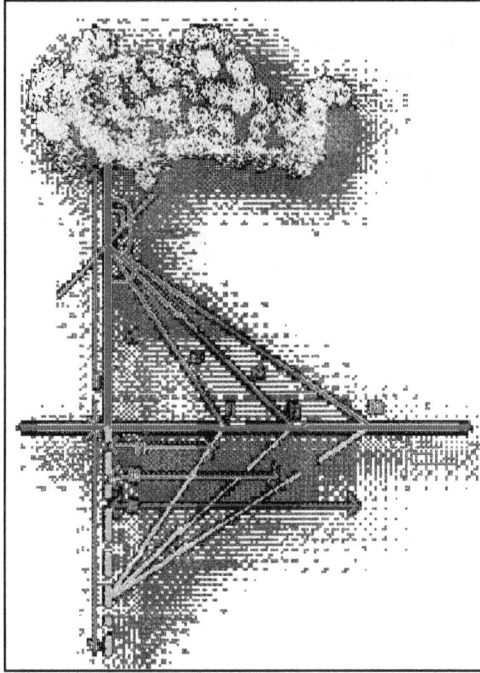

Figure 23.1: Geometry for Electric Field Computation

$$\vec{A} = \frac{\mu^c}{4\pi} \int \frac{I(z',t')\overline{dZ'}}{r}$$

(4)

The lightning induced potential developed on the earth surface forces the lightning current to traverse inside the earth surface. The movement of current inside the earth is governed by the intensity of the current and characteristics of the earth media. This feature basically compelled to assume the earth sub-surface to be a horizontally stratified structure having uniform thickness and possessing homogeneity and isotropic nature.

The information gathered through the lightning induced electric field is incorporated into resistivity kernel function, which essentially depends on resistivity and thickness of sub-surface layers (Koefoed, 1976) and is defined as:

$$K(\lambda_j) = \sum_{i=1}^{p} \sum_{j=1}^{m} w_i e^{-\epsilon_i \lambda_j}$$

(5)

where,

      p: 1,2,3,4....and,

      m: 1,2,3

This exponential function has its values dependent on the lightning electric field generated on the earth surface (Okolie *et al.*, 2008). In order to delineate information regarding the resistivity variation of sub-surface layers a suitable inversion technique is intended to be employed. A resistivity transform function is developed using equation (5) for the inversion process (Anders and Auken, 2004; Drahor *et al.*, 2007). Once this function is resolved it is compared with the function evolved through the assumed layer parameters.

For the purpose an interactive iterative scheme is developed as:

$$T_L(p_q, P) = T_m(p_q, P^0) + \sum_{q=1}^{2M-1} (P_S - P^0)\frac{\partial T_q}{\partial P_s} \tag{6}$$

where,

      N: No. of layers in the assumed model

      p_q: Known Parameters

      P_s: Unknown layer parameters

      P^0: Initial values of the assumed parameters

      T_L: Resistivity transform for the lightning data

      T_m: Resistivity transform for the assumed model

The yielded results are compared with the data generated through the lightning parameters (electric field).With this exercise it is easier to reconstruct the real arrangement of different layers at different depths below the surface of earth. This procedure eventually proves beneficial in constructing a finale geological structure of earth sub-surface.

Conclusion

The present paper tends to discuss the mathematical analysis of lightning return-stroke on the ground. The modeling has been attempted through analyzing the propagation effects of lightning electromagnetic field along a finitely conducting ground. The modeling is suitably and straight away employed in interpreting resistivty variation beneath the earth surface. The synthesis essentially involves the assumption of a finite ground having a three layer stratified structure. It has been shown that the adopted methodology is simple and robust yielding appropriate results.

References

Anders, V., C. and Auken, E., Optimizing a layered and laterally constrained 2-D inversion of resistivity data using Broyden's update and 1-D derivatives, 56, 247-261, 2004.

Bruce, C. E. R. and Golde, R. H., The lightning Discharge, J. Inst. Elec. Engrs., 88, 487-505, 1941.

Cooray, V., On the concepts used in return-stroke models applied in engineering practice, Trans. IEEE (EMC), 45, 101-108, 2003.

Divya and Rai, J., Calculation of electric field from lightning above finitely conducting ground, Ind. J. Radio and Space Phys., 15, 96, 1986.

Delfino, F., Procopio, R. and Rossi, M., Lightning return-stroke current radiation in presence a conducting ground: 1. Theory and numerical evaluation of the electromagnetic fields, J. of Geophys. Res., 113, 2008.

Drahor, M. G., Gokturkler, G., Berge, M. A., Kurtulmu, T. O., and Tuna, N., 3-D Resistivity Imaging from an Archaeological Site in South-Western Anatolia: A Case Study, Near Surface Geophys., 5, 195-201, 2007.

Koefoed, O., Progress in the direct interpretation of resistivty sounding: an algorithm, Geophys. Prospect. (24), 617-632, 1976.

Kumar, K., Rai, J., Singh, V. and Niwas, Sri, Study of VLF radiation from lightning above finitely conducting multilayered earth, J. Meteor. Soc., 91, 195, 1990.

Okolie, E. C., Egbai, J. C. and Oseji, J.O., Comparative Investigation Strata and Groundwater Distribution in Overokpe Array, Nig. J. of Sc. And Environ., 7, 91-98, 2008.

Pathak, P.P., Geophysical prospecting with lightning radiation, J. Natural Phys. Sci., 4, 69, 1990.

Shoory, A., Mimouni, A., Rachidi, F., Corray, V., Moini, S. and Sadeghi, S.H.H., Validity of simplified approach for the evaluation of lightning electromagnetic fields above a horizontally stratified ground, IEEE Trans. on electromagnetic compatibility, 52, 657-663, 2010.

Thottappillil, R. and Rakov, V.A., on different approaches for calculating lightning electric fields, J. Geophys. Res., 106, 14191-14205, 2001.

Thottappillil, R., Rakov, V. A., Review of three equivalent approaches for computing electromagnetic fields from an extending lightning discharge, J. of Lightning Research, (1), 90-110, 2007.

Wait, J.R., Image theory of a quasistatic dipole over dissipative half space, Electronics Lett., 5, 281-282, 1969.

2013, Environmental Technology

Editors: D.R. Khanna, A.K. Chopra, Gagan Matta, R. Bhutiani & Vikas Singh

Published by: DAYA PUBLISHING HOUSE, NEW DELHI

Pages 203–209

Chapter 24

A Spatial Model for Socio-Economic Zoning of Flood Plains

Om Prakash Dubey and Walmi

U.P. Water and Land Management Institute,
Walmi Bhawan, Utretia, Lucknow

ABSTRACT

The main problems faced during the monsoon in India are flooding, drainage congestion and bank erosion. The flood plains are the playground of the river and should be left as such without any human interference. Alluvial rivers are of meandering or braided form and decide their own shapes normally having a water way related to their nominal discharge. Very roughly, the width of these "playground" are four to six times such a waterway width for the meandering rivers and even larger for the braided rivers. As a result, a very large part of the alluvium would have to be left alone without human settlements. This is clearly not practicable and pragmatic approach of a management. This pushes the need for flood hazard mapping and zoning. In this study, an attempt has been made to develop a simple approach for flood plain zoning. Data for the analysis has been generated by synergistic use of remote sensing, field visit and ancillary data. Data analysis has been carried out in GIS environment.

Introduction

Flood hazard mapping and zoning techniques have development over the period as reflected from the pioneering works of several individuals and organizations including Ellies 1969, Sheoffer, 1964, U.S. Army Crop of Engineers, 1973, Sheoffer Ellies and Speecker 1970, Weolman 1971 and Dingonar 1975. Several environmentalists like Hack and Goodlett 1960, Sigafors 1964 and Rechardson 1968 have also devised alternative processors for flood hazard mapping. India being the second largest flood prone country in the world, the need for flood inundation as well as socio-economic zones in flood prone areas is most essential for the country's economic development and environmental management. After the unprecedented flood of 1954, the government of India took several initiatives and constituted a number of committees to study the problem of land in the floods. The some important initiatives are policy statement 1954, high level committee on floods 1957, policy statement of 1958 Ministerial Committee on flood control 1964, Ministers Committee on floods and flood relief 1972, Working groups on flood control for five years plan, *Rashtriya Barh Ayog* 1980, National Water policy 1987, National Commission for Integrated Water Resources Development Plan 1996. Historical flood damage data 1953-2000 brings out the fact that during this period the area affected by floods remained more or less constant, whereas the population effected and crop damage are showing an increasing trend.

There is a general perception that in spite of spending large amount of money for various flood management works in the country, the flood damage is increasing. Further, the shape of research in this area of vital human concern is rather slow in the country. In this regard mention may be made of the work of Kayastha and Yadava 1977 Chakraborty 1979, 1991, Goswami 1991 and Rango and Enderson 1994. The use of remote sensing techniques particularly those that are satellite based has significantly improved the quality and coverage of flood mapping in the country in recent years. Large scale investment in flood protection, huge recurring damages, gradual and steady increase of population pressure on land forces demand for objective flood plain zoning considering the constraints and aspiration of land users on one hand and flood disaster on the other hand. In this study an attempt has been made to develop an analytical object oriented model in GIS for flood plain zoning. Inputs to the model were extracted from synergistic use of remote sensing data and conventional data. It is expected that model may be used for utilization of flood plains in such a manner so as to reduce the flood damage and disaster. It is an established fact that flood hazard and its management in an area depends upon several geo-parameters pertaining to above the surface, on the surface and subsurface of that area.

These parameters include rainfall, ground elevation, land cover drainage density, texture of surface material, ground slope in addition to social and economic factors (c.f Le Grand 1967, Abyakar and Mansi 1974, Bouwer, 1978, Kriz 1971, Reeves *et al.*, 1978, Turc 1977, Dubey *et al.*, 1984 and Krishnamurthy *et al.*, 1996). This pushes the flood plain zoning program multivariate geomatical study. The techniques commonly adopted for the analysis of multivariate data are Cluster Analysis, Discriminant Analysis, Multivariate Regression and Multivariate Optimization (Dunteman, 1984). In this study data mixing or combination (extracting components) that involves the selection of a set of weights to the dependent variables has been adopted for data integration.

The weights have been assessed in such a manner that the combination becomes representative and at the same time it has maximum correlation with the independent parameter, the flood hazard in the present investigation. One of the most widely used statistical techniques in geomatics is the analysis of the variance. Wide application of this technique lies in the fact that the variance of a composite is equal to the sum of the individual variances of the in put parameters. This implies that if two or more unrelated factors introduce variability in a set of observations, the total variability can be represented in to individual portions that add up to the total. These portions can then be evaluated to assess their relative contributions to the total variability in the observed parameter (Dunteman 1984). The linear mixing mode is being described below.

Linear Mixing Model (LMM)

Linear mixing modeling is a branch of statistical science (Wang 1990, Maselli *et al.*, 1996, Bryant 1996, Kant and Badrinath, 1998). It is a method of analyzing a set of observations (obtained from a given sample) from their inter correlation to determine whether the variations can be accounted adequately by a number of basic categories smaller than that which the investigation was started (Fruchter, 1967). Let us consider a multivariate system consisting of 'p' responses described by the observable random variables $X_1, X_2, X_3. X_p$. The observable random vectors have mean x and co variance 'S'. The Linear Mixing Model (LMM) postulates that 'X' is linearly dependent upon few unobservable variables $F_1, F_2, \ldots\ldots F_m$ called linear composite (LC) and additional source of variations $e_1, e_2, \ldots. e_p$ called specific factors. The General LMM involving y as dependent variable and p independent variables, $x_1, x_2, x_3\ldots x$ p can be written as:

$$y = b_0 + b_1 x_{11} + b_2 x_{12} + \ldots\ldots\ldots\ldots + b_n x_{1n} + e \qquad (1)$$

Hence LMM in matrix form may be written as:

$$X - x = L F + e$$

$$Y = B X + E, \qquad (2)$$

Where Y is a column matrix representing the dependent variable, B is a vector representing the weights to the dependent variable, X is a matrix representing the independent variables, and E is a matrix representing the random observational error. It is worth mentioning here that there is no unique weight matrix B that determines the model. It can be easily proved that if P is any orthogonal matrix then a modified weight matrix B_m, which is equal to the multiplication of matrices B and P, that is, $B_m = B P$, also represents the model.

Where, (X-x) is a vector having p elements containing deviations of observed variable X and its mean value x, L is matrix of LC loading having p rows and m column, C is vector of Composite having m rows and e is error vector having p elements. From the above equation it is evident that "p" deviations $(X_1 - x_1) \ldots. (x_p - x_p)$ are expressed in terms of (p + m) random variables, $c_1, c_2, \ldots. c_m, e_1 \ldots. e_p$. With so many unobservable quantities a direct solution of LMM from the observations on $x_1, x_2 \ldots. x_p$ is difficult. However, with the help of following assumptions about the random vectors c' and e', the model reduces to simple and easy form. These assumptions are (1) Original variables are linearly related. (2) Common composite 'c' and unique factors 'e' have mean zero and standard deviation unity. (3) Common factor 'c' and unique factor 'e' is independent. The LMM proceeds by imposing conditions that allow one to uniquely estimate the loading and the specific variance matrix. The loading matrix is then rotated, where the rotation is determined by some, 'ease of interpretation', method. Once the loading and the specific variance matrix are obtained composites are identified and estimated values for the composites themselves (called composites scores) are frequently constructed. (Johnsons, and Wichern, 1988). In LMM method composites are determined so as to account for maximum variance of all the observed variables. The residual terms (*i.e.* specific factors e_i) are assumed to be small in this method. (Joreskog, Klovan and Reyment; 1976)

In the present study the problem has been formulated as, the determination of weight vector, $b' = [b_{11}, b_{12}...b_{ip}]$, such that the variance of: the composite ($b'x = b_{11}x_1 + b_{12}x + ... + b_{ip}x$), which is represented by the matrix $b'Vb$, is maximized. V_{pxp} is a covariance matrix of the observed independent variables. Maximization has been carried out subject to the constraints described in equation (3). The objective function has been written in equation 4.

$$b'b = \sum_{i=1}^{p} b_{li}^2 = 1 \qquad (3)$$

$$y = b'Vb - \lambda(b'b - 1) \qquad (4)$$

Where y is the function to be maximized, I represent the identity matrix and l a Lagrange multiplier. The condition, (b'b-1) reflects the condition that b'b-1 = 0. This condition ensures a unique solution. The vector b' has been obtained by maximizing the above-mentioned objective function. Stationary points have been obtained by differentiating equation (4) with respect to b' and equating to zero.

$$\frac{\delta y}{\delta b} = 2Vb - 2\lambda b = 0$$

$$(V - \lambda I)b = 0 \qquad (5)$$

Since $b \neq 0$; the matrix $(V - \lambda I)$ must be singular, that is, $(V - \lambda I) = 0$; Pre multiplying equation (5) by b' yields.

$$b'(V - \lambda I)b = 0, \text{ or}$$

$$b'Vb = - \qquad (6)$$

Equation (6) suggests that b' may be obtained by maximizing $b'Vb = \lambda$, that is the variance of the composite. The vector, b, associated with the largest root and other roots can be evaluated then after. The elements of the vector are the weights to the variable. Literature shows that it is a common practice to consider the sample correlation matrix instead of the sample covariance matrix while calibrating the model. Keeping this in view in the present study, sample correlation matrix has been considered in order to determine the weights.

Database

Present study has been carried out in a part of Indo-Gangetic plain, covering about 100 sq. km in Haridwar District of Uttrakhand State. The area mainly consists of unconsolidated deposits. The soil is generally coarse grained and aquifers are generally water table type. As mentioned earlier that the flood plain and associated damage depends upon several above surface, surface and sub surface geo-parameters. In the present study land capability, percentage of *kharif* agriculture, agricultural intensity, population density, density of domestic animals, density of poultry units, density of agricultural labor has been assumed to be the factors influencing the flood damage. Following data sets have been used. Topographic map of the area at 1:250,000 and 1:50,000 scale. Geo Coded Thematic Mapper, False Color Composite (TM - FCC) imagery at 1:50,000 scales. Ancillary data related to social and economic parameters as mentioned above.

Data Analysis and Results

First of all the area was decomposed in to geomorphic units namely Upper Bazada, Lower Bazada, Alluvial area, Flood plain Active and flood plain older based on image elements on satellite data. Socio-economic parameters as mentioned above namely, density of agricultural labor, land capability, population density, agricultural intensity, percentage of *kharif* land, animal density, poultry density, and house have been collected through field traversing. Preliminary analysis revealed that in general study area receives about 1000 mm annual rainfall. During 1972 to 1999 land use of the area remained more or less constant except seasonal variation and increase in cropping intensity. Prevailing soil characteristics favors groundwater recharge.

Geo parameters considered for the study show large variation. This observation has been interpreted that combination or composite of the geo parameters may act as a good indicator for the flood plain damage. Encouraged by above findings the database was put to detailed analysis. As mentioned earlier in the present study correlation matrix (Table 24.1) has been used as a basis for the study. Then after latent roots (Table 24.2) of the correlation matrix was calculated. Analysis of Table 24.2 reveals that latent roots are, 6.4608, 0.4342, 0.0639, 0.0371, 0.0028,.0012, and 0.0002 respectively.

Table 24.1: Correlation Matrix

Variable	1	2	3	4	5	6	7	8
1	1.000							
2	0.796	1.000						
3	0.728	0.881	1.000					
4	0.864	0.950	0.965	1.000				
5	0.916	0.963	0.876	0.969	1.000			
6	0.718	0.866	0.885	0.896	0.813	1.000		
7	0.777	0.871	0.969	0.958	0.866	0.961	1.000	
8	0.717	0.809	0.899	0.937	0.830	0.853	0.961	1.000

Further, the first latent root accounts for 92.3 per cent and the second latent root accounts for 6.3 per cent of the total variation in the data. Collectively these two latent roots account for 98.6 per cent of the total variation in the data. Remaining roots accounts for only 1.4 per cent of the total variation in the data. Therefore in the present study, only two linear composites considering annual rainfall, percentage vegetal cover, drainage density, ground elevation, land slope, and grain size have been evaluated and considered for further analysis. The weights to the independent variables for the linear composite (C1) and linear composite (C2) have been calculated using equation (6).

Table 24.2: Latent Roots and their Contribution

Linear Combination	Latent Root	per cent	Cumulative per cent
C1	6.4608	92.30	92.30
C 2	0.4342	6.30	98.60
C 3	0.0639	0.92	99.52
C 4	0.0371	0.42	99.94
C 5	0.0028	0.04	99.98
C 6	0.0012	0.02	100.00

A study of the Table 24.2 reveals that for the first linear composite, C1 which accounts for about 92 per cent variation in the data, is more or less equally loaded for all the variables. Whereas, the second linear composite C_2, which accounts for about 6 per cent is heavily loaded with land slope. Numerical values of the weights for the different parameters are tabulated in Table 24.3 have been used to calculate the total score pertaining to a particular land unit. For the different land units in the study area total scores have been calculated.

Table 24.3: Weights for Linear Combinations (C_1)

Parameters	Weights
Agricultural Labour	0.859
Land Capability	0.945
Population Density (no./ha)	0.952
Agricultural Intensity	1.0
Kharif Cultivation (per cent of total area)	0.958
Animal density (no./ha)	0.925
Poultry Density (no./ha)	0.975
House Density (no./ha)	0.926

The result of the above study was tested in a part of the study area. Total score for each land unit in the test area was determined as described above. Statistical analysis of the totals scores was carried out in order to determine the mean score (μ), and standard deviation (σ). Based on the total score the groundwater availability in an area was categorized in to three classes, namely A, B and C. The categories A and C represent the maximum and minimum groundwater availability areas respectively. Category B represents the moderate availability of the groundwater. A land unit is assigned groundwater availability code A if its total score (Ts) is more than mean score plus standard deviation (Ts > ($\mu + \sigma$)).

It is assigned code B if total score lies between mean score minus standard deviation and mean score plus standard deviation (($\mu - \sigma$) > Ts > ($\mu + \sigma$)). It is assigned a code C if its totals score lies less than the mean score minus the standard deviation (Ts < ($\mu - \sigma$)). Based on the present study it can be concluded that the proposed parameter weights (Table 24.3) can be used to delineate the flood plain zones. Proposed weights can also be used for data analysis in Geographical Information System (GIS) environment. This will help in data compression and efficient data analysis.

Acknowledgements

This research work is a small part of a research program for Bazada Land funded by MoWR through INCOH. The authors are highly thankful to the MoWR, INCOH in particular to Shri Masood Hussain and Dr. Ramakar J Jha, for their help and support in conducting the research program. Authors are also thankful to the organizations and individuals for their help in data sharing and data analysis.

References

Dunteman, G.H. (1984). Introduction to Multivariate Analysis, Sage Publications, Boverly Hills, London. 40 - 100.

Dubey, O.P., Srinivas, A.K., Awasthi (1984). Analysis of Remote Sensed Data for Groundwater Studies of Piedmont Zone. Proceeding. V Asian Conference on Remote Sensing. E – 6 – 1 to E - 6 - 10.

Dubey, O. P. (1991). Estimating soil Grain Sizes from Reflection Data, IJSC, 19, (3), 29-37.

Krishnamurthy, J., Venkatesa Kumar, N., Jayaraman, V. and Manivel, M. (1996). An Approach to Demarcate Groundwater Potential Zones through Remote Sensing on a Geographic Information System, International Journal of Remote Sensing, 17, (10), 1867-1884.

Kriz., H. (1971). Relations Between the Abstraction of Groundwater and the Elevations of the Groundwater Level, J. Hydrology. 13 254-262.

LeGrand H.E. (1967). Groundwater of the Piedmont and Blue Ridge Provinces in the Southeastern States. Geological Circular 538, USGS.

Orlov, D.S. (1966). Quantitative Patterns of Light Reflection on Soils (in Manual of Remote Sensing, Vol 2, American Society of Photogrammetry, N.Colwell (ed) pp.2211-2228.

Orlov, D.S. (1969). Quantitative Laws of Reflection of Light by Soils. Soviet Soil Science, (11): 84 - 95.

Reeves, M.J., Parry, E.L. and Richardson, G. (1978). Preliminary Evaluation of Groundwater Resources of the Western Part of the Vale of Pickering, Q.J. Engg. Geol. (11): 253-262.

Turc, G. (1977). Two-dimensional Searches for Highlighted Well Sites, Groundwater, 15, (4), 269-275.

2013, Environmental Technology
Pages 211–216
Editors: D.R. Khanna, A.K. Chopra, Gagan Matta, R. Bhutiani & Vikas Singh
Published by: DAYA PUBLISHING HOUSE, NEW DELHI

Chapter 25

Modeling Groundwater Pollution Potential in Rural Watersheds

Om Prakash Dubey and Walmi

U.P. Water and Land Management Institute,
Walmi Bhawan, Utretia, Lucknow

ABSTRACT

In order to fulfill growing needs, pollutants are being increasingly added to the groundwater system through various human activities and natural processes. For optimal developmental activities and sustained agricultural growth, it is essential to assess the existing and anticipated level of groundwater pollution (GWP). In this study, an attempt has been made to represent the GWP system by Factor Analytic Model (FAM). Input to the model has been given through synergistic use of remote sensing and ancillary data. The model has been validated in a part of Hardwar district, Uttarakhand state.

Background

In order to fulfill growing needs, pollutants are being increasingly added to the groundwater system through various human activities and natural processes. Applications of fertilizers and pesticides to enhance crop production have become a common practice. In case fertilizer application exceeds the plant uptake, the residual joins the water table. This increases nitrate concentration in groundwater. Similarly, excess applications of pesticides that are complex organic chemicals may have adverse health effects.

Long-term use of saline irrigation water combined with poor management and adverse climatic conditions for example, low rainfall and high evapotranspiration, leads to accumulation of salts in the root zone. Poor agricultural practice results in a loss of crop yield and deterioration of soil structure. Poorly designed and improperly managed waste disposal sites contribute significant amount of leachate. This leachate may affect the water quality.

Improperly designed and maintained septic tank becomes a threat to groundwater quality. Disposal of waste through wells adds pollutant to the groundwater and also accelerate their movement towards a production well. The major pollutant is industrial waste that includes heavy metals, toxic compounds and radioactive material. Another significant source of metal contaminants are tailing produced at mining sites. In India, even during ancient time, well-defined legislation was in existence to control the water pollution.

In recent years, land and water sectors were put to stresses in order to meet the demand of the growing population. Groundwater is more dependable source of water as compared to surface water (Viverkar 1999). Gradually, quality of water in addition to quantity is gaining importance in the selection of suitable sites for the groundwater development. Groundwater Pollution Vulnerability (GPV) of any given geographical location depends upon a wide range of above surface, surface and subsurface environmental parameters (Brown 1972, Jackson1986).

Evaluation of GPV involves decision-making keeping in view multiple interwoven criteria (ESCAP 1996). Generally, data required for GPV studies are either not available or not sufficient. Collection, storage, and processing of required data is difficult, costly, and time consuming. As a result GWV studies are generally based on questionable data.

Remotely sensed data has proven capability in providing many above surface, surface and sub surface characteristics of a land unit. Synergistic use of remote sensing and ancillary data can be made for the development of the database required for GPV (Daniel *et al.*, 1994, ESCAP 1996). A suitable Geographical Information System (GIS) can be used to store, process and retrieve the developed database. GPV mapping is a complex system (Biawas 1971, Hamil *et al.*, 1996). In this study an attempt has been made to represent the GWP system by Factor Analytic Model (FAM).

The Analytical Model

In this study a factor analytic model (FAM) has been developed. The FAM involves decomposing (Satty 1988, Mendoza 1997) the complex groundwater pollution system in to a number of simpler components forming a cascade. At each cascade level the decision can be taken in simpler manner. The decision process moves from one cascade to another to arrive at the final decision. FAM has been used to develop a decision support system for weighting a particular land characteristic keeping in view its GPV.

The decision cascade process starts from the lowest cascade level and progressively moves upwards until final decision is made. At each level pair wise comparisons are made between factors at that level. These comparisons lead to priority vectors that are propagated up the cascade to arrive at a final priority vector (Table 25.1). The FAM finally ranks a land unit in to a predefined GPV Class, based on its attributes.

Table 25.1: Variable Loading (RIW)

Goal		Level I		Level II		Level III	
	Feature	RIW	Feature	RIW	Feature	RIW	
GWPP	Surface	(0.65)	Land use	(0.44)	Agriculture	0.40
					Water Body	0.25
					Barren Land	0.18
					Thin forest	0.10
					Thick forest	0.05
					Settlement	0.02
			Land Slope	(0.26)	Low Slope	0.72
					Mild Slope	0.21
					Milder Slope	0.07
			Distance from Paleo Channel	(0.18)	Less than 50 m	0.90
					More than 50 m	0.10
			Distance from Flood Plain	0.04	Up to 50m	0.90
					More than 50 m	0.10
			Soil	0.06	Sand	0.56
					Sandy loam	0.27
					Loamy sand	0.13
					Clay	0.04

Contd...

Table 25.1–Contd...

Goal						
			Level			
	Level I		Level II		Level III	
Feature	RIW	Feature	RIW	Feature	RIW	
		Distance from Urban areas	0.02	Less than 0.5 km	0.65	
				0.5km–1.0m	0.28	
				More than 1.0 km	0.07	
Sub Surface	0.24	Aquifer Media	(0.5)	Sand and Boulder	0.63	
				Sand Boulder and Clay	0.28	
				Sand and Clay	0.09	
		Permeability in Vertical Direction	(0.5)	HighLow	0.900.04	
Groundwater	(0.11)	Groundwater Depth	(0.60)	< 5m	0.73	
				5-15m	0.19	
				> 15m	0.08	
		Rainfall Recharge	(0.32)	High	0.65	
				Medium	0.28	
				Low	0.07	
		Water Quality	(0.08)	SAR Value Low	0.75	
				SAR Value High	0.25	

Decision cascading has been carried out in the following steps.

☆ Step (1): The decision making process is decomposed in to a set of cascades. The goal of the analysis is at the top level. The elements of the lower level include the attribute such as objectives perhaps even more redefined attributes follows at the next lower level – until the last level.

☆ Step (2): In the second phase, pair wise comparisons of the attributes or elements at a particular cascade level relative to their contribution or significance to the elements of the next higher cascade level is made. This phase constitutes the evaluation (qualitative) or assessment (quantitative) of the decision making process. Specifically the input matrix of pair wise comparisons expresses the relative of influence of an element over the others.

☆ Step (3): In the third phase, the pair wise input matrix is decomposed spectrally. Spectral decomposition provides an estimate of the relative influence weight (RIW) of the elements at a particular cascade.

☆ Step (4): Groundwater Pollution vulnerability (GPV) of a land unit can be determined by RIW of the above surface, surface, and subsurface parameters influencing the GPP.

The FAM is mathematically sound but pair wise comparison is highly subjective. In order to get optimal results with minimum subjectivity the FAM was further modified to accommodate multi criteria through linear mixing modeling (LMM), a branch of statistical science (Wang 1990, Maselli *et al.*, 1996, Bryant 1996, Kant.and Badrinath 1998). LMM is a method of analyzing a set of observations for their inter correlation to determine whether the variations can be accounted adequately by a number of basic categories smaller than that which the investigation was started (Fruchter, 1967). Let us consider a multivariate system consisting of 'p' responses described by the observable random variables X_1, X_2, X_3, X_p. The observable random vectors have mean x and co variance 'S'. The LMM postulates that 'X' is linearly dependent upon few unobservable variables $F_1, F_2, \ldots\ldots F_m$ called linear composite (LC) and additional source of variations $e_1, e_2, \ldots. e_p$ called specific factors. Hence LMM in matrix form may be written as:

$$(X - x) = L F + e$$

Where, (X-x) is a vector having p elements containing deviations of observed variable X and its mean value x, L is matrix of LC loading having p rows and m column, and e is error vector having p elements. With so many unobservable quantities a direct solution of LMM from the observations on $x_1, x_2 \ldots. x_p$ is difficult. However, with the help of following assumptions about the random vectors, the model reduces to simple and easy form. These assumptions are (1) Original variables are linearly related. (2) Unique factors 'e' has mean zero and standard deviation unity. (3) Factors and unique factor 'e' is independent.

The LMM proceeds by imposing conditions that allow one to uniquely estimate the loading and the specific variance matrix. The loading matrix is then rotated, where the rotation is determined by some, 'ease of interpretation', method. Once the loading and the specific variance matrix are obtained composites are identified and estimated values for the composites themselves (called composites scores) are frequently constructed (Johnsons, and Wichern, 1988). Composites are determined so as to account for maximum variance of all the observed variables. The residual terms (*i.e.* specific factors e_1) are assumed to be small in this method. (Joreskog, Klovan and Reyment; 1976).

Min $\qquad [\{(X-x) - LF\}^T \{ (X-x)-LF\}]^{-1}$

Subject to $\quad f_i = 1, f_i > 0$

FAM can be calibrated using historical database. Weights or membership function at each cascade level for different land characteristics can be developed. The importance of a particular land characteristic is decided on the basis of a linguistic measure of importance. A comparison is made between various themes, land elements on a common scale and a confusion matrix representing their relative importance is developed. The confusion matrix can be decomposed spectrally in to components. The vector corresponding to this component represents the weights to different land characteristic that were considered influencing the decision. For data mixing RIW at different cascade level is shown in the Table 25.1 as an example. These weights can be used to map the GPV of an area.

Conclusion

For sustainable developments of a region, reliable estimate of groundwater quality is of paramount importance. Generally sufficient data required for groundwater pollution vulnerability are not available for Indian watersheds. Satellite data can be analyzed to generate database required for GPV studies. Generated database can be put to analysis for extracting the most influential composite and subsequently the variable loading. Using the proposed FAM the study area was classified into different classes in terms of their potential to pollute the groundwater. The model efficiency was tested by carrying out field surveys and found to above 80 per cent. The model can be used for evaluating the GPP in any area after calibration. The added advantage of the proposed approach is that it compresses the data up to 70 per cent that helps in efficient analysis and prediction.

2013, Environmental Technology   *Pages 217–229*

Editors: D.R. Khanna, A.K. Chopra, Gagan Matta, R. Bhutiani & Vikas Singh

Published by: DAYA PUBLISHING HOUSE, NEW DELHI

Chapter 26

Carbon Credit and Climate Change: A Biggest Threat to India

Kumkum Pandey and Deepa Vinay

Department of Family Resource Management, G.B.P.U.A. and T.,
Pantnagar, Uttarakhand – 263 145

ABSTRACT

Carbon Credits are a tradable permit scheme under United Nations Framework Convention for Climate Change which gives the owner the right to emit one metric tonne of CO_2 equivalent. They provide an efficient mechanism to reduce the green house gas emissions by monetizing the reduction in emissions. Rural India is facing the biggest threat of climate change. Everyday activities like driving a motorbike, air conditioning, heating and lighting houses consume energy, industrial processes and fossil fuel combustion produce emissions of greenhouse gases, which contribute to climate change. When the emissions of GHGs are rising, the Earth's climate is affected, the average weather changes and average temperatures increase. Any sharp rise in sea level could have a considerable impact on India. The United Nations Environment Programme included India among the 27 countries that are most vulnerable to a sea level rise. In agriculture and forestry different sources and sinks release, take up and store three types of GHGs: CO_2, methane and nitrous oxide. Many agricultural and forestry practices emit GHGs to the atmosphere. Addressing climate change is not a simple task. To protect India, their economy, and land from the adverse effects of climate change, India must reduce emissions of carbon dioxide and other greenhouse gases. To achieve this goal the concept of Clean Development Mechanism has come into vogue as a part of Kyoto Protocol for which objective is the "stabilization of greenhouse gas concentrations in the atmosphere at a level that would prevent dangerous anthropogenic interference with the climate system".

Keywords: Carbon credit, Climate change, Greenhouse gases, Solar energy and Global warming.

Carbon Credit

Carbon Credits are a tradable permit scheme under United Nations Framework Convention for Climate Change which gives the owner the right to emit one metric tonne of CO_2 equivalent. They provide an efficient mechanism to reduce the green house gas emissions by monetizing the reduction in emissions. Sustainable and clean renewable energy systems such as hybrid solar/wind electric generators can be used to eliminate or reduce carbon dioxide emissions by replacing old diesel, oil, gas or coal fired electric generators which emit greenhouse gases that produce global warming. Carbon sequestration credits or offsets are calculated by the amount of carbon emissions that would have been emitted if a diesel or other traditional polluting electric generator was used to produce the same amount of electricity. Companies and electric utilities in countries can buy these emission reduction carbon credits to replace the emissions from their coal burning electric power plants to meet regulatory requirements. Carbon credit leads to various benefits like- provide carbon funds for RE project implementation, reduce poverty and provide offgrid electricity, provide employment for carbon sequestration and Boost economic development in rural communities. By, switching to Clean Development Mechanism Projects, India has a lot to gain from Carbon Credits: a) It will gain in terms of advanced technological improvements and related foreign investments. b) It will contribute to the underlying theme of green house gas reduction by adopting alternative sources of energy. c) Indian companies can make profits by selling the CERs to the developed countries to meet their emission targets.

Climate Change

India, a large developing country (1.4 billion by 2040), is likely to be impacted more due to climate change. The rural climatic sensitive sectors are agriculture, forests, fisheries etc. Hence, climate change is a very critical issue for India more than other countries and needs due attention. Due to climate change there will be impact on agriculture sector with reduce food production, effect on water resources due to change in stream flow, increase drought incidence and retreat of glaciers and untimely rains. There will be effect on forest ecosystems, such as loss of biodiversity, change in forest types and shift in forest boundaries. Climate change will affect coastal zones and there will be increase in sea levels and increase in cyclones and hurricanes. There will be impact on health and diseases. There is need to address to climate change through mitigation and all activities be aimed at reducing Green Houses Gases (GHG) or carbon intensity of goods and services, reducing CHG emissions(GHGs) and increase removal of CO_2 from the atmosphere.

A National Action Plan on Climate change just been released by Prime Minister of India, mandates setting up of energy benchmarks for each sector and allows trade in energy saving certificates. This is expected to kick start a domestic trade in energy just as the world trade in carbon emission certificates. It is proposed to save 10,000 MW by the end of 2012 through energy efficiency measures. The key demands of the India's National Action Plan as climate change are, 1) Solar energy boost (1000 MW) solar power by 12th Plan). 2) Steel, power and textile industries to trade in energy efficiency targets. 3) Minimum target of 5 per cent renewable energy for power grids to procure. 4) Nuclear power plant on climate mitigation package. 5) Critical data to be digitized: sharing and access made easier. 6) recycling from automobiles at the end of their life. 7) 5000 MW thermal plants to be closed by 11th Plan. Additional 10,000 MW to be shut or overhauled by 12th Plan. The ICAR has also launched a Network Project on "Climate Change" involving 15 research institutes and State Agricultural Universities for conducting critical research on crops, livestock and fisheries. ICAR has also entered into agreement with the International Centre for Research on Agro-forestry (ICRAF) for collaborative research on farm forestry aimed specifically at dealing with climate change.

The Link between Climate Change, GHG Emissions, Agriculture and Forestry

Climate change is one of the biggest threats we face. Everyday activities like driving a car or a motorbike, using air conditioning and/or heating and lighting houses consume energy and produce emissions of greenhouse gases (GHG), which contribute to climate change. When the emissions of GHGs are rising, the Earth's climate is affected, the average weather changes and average temperatures increase.

In agriculture and forestry different *sources* and *sinks* release, take up and store three types of GHGs: carbon dioxide (CO_2), methane (CH_4) and nitrous oxide (N_2O). Many agricultural and forestry practices emit GHGs to the atmosphere (FAO, 2009). Figure 26.1 shows the main *sources* of agricultural GHGs: for example, by using fertilizers N_2O is released from the soil and by burning agricultural residues CO_2 levels rise. CH_4 is set free in the digestion process of livestock, as well as if rice is

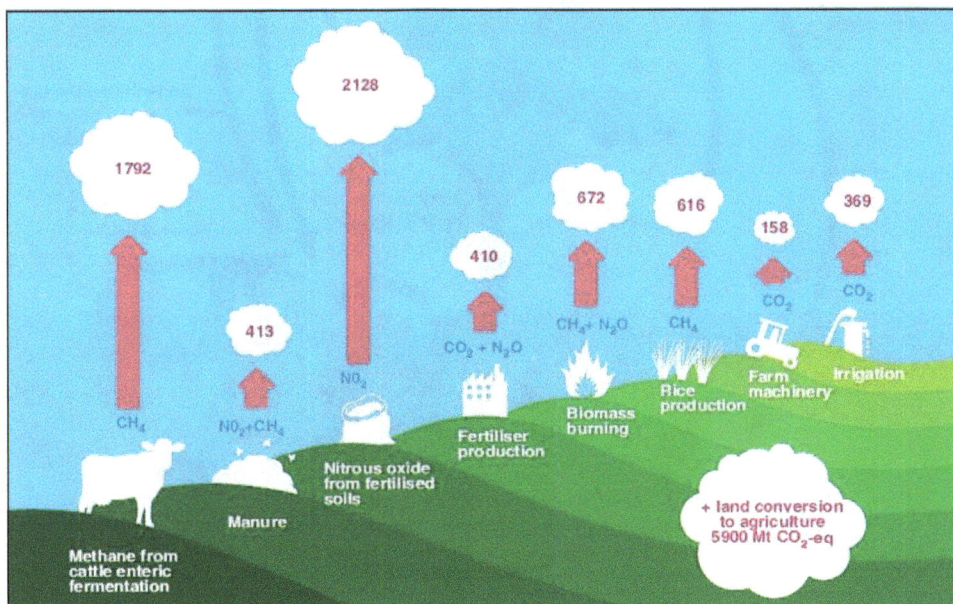

Figure 26.1: Sources of Agricultural GHGs in Megatons (Mt) CO_2-eq
Source: Greenpeace International, 2008

grown under flooded conditions. When land is converted to cropland and trees are felled, a source of CO_2 emissions is created.

Agriculture is an important contributor to climate change, but it also provides a *sink* and has the potential to lessen climate change. Figure 26.2 shows the components of the land carbon cycle: carbon is stored – sequestered - above-ground by plants, crops and trees, and below-ground in the soil and roots. *Carbon sequestration* means that carbon dioxide is captured from the atmosphere through photosynthesis by the tree or plant to store it as cellulose in its trunk, branches, twigs, leaves and fruit and oxygen is released to the air in return. Also the roots of the trees and plants take up carbon dioxide. Decomposing organic materials increase the amount of carbon stored in the soil, which is higher than the total amount in the vegetation and the atmosphere. Animals breathe in oxygen and breathe out CO_2 and through their faeces carbon and N_2O is released to the soil.

To slow down climate change impacts, the emissions of GHGs need to be reduced immediately. As explained above several activities in agriculture and forestry

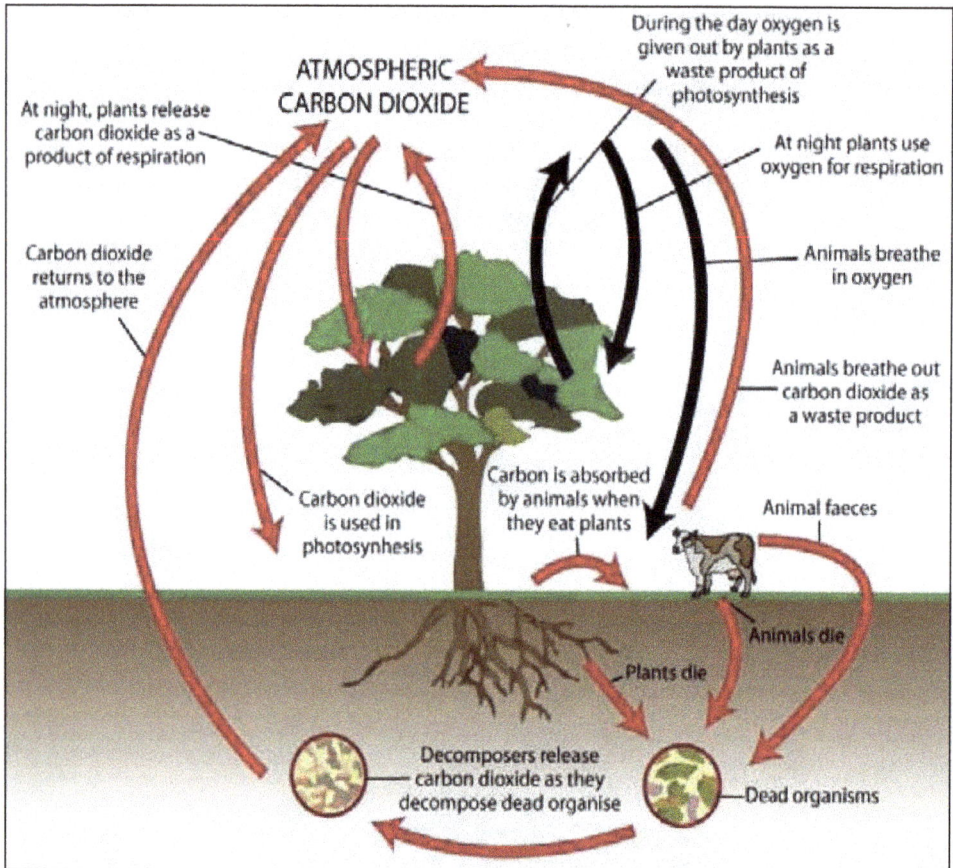

Figure 26.2: Carbon Cycle at Plot Level
Source: www.energex.com.au/switched_on/being_green/being_green_carbon.html

---

IN THE AGRICULTURE, FORESTRY AND OTHER LAND USE (AFOLU) SECTORS DIFFERENT TYPES OF ACTIVITIES CAN HELP TO REDUCE OR AVOID EMISSIONS, OR INCREASE THE REMOVAL OF GHGS:

- **Forestry activities** such as afforestation and reforestation, sustainable forest management, agroforestry, avoided deforestation/reducing emissions from deforestation and forest degradation (REDD);
- **Agricultural activities** such as cropland and grazing land management, livestock management (improved feeding practices), peatland management and manure management;
- **Energy** activities such as increasing the energy-efficiency at household or community level, sustainable biofuel production, and the employment of Integrated Food and Energy Systems;
- **Biodiversity** enhancing activities such as watershed and soil management, biodiversity conservation.

Box 26.1
Source: Baalman, P. and B. Schlamadinger, 2008

contribute to GHG emissions. Changing these, and switching to new sustainable land management practices (Box 26.1) can support the uptake and the reduction of GHGs. Some agricultural activities can increase the amount of organic matter and carbon in the soil by using cover crops or reduce the emissions of methane through improving feeding practices. Sustainable forest management can avoid the destruction of forests and the release of CO_2, and planting new trees sequesters more CO_2.

Kyoto Protocol

Kyoto Protocol was adopted in the third conference of the parties of the UNFCCC (COP_3) in Kyoto, Japan on, 11th December 1997. 197 countries signed the document first. India became a signatory in 2003. Kyoto Protocol of 1997 contrives and suggested reduction of emission of greenhouse houses into the atmosphere and fixed individual quotas for each of the ISO member countries. Said quota mandates the tolerable. The developed countries are supposed to reduce their carbon emissions to the tune of 20 per cent by 2012. The Kyota Protocol is designed to cut greenhouse gas emissions by making the polluter start paying for *climate change. Kyoto Protocol includes three flexibility mechanism for reduction of GHGs:*

1. The Kyoto protocol's clean development mechanism (CDM) allows developed countries to gain emissions credits for financing environmentally friendly projects based in developing countries.
2. A country can also earn emissions credits something called joint implementations, which allows a country to benefit by carrying out something like a reforestation project in other industrialized country or "economy in transition"

3. The third mechanism is carbon trading. It allows countries to buy emission credits from countries that do not need them to stay below their emission quotas.

Clean Development Mechanism (CDM)

The CDM is one of the three flexible mechanisms established under the Kyoto protocol. It allows developed countries to invest indirectly in GHG emission reduction projects in developing countries, by buying tradable Certified Emission Reduction (CERs). Since developing nations generate lower gas emissions and also need to develop hence, an acceptable rating system need to be formulated. The carbon credit was introduced in the mainstream to precisely to address this disparity. Four of the 25 EU states – the Czech Republic, Greece, Italy and Poland have not joined the system at all, and the UK is resquabbling with European Commission over its emission allowance under the scheme. No country wants to put its business a disadvantage so all played safe, erring on the side of generosity when setting emission quotas for industry.

Carbon Credits are tradable permit bonds set with a signed monitory value that have been devised to implement a global cut back on gas emissions. Each carbon credit gives the owner the right to emit one ton of greenhouse gas into the atmosphere. If an individual or a company exceeds the set credit quota, the purchase of additional bonds equivalent to the exceeded value becomes a form of penalty. Those companies which have gone below emission levels can be paid for the effort. Each country that has signed on to Kyoto has its own target for slashing CO_2 emission. Countries that cut their emission of greenhouse gases get credits for their efforts: one credit for each ton of reduced CO_2 emissions. Under the Kyoto Protocol, companies that would fail to meet the quota or carbon caps are required to undertake greenhouse reduction projects in other countries where costs are deemed much lower in order to compensate for the excess generated in their locality. This also helps in developing country to get new technology.

The countries should join hand in undertaking joint projects to reduce CO_2 emissions and develop new and efficient technologies for use in urban as well as rural areas. We need to evolve varieties and genomes which will be able to bear the rise in temperature, drought, floods etc. The organic farming needs to be promoted instead of use of chemicals as fertilizer factories have been found to be the main contributor to global warming and burning of straw needs to be curbed Burning of fossil fuels that release heat trapping gases are responsible for global warming that may disrupt water and food supplies with even more droughts, floods and heat waves. CO_2 levels are at about 300 ppm. Global warming has resulted in change in climate all over the world and concern is felt due to melting of the Glaciers. It is estimated that globally, 30 per cent of the Earth's species could disappear, if temperature rise by 4.5° F(2.2° C) and upto 70 per cent, if the temperature rise by 6.3° F(3° C).

Greenhouse Gases with Different Global Warming Potential vs CO_2

☆ CO_2 - Carbon dioxide - 1

☆ CH_4 - Methane - 21

☆ N_2O - Nitrous oxide - 310

☆ PFCs - Perfluorocarbons - 6,500-9, 200

☆ HFCs - Hydrofluorocarbons - 11,700

☆ SF_6 - Sulphur hexafluoride - 23,900

The hardest hit will be plants and animals in cooler climate or at higher elevations and those with limited ranges or tolerance for temperature change. Global carbon emission is about 8 billion tonne per year and by 2057 it is expected to be 16 bt/year. It is difficult to track emission values and difficult to know if a particular company exceeded its quota or achieved reduced emissions. In Agriculture it is much more difficult to keep such record and also to fix emission reduction quota. However, for industries a system known as the CO_2 calculator has been formulated to estimate the gas emitted in the atmosphere. The combustion of fossil fuels borne out of the use of vehicles powered by gasoline and the production run of gas, oil and coal fired power plants, tops the list of atmosphere pollutants. The major pollutant industries are *cement, steel, textile, air conditioners, refrigerators and fertilizer.*

India has already initiated integration of CDM in National Policy by establishing an Inter Ministerial Committee on Climate change. A National CDM Authority has been constituted by legislation passed by Cabinet Committee. The 42 SD (Sustainable Development) criteria set by GOI are also approved by Inter Ministerial Committee and all projects have to comply to their criteria to get host country's approval. The SD criteria for CDM projects are zonal, economic, environmental and technological well being. India established CDM authority in December 2003 and 1st project on CDM was registered in March 2005 and on October 2005, first project received CER's. Out of 818 CDM project sanctioned, 283 (35 per cent) are located in India, 123 project (15 per cent) in China, 3 per cent in Chile, 12 per cent in Mexico, 13 per cent in Brazil and 22 per cent (other).

India's CDM Projects

While laying more emphasis on protection of environment, India's implementing several environmental friendly projects for future generations with aim to reduce CO_2 emissions substantially. Two largest projects, the Sasan and Krishnapatnam UMPPs, with a total capacity of 7920MW, have been registered with the CDM - Executive Board making us the largest supplier of CERs among Indian power generation companies. The two projects would together generate 3.48 million CERs every year and others are Jharkhand Integrated Power Ltd., Chitrangi Power Ltd., Samalkot Power Ltd., Rajasthan Sun Technique Energy Private Ltd

India gets the World's Largest Carbon Credit Project

India has bagged the world's largest carbon credit project that will help replace 400 million incandescent light bulbs with energy saving CFL bulbs at dirt-cheap prices in a year while preventing 40 million tonnes of carbon from entering the atmosphere annually. The project, which will allow the government, investors, discoms and CFL manufacturers to sell CFLs at Rs 15 each, instead of the Rs 100 they currently cost on average, has been approved by CDM. The mammoth size of the project can be gauged from the fact that the world's second largest CDM project earns only about 1.5 million credits a year in comparison. "Almost half the households in India will immediately benefit from the scheme and as other areas get electrified, those villages will get added on. There are roughly 400 million light points at present in the country that we will provide the subsidized CFL bulbs for the Bureau of Energy Efficiency, which is the nodal agency for the grand project. The scheme – called Bachat Lamp Yojna – works like this. For every ten bulbs that consumers use for a year, a *tonne of carbon* is prevented from escaping into the atmosphere as CFL bulbs use substantially less power than incandescent ones. For every tonne of carbon saved, the Bureau of Energy Efficiency, acting as the anchor, gets a carbon certificate from the UN, which it then hands over to the investor. In anticipation of the project clearance, it has begun distribution of the energy saving bulbs to consumers. States like *Kerala, Uttarakhand and Punjab and cities like Hyderabad* are rolling out the scheme within a month. Buoyed by the speed at which they have been able to get the green light for the project from UN, BEE has begun discussions on the next quantum leap which could save even more power for the country and generate a substantially higher number of carbon credits – putting Indian households on to LED lamps.

The Carbon Market

The carbon credit market is the fastest growing market in the world. The volume of carbon credits sold by developing countries (India) doubled between the year 2003 and 2004 and tripled by 2005. In 2006 alone, carbon transactions worth USD 30 billion (INR 1.19 trillion) were conducted globally, transferring some USD 5 billion from the countries of the global north to the global south. Out of the total number of carbon contracts signed in the world till date, India has the second largest portfolio with a market share of 12 per cent, trailing only behind China, which has a whopping market share of 61 per cent. India has a lot to look up to and aim for as far as the carbon credit market is concerned.

The international carbon market is large and there is a big demand from CERs. The CER issued for a project activity may be traded, in a similar way as company shares, in the market. During 2006, the carbon market worldwide was worth $ 22.5 billion (Rs 88,000 crore)1 and transactions of about 1.6 billion tones of CO_2 equivalent (CO_2 e) took place. The carbon market is expected to grow significantly in 2007, possibly up to 50 per cent. India is a key player in the carbon market. Out of over 2000 CDM projects under development all over the world, the highest number of projects (about 650) is located in India. However, in terms of the actual volume of carbon credits or Certified Emission Reductions (CERs) traded, India ranks second with a current potential of 323,000 CERs by 2012. India has developed a large number of

CDM projects which works in different areas include- renewable energy (hydro, wind, solar, biomass, etc.), energy efficiency, demand-side management, recovery of methane (landfill, coalmine, wastewater,etc.), destruction of HFCs (HFC-23) and NO_2, SF_6 (electricity/transmission/distribution lines), fuel switching, transportation, a forestation and reforestation.

Also, UNFCCC has recognized the need to simplify the procedures to promote small-scale CDM projects. Energy efficiency improvements projects that reduce energy consumption upto the equivalent of 15 GWh or renewable energy projects with a capacity of upto 15 MW or any other projects that both reduce emissions and directly emit less than15 kilotons of CO_2 annually qualify to be in the small-scale project category. Bundling of several project activities to form a single CDM project activity can is possible, provided the project activities in the sub-bundle belong to the same type.

Growth of Carbon Market in India Depends on Governement Strategy

The growing pressure on countries to address climate change (*World Bank, 2007*) has given rise to a multi-million dollar international market for buying and selling emissions of greenhouse gases. The Kyoto Protocol expires in 2012, and international talks have already begun to decide the shape of a new treaty that will succeed it. After 2012, the carbon market is expected to expand exponentially. Researches indicated that it could grow to $100 billion annually, becoming a significant source of foreign capital flows. The enormous potential of carbon financing for Indian enterprises, the benefits have so far largely been availed of by small and medium enterprises (SMEs). Public Sector Units (PSUs) have mostly remained away, in large part due to lack of knowledge. For India to cash in on the vast potential of the carbon market, a government strategy needs to be devised to derive the maximum benefits from it and address current market failures. The country also needs to build the capacity of its Public Sector Units (PSUs) to avail of carbon finance. This can be done by systematically screening massive infrastructure and urban development projects to see if they are eligible for such finance. There needs to be a pilot project to demonstrate how the carbon market can catalyze investments in renewable energy to benefit the 400 million people in India's rural areas.

India sellers don't know how to access buyers from industrialized countries, as the majority of transactions are done on a bilateral basis. Although there is an interest from many players in India to launch a carbon trading platform – which would enable sellers to obtain bids on their carbon credits through public trading, much like the stock market – they have been unable to do so due to a lack of regulatory clarity. India should also consider the establishment of a carbon fund, aimed at accelerating the capacity to develop carbon finance opportunities along the lines of the China CDM has a significant role to play as well. Many of the Indian projects are currently driven by SMEs and stand disadvantaged because each project generates a small quantity of carbon credits. Hence, the private sector needs to build its expertise to club small projects together in order to improve their market access. In the end, the growth of the carbon market will largely depend on the realization that the market

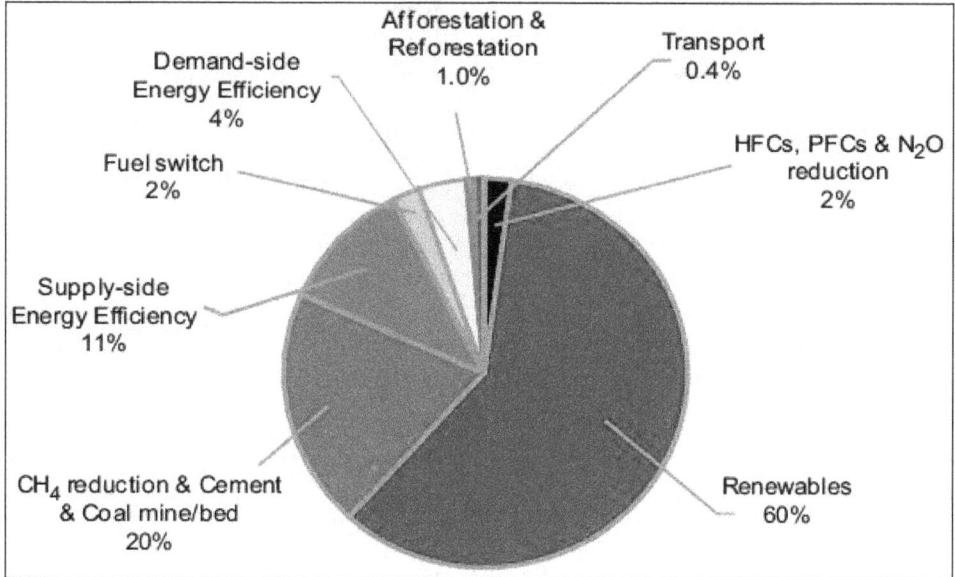

Figure 26.3: Percentage of CDM Projects in each Category
Source: UNEP Risoe, 2010

can assist India in achieving low carbon growth, and the consequent development of a strategy to cash in on the opportunities this offers.

Criticism and Controversies

Questions have been raised over carbon crediting. Some argue that it offers an easy way to avoid necessary changes in the culture of consumption. While there is some truth in this, it is also clear that awareness of climate change will not transform economies overnight. As a stepping stone on the way to a cleaner future, carbon markets can be considered a good first step. Other criticisms have been aimed at certain types of project, in particular at reforestation. The principle behind reforestation is that trees absorb carbon dioxide from the atmosphere as they grow. However in practice it is hard to predict how long trees will remain standing before they are felled or burnt. Although Carbon Catalog includes reforestation projects, their effectiveness is a subject of controversy.

There are other reasons why some specific projects have been criticized. Some lack additionality, meaning that they would still have occurred without the support of carbon funding. Supporting such a project is not a valid way to neutralize carbon emissions. In other cases, providers have sold the same tonne of carbon more than once, or money has failed to reach the project as promised. Several certification standards have emerged to address these issues, and these are listed in Carbon Catalog alongside projects which apply them. The aim of Carbon Catalog is to increase the level of trust and information in the carbon market. If chosen and researched carefully, we believe carbon credits can make a positive contribution to the well-being of the planet.

Future Outlook

The rising pressure on countries to address climate change has paved the way for the rise of a multimillion dollar international market for buying and selling emissions of greenhouse gases. Ever since its establishment in 2001, the carbon market has captured the attention of Indian entrepreneurs. Majority of projects selling carbon credits so far include renewable energy (such as wind power, biomass cogeneration and hydropower), energy efficiency measures in several sectors (such as cement, petrochemicals and power generation) as well as the reduction of industrial gases that contribute to climate change. The market for carbon credit deals in India is expected to be around USD 6 billion (INR 236.4 billion) in the next four years. About 34 per cent of the total numbers of CDM projects that have been approved are from India. Environmental finance as an asset-class is pegged at USD 1 trillion globally by 2012. Indian CERs are predominantly bought by European and Japanese companies. With Australia ratifying the Kyoto Protocol, strong demand is expected from that region. Recently, the ITL (International Transaction Log) was awarded to the Kyoto Treaty by UNFCCC, to help transfer the CDM registry to a national level. As a matter of fact, the price of CERs is bound to rise with the ITL in place. CERs were priced at around USD 17 (around INR 672) as on March 2007 as compared to European Union Allowances or EUAs, which are the regional equivalents of CERs, selling as high as around USD 30 (around INR 1186). Due to lack of transfer facilities of Asian CERs to the national registry of a European buyer, the European buyers ended up paying a higher price for locally available EUAs. This price difference between CERs and EUAs is expected to lessen thereby offering better deals to Indian players.

Conclusion

There is a great opportunity awaiting India in carbon trading which is estimated to go up to $100 billion by 2010. In the new regime, the country could emerge as one of the largest beneficiaries accounting for 25 per cent of the total world carbon trade, says a recent World Bank report. The countries like US, Germany, Japan and China are likely to be the biggest buyers of carbon credits which are beneficial for India to a great extent. The Indian market is extremely receptive to CDM. Having cornered more than half of the global total in tradable certified emission reduction (CERs), India's dominance in carbon trading under the CDM of the UNFCCC is beginning to influence business dynamics in the country. India Inc pocketed Rs 1,500 crores in the year 2005 just by selling carbon credits to developed-country clients. Various projects would create up to 306 million tradable CERs. Analysts claim if more companies absorb clean technologies, total CERs with India could touch 500 million. Of the 391 projects sanctioned, the UNFCCC has registered 114 from India, the highest for any country. India's average annual CERs stand at 12.6 per cent or 11.5 million. Hence, MSW dumping grounds can be a huge prospect for CDM projects in India. These types of projects would not only be beneficial for the Government bodies and stakeholders but also for general public.

Abbreviations

CDM: Clean Development Mechanism

UNFCCC: UN Convention on Climate Change

CERs: Certified Emission Reduction

ITL : International Transaction Log

UN: United Nations

SMEs: Small and Medium Enterprises

PSUs: Public Sector Units

GHG: Greenhouse Gases

ICRAF: International Centre for Research on Agro-forestry

References

Baalman, P. and B. Schlamadinger. 2008. Scaling Up AFOLU Mitigation Activities in Non-Annex I Countries. Working Paper. A report by Climate Strategies and GHG Offset Services for the Eliasch Review.

Bulletin of NSWAI-ENVIS is published by the National Solid Waste Association of India in Mumbai, 7th issue - february, 2007.

M. Poffenberger. Community Forestry in Northeast India: Recommendations for Action. Community Forestry International, Santa Barbara, California, United States.

FAO. 2009. Food Security and Agricultural Mitigation in Developing Countries: Options for Capturing Synergies. Rome, Italy. www.fao.org/docrep/012/ i1318e/i1318e00.pdf.

FAO. 2010. Making the Step From Carbon to Cash – A Systematic Approach to Accessing Carbon Finance in the Forest Sector. Forest and Climate Change Working, 10 pp.

Greenpeace International. 2008. Cool Farming: Climate change impacts of agriculture and mitigation potential.

IPCC. 2007. Changes in Atmospheric Constituents and in Radiative Forcing. In: Climate Change 2007: The Physical Science Basis. Contribution of Working Group I to the Fourth Assessment Report of the Intergovernmental Panel on Climate Change. Cambridge University Press, Cambridge, United Kingdom and New York, NY, USA.

Ministry of Environment and Forests. 2001. State of Forest Report, 2001. Ministry of Environment and Forests, Government of India, Dehadrun, India. Available at envfor.nic.in/nfc/fc-stat.htm.

Planning Commission. 2003. Report of the Working Group on National Action Plan for Operationalising Clean Development Mechanism (CDM) in India. Government of India, Delhi, India. Available at planningcommission.nic.in/ reports/genrep/fin_CDM.pdf.

Roy, P.S. and P.K. Joshi. 2002. Forest Cover and Assessment in North East India: Issues and Policies. Indian Institute of Remote Sensing (NRSA), Dehradun, India.

World Bank (2007). Report on Growth of Carbon Market in India Depends on Government Strategy, New Delhi.

www.cd4cdm.org/Guidebooks.htm

http://edugreen.teri.res.in

http://www.biobin.net

http://www.ccc.govt.nz

2013, Environmental Technology

Editors: D.R. Khanna, A.K. Chopra, Gagan Matta, R. Bhutiani & Vikas Singh

Published by: DAYA PUBLISHING HOUSE, NEW DELHI

Pages 231–237

Chapter 27

Electric Field and Heat Induced in Blood Due to Radiation from Radio Antenna

*Hemendra Tripathi[1], Devendra Singh[2],
Vijay Kumar[3] and P.P. Pathak[1]*

*[1]Department of Physics and [2]Department of Physics, FET,
Gurukula Kangri University, Haridwar – 249 404
[3]Department of Physics, Graphic Era University,
Dehradun – 248 002*

ABSTRACT

Deposition of energy in consequence of induced electric field due to radio frequency radiation emitted by radio broadcasting antenna is calculated. The high frequency electromagnetic field transmitted by antenna is penetrated and absorbed in the biological body which causes the thermal effect and tissue gets heated. Induced electric field and specific absorption rate in blood at the different distances from the broadcasting antenna are assessed. Variation in the depth of blood is made from 10 μm to 20 μm. The comparison is made with the international guidelines of WHO and ICNIRP used to measure the amount of energy deposition in the tissues due to radio frequency antenna of high power. The thermoregulatory mechanism of the body which maintains the body temperature at a set point is also discussed. The incapability of various thermal sensors to initiate mechanisms of heat loss due to unavailability of sufficient mechanisms to carry away all of the unwanted heat during the gain of heat energy deposited through broadcasting antenna may cause the damage of tissues which would be harmful for human being. The recommendations for installation of high power transmitters of radio frequency radiation must be simple and followed carefully.

Keywords: Induced electric field, Specific absorption rate, Thermoregulatory mechanism, Broadcasting antenna.

Introduction

In present times of faster communication, we are exposed to many types of radiation from all types of communication systems like TV, radio and mobile telephony etc. Generally speaking, some radiation penetrates and is being absorbed by the human body when it is exposed to it. Depending on the frequency of radiation, the human body interacts with such a field via induced currents and thermal effects. The field generated within the body, the so-called internal field, is determined by the amount by which a body is influenced by electromagnetic fields. The EMF can be characterized by several different parameters (field strength, field direction, field orientation, field complexity and so on) (Gandhi and Mohammad, 2008). The quantity used to measure how much RF energy is actually absorbed in a body is called the specific absorption rate (SAR). This SAR also varies with the dimension of tissues (Dein and Amr, 2010). The absorbed microwave energy produces molecular vibration and converts the energy into heat. When the rate of energy absorption is high, it produces heating in living tissues (Ozen *et al.*, 2008). The heat generated in the medium is proportional to the absorbed power. The biological effects of radiofrequency energy depend on the rate at which power is absorbed (Osepchuk and Petersen, 2003). If the organism cannot dissipate this heat energy as fast as heat is produced, the internal temperature of the body will rise.

Exposure from TV and Radio transmitters have been studied by Joseph and Martens (2006) and Sirav and Seyhan (2009). Pathak *et al.* (2003) and Kumar *et al.* (2008) assessed the temperature change in tissues per second by taking the TV transmitters of high frequency radiation as a point source. Due to exposure to microwave radiation many health risks have been studied including brain tumors, acoustic neuroma, leukemia and testicular cancer, sleep disturbances and headache (Johansen, 2004; Takebayashi *et al.*, 2006, 2008; Hardell *et al.*, 2007). Wessapan (2011) calculated the local SARs and temperature increase of human model for various operating frequencies. also focusing on the interaction between electromagnetic field and organs in the human trunk.

In this manuscript the specific absorption rate as a result of induced electric field is assessed due to the radio frequency antenna of All India Radio, situated in Mussoorie (Uttarakhand), India. The calculation of induced electric field and SAR in blood is made by taking the transmitter of finite length dipole antenna. The frequency used for the calculation is 102.1 MHz of FM (rel.) broadcasting and power of the antenna is 10 kW. The calculation is made at different distances from the antenna and at different depth inside the blood.

Public access to broadcasting antennas is normally restricted so individuals cannot be exposed to high-level fields that might exist near antennas. Ambient RF radiation levels in inhabited areas near broadcasting facilities must be typically well below the exposure levels recommended by current standards and guidelines. Therefore, precautions must be taken to ensure that maintenance personnel are not exposed to unsafe RF fields.

Thermoregulation

Thermoregulation of the human body in the presence of RF fields follows the laws of thermodynamics in which heat and work are balanced and conserved. The whole gain energy becomes equal to the spend energy by the body. Thus there is no excess temperature in the body in this natural process. More specialized adaptation of the relevant equations has been developed by thermal physiologists. The balance of heat energy is expressed by the following equation;

$$M \pm W = E \pm R \pm C \pm D \pm S \tag{1}$$

where,

M is the rate at which thermal energy is produced through metabolic processes, W, the rate at which the work is produced, E, the rate of exchange with the surroundings via evaporation, R, the rate of heat exchange with the surroundings via radiation, C, the rate of heat exchange with the environment via convection, D, the rate of heat exchange with the surroundings via conduction and S represents the rate of body heat storage.

The ability of various thermal sensors to initiate mechanisms of heat loss, during the gain of heat energy deposited by the dissipation of electrical currents deposited by RF energy depends on the availability of sufficient mechanisms (vasomotor, sudomotor in addition to convection and radiation) to carry away all of the unwanted heat. Without this, cooling cannot occur and the core body temperature will continue to rise.

Thus, when electromagnetic radiation is penetrated inside the body, the energy is absorbed by the tissues of the biological material. It works as a source of production of extra energy inside the body.

$$M \pm W + E_R = E \pm R \pm C \pm D \pm S \tag{2}$$

where,

E_R Energy due to electromagnetic radiation of broadcasting antenna. Above equation becomes unequilibrium because production of energy becomes greater to the energy inside the body. This excess energy may increase the temperature of the tissue and may harmful in much other way for tissue life.

Material and Methods

As we want to access the effect of broadcasting antenna, whose power P is the only parameter we know, we have to relate the electric field (more generally E_{rms}) to the power of transmitters which is r distance far from the biological body (Prasad, 1999).

$$E_{rms} = \frac{\sqrt{90P}}{r} = 9.487\frac{\sqrt{P}}{r} \tag{3}$$

When a human body is exposed to the EM wave of electric field E_{rms}, it penetrates into the body. It results into inside or induced field E_i at a given depth z given by Polk (1996).

$$E_i = E_{rms} \exp\left(\frac{-z}{\delta}\right) \tag{4}$$

where,

δ is skin depth, which is the distance over which the field decreases to 0.368 of its value just inside the boundary, given as

$$\delta = \frac{1}{\omega\sqrt{\left\{\frac{\mu\omega}{2}\left[\left(1+p^2\right)^{\frac{1}{2}}-1\right]\right\}}} \tag{5}$$

where,

μ is permeability of body material and ε its permittivity, $p = \frac{\sigma}{\varepsilon\omega}$, σ being its conductivity.

It is generally accepted that the specific absorption rate (SAR) is the most appropriate metric for determining electromagnetic exposure, i.e. the mass averaged rate of energy absorption in tissue, is related to the induced electric field E_i (V/m) can be determined at any point from the relation (Hirata et al., 2008).

$$SAR = \frac{\sigma E_i^2}{\rho} \tag{6}$$

where,

σ is the conductivity of the tissues for which the calculation is made and ρ is their mass density.

Calculation

This electromagnetic wave penetrates inside the body. Thus, the induced electric field at different distances in the blood of the human body at different depth is numerically evaluated in Table 27.1. The induction of field causes the deposition of energy in the form of specific absorption rate (SAR). The values of SAR corresponding to induced electric field are also calculated in Table 27.2. The conductivity of the body tissues is taken from Gabriel et al. (1996 a, b, c).

Results and Discussion

Tables 27.1 and 27.2 show the induced electric field and SAR at different distances from the broadcasting antenna in the blood. The depth of penetration ranges from 10 μm to 50 μm. The broadcasting antenna of power 10 kW is used for calculation which is running at radio frequency i.e. 102.1 MHz to broadcast the signals. The values of SAR are above up to the distance 110 m than the value recommended by WHO and ICNIRP guidelines. However, the corresponding values at a distance of 120 m are below the guidelines but, very near to the limiting point. Table 27.2 represents that the variation in SAR on increasing the depth, is not very large after 70 m and it become approximately constant after 120 m.

Table 27.1: Induced Electric Field (V/m) at Different Depths in Blood at Different Distances Due to Radiation from All India Radio Antenna of Power 10kW at 102.1 MHz

Distance (m)	10 μm	20 μm	30 μm	40 μm	50 μm
10	94.8522	94.8344	94.8166	94.7988	94.7810
20	47.4261	47.4172	47.4083	47.3994	47.3905
30	31.6171	31.6111	31.6052	31.5993	31.5933
40	23.7135	23.7091	23.7046	23.7002	23.6957
50	18.9704	18.9669	18.9633	18.9598	18.9562
60	15.8090	15.8061	15.8031	15.8001	15.7972
70	13.5505	13.5479	13.5454	13.5428	13.5403
80	11.8568	11.8545	11.8523	11.8501	11.8479
90	10.5390	10.5370	10.5351	10.5331	10.5311
100	9.4852	9.4834	9.4817	9.4799	9.4781
110	8.6234	8.6218	8.6201	8.6185	8.6169
120	7.9045	7.9030	7.9015	7.9001	7.8986

Table 27.2: Specific Absorption Rate (W/kg) at Different Depths in Blood at Different Distances Due to Radiation from All India Radio Antenna of Power 10kW at 102.1 MHz

Distance (m)	10 μm	20 μm	30 μm	40 μm	50 μm
10	10.4961	10.4922	10.4883	10.4843	10.4804
20	2.6240	2.6231	2.6221	2.6211	2.6201
30	1.1662	1.1658	1.1653	1.1649	1.1645
40	0.6560	0.6558	0.6555	0.6553	0.6551
50	0.4198	0.4197	0.4195	0.4194	0.4192
60	0.2916	0.2915	0.2914	0.2912	0.2911
70	0.2142	0.2141	0.2141	0.2140	0.2139
80	0.1640	0.1639	0.1639	0.1638	0.1638
90	0.1296	0.1295	0.1295	0.1294	0.1294
100	0.1050	0.1049	0.1049	0.1048	0.1048
110	0.0868	0.0867	0.0867	0.0867	0.0866
120	0.0729	0.0729	0.0728	0.0728	0.0728

Conclusion

However, while technologies have been a vital part of our daily and we cannot avoid this, nor would we wish to, because technology makes our lives healthier, wealthier and safer but these radiations penetrate the tissues and may heat the tissue due to long term exposure, to its extent to bear which would be harmful for our health. Therefore, the international guidelines for installation of high power radio frequency transmitters must be made very clear and followed strictly. The installation of high power broadcasting antenna must be kept at least 120 m away from the high populated area. The public should also be made aware of the use of radio frequency sources and the minimum distance which must be kept by them to live safer and healthy.

References

Dein A.Z.E. and Amr A. 2010, "Specific Absorption Rate (SAR) Induced in Human Heads of Various Sizes When Using Mobile Phone", *Proceeding of the World Congress on Engineering (WCE), London (U.K.)*, 1, June 30 – July 2.

Gabriel C., Gabriel S. and Corthout E. 1996, "The Dielectric Properties of Biological Tissues: I. Literature Survey", *Phys. Med. Biol.* 41, 2231-2249.

Gabriel S., Lau R.W. and Gabriel C. 1996, "The Dielectric Properties of Biological Tissues: II. Measurements in the Frequency Range 10 Hz to 20 GHz", *Phys. Med. Biol*, 41, 2251-2269.

Gabriel S., Lau R.W. and Gabriel C. 1996, "The Dielectric Properties of Biological Tissues: III. Parametric Models for the Dielectric Spectrum of Tissues", *Phys. Med. Biol.* 41, 2271-2293.

Gandhi F.M. and Mohammad M.A.S. 2008, "Thermal Effects of Radiofrequency Electromagnetic Fields on Human Body", *Journal of Mobile Communication*, 2(2), 39-45.

Hardell L., Carlberg M., Ohlson C.G., Westberg H., Eriksson M. and Hansson Mild K. 2007, "Use of Cellular and Cordless Telephones and Risk of Testicular Cancer", *Int J Androl*, 30(2), 115-122.

Hirata A., Shirai K. and Fujiwara, O., 2008, "On Averaging Mass of SAR Correlating with Temperature Elevation Due to a Dipole Antenna", Progress in Electromagnetics Research, PIER, 84, 221-237.

Johansen C. 2004, "Electromagnetic Fields and Health Effects – Epidemiologic Studies of Cancer, Diseases of the Central Nervous System and Arrhythmia – Related Heart Disease", *Scand J Work Environ Health*, 30 (Suppl1), 1-30.

Joseph W. and Martens L. 2006, "Reconstruction of the Polarization Ellipse of the EM Field of Telecommunication and Broadcast Antennas by a Fast and Low-Cost Measurement Method", *IEEE Trans Electromagn. Compat*, 48 (2), 385-396.

Kumar V., Vats R.P. and Pathak P.P. 2008, "Harmful effects of 41 and 202 MHz radiations on some body parts and tissues", *Indian Journal of Biochemistry and Biophysics*, 45, 269-274.

Nielsen J.B., Elstein A., Hansen D.G. and Kildemoes H.W., I.S. Kristiansen and Stovring H. 2010, "Effect of Alternative Styles of Risk Information on EMF Risk Perception", *Bioelectromagnetics*, 31, 504-512.

Osepchuk J.M. and Petersen R.C. 2003, "Historical Review of RF Exposure Standards and the International Committee on Electromagnetic Safety (ICES)", *Bioelectromagnetics Suppl.*, 6, S7-S16.

Osepchuk J.M. and Petersen R.C., 2008, "Safety and Environmental Issues", RF and Microwave Applications and Systems, Taylor and Francis Group, LLC, Chapter 21, 21.1-21.21.

Ozen S., Helnel S. and Cerezci O. 2008, "Heat Analysis of Biological Tissue Exposed to Microwave by Using Thermal Wave Model of Heat Transfer (TWMBT)", *Burns*, 34, 45-49.

Pathak P.P., Kumar V. and Vats R.P. 2003, "Harmful Electromagnetic Environment Near Transmission Tower" *Indian J. Radio Space Phys.* 32, 238-241.

Prasad K.D. 1999, "Electromagnetic Waves", in Electromagnetic Fields and Waves, First Edition, Satya Prakashan, pp 425-520, New Delhi.

Sirav B. and Seyhan N. 2009, "Radio Frequency Radiation (RFR) from TV and Radio Transmitters at a Pilot Region in Turkey", *Radiat. Prot. Dosimetry*, 136 (2), 114-117.

Stuchly M.A. and Stuchly S.S. 1996, "Experimental Radio Wave and Microwave Dosimetry, " in C. Polk and E. Postow (eds.), Handbook of Biological Effects of Electromagnetic Fields, Second Edition, Boca Raton, CRC Press, 295-336.

Takebayashi T., Akiba S., Kikuchi Y., Taki M., Wake K., Watanabe S. and Yamaguchi N. 2006, "Mobile Phone Use and Acoustic Neuroma Risk in Japan", *J Occup Environ Med*, 63(12), 802-807.

Takebayashi T., Varsier N., Kikuchi Y., Wake K., Taki M., Watanabe S., Akiba S. and Yamaguchi N. 2008, "Mobile Phone Use, Exposure to Radiofrequency Electromagnetic Field and Brain Tumour: A Case-Control Study", *Br J Cancer*, 98, 652-659.

Wessapan T., Srisawatdhisukul S., Rattanadecho P. 2011, "The Effects of Dielectric Shield on Specific Absorption Rate and Heat Transfer in the Human Body Exposed to Leakage Microwave Energy", International Communications in Heat and Mass Transfer 38, 255–262.

Wideman P.M. and Schutz H. 2008, "Informing the Public About Information and Participation Strategies in the Siting of Mobile Communication Base Stations: An Experimental Study", *Health Risk Soc.*, 10, 517-534.

2013, Environmental Technology
Editors: **D.R. Khanna, A.K. Chopra, Gagan Matta, R. Bhutiani & Vikas Singh**
Published by: **DAYA PUBLISHING HOUSE, NEW DELHI**

Pages 237–244

Chapter 28

Biopesticides: An Alternative of Chemical Pesticides to Minimize Environmental Pollution

Abhishek Soni[1], Gangaram Masar[1], R.K. Kaurav[1],
Neetu Arya[2] and R.C. Saxena[2]

[1]Department of Zoology,
Bherulal Patidar Govt. P.G. College, Mhow, M.P.
[2]Pest Control and Ayurvedic Drug Research Lab
S.S.L. Jain P.G. College, Vidisha, M.P.

Introduction

The plant kingdom particularly the Indian flora is a rich source of safer chemicals which could be exploited for insect pest control. The use of biopesticides or the Natural Plant Product is quite obvious because of the several advantages over the synthetic chemicals such as:

1. Biopesticides are easily available and the growth of the plant in forest and wild areas makes it economical also for the use.
2. Biopesticide of the plant origin possess at least or no mammalian toxicity and thus constitute no health hazard.
3. There is no risk of pest resistance to the natural products.
4. These is minimal risk of resurgence.
5. Surface persistency lasts for the longer period of time.
6. Natural Products are less expensive and their use as biopesticides would not disturb the ecosystem.

7. The most beneficial effects of the natural product is that there is no adverse effect on plant growth, the cooking quality of the food grains and even no loss in the germinating property of the stored grains.

Review of Literature

Phytochemicals derived from various botanical sources have provided numeral beneficial sources of bioinsecticides during the last two decades. There are much reports available on the natural products for vector control programme. But the literature is quite scanty on the mode of application of JHA and other biocidal compounds on stored grain pests.

The isolation and identification of natural biocidal compounds from natural products of plant origin gained momentum during the last two decades. When Bowers *et al.* (1966); Staal (1967); Slama (1973); Brooks (1979); Schooneveld (1979) and Bowers and Nishida (1980) have isolated these chemicals from natural resources of plants. Recently, Krishna *et al.* (2008) and Diwan and Saxena (2009-10) have also reported biopesticidal activity against insect pest of crops as well as human health hazards. They have also suggested biopesticides as suitable alternative to the chemical pesticides.

Materials and Methods

Laboratory Culture of the Test Insects

The rust red flour beetles, *Tribolium castaneum* (Herbst) and confused flour beetles, *Tribolium confusum* (Duval) is cosmopolitan pests of stored grains, specially the milled products all over the globe. The present investigation aims at determining the effect of alkaloids in the development of *Tribolium castaneum* and *Tribolium confusum* in the laboratory in a sterilized glass jars (10cm×15cm) containing wheat flour, rice and maize flour.

The culture jars and wheat flour were first sterilized at a temperature of 60°C for 36 hr after cooling the flour, the powdered yeast was added to it at the rate of 5 percent (w/w). About the ¾ of the jars were filled with yeast mixed wheat flour and 100 adults, taken from the parental stock were released in each jar. The jars were covered with muslin cloth and kept in wooden cupboards in dark at a temperature 30±2°C and RH 70±5 per cent.

Bioassays Methodology

The experimental bioassays were performed using the two plant alkaloids obtained from *Annona squamosa* and *Argemone maxicana* using laboratory cultured parental stock adults and larvae of *Tribolium castaneum* and *Tribolium confusum*.

Observations and Results

Result mentioned in Tables 28.1 and 28.2 report the angular transformed value for the different alkaloidal compound of both *Annona squamosa* and *Argemone maxicana* for the larvae as well as the adult beetles of *Tribolium castanerm* and *Tribolium confusum* in Table 28.1 and 28.2, the larvicidal activities are reported. From Table 28.1, it appears that the activity is maximum in *Annona* followed by *Argemone*. Similarly in Table 28.2, also the activity to the adult beetles falls in the same order. This is also quite true in case of *Tribolium confusum* where higher concentration completely shows feeding deterrent activity (500 ppm).

Table 28.1: Larvicidal Action of Plants Extract Against *Tribolium castaneum*

Concentration in (ppm)	Different Plant Extracts, Mortality in Percentage and Arc Sign Value	
	Annona squamosa	Argemone maxicana
2000	48 (7.00)	40 (6.40)
3000	62 (7.93)	53 (7.34)
4000	75 (8.71)	66 (8.18)
5000	88 (9.43)	80 (9.00)

Values in the paranthesis are arc $\sqrt{n+1}$ control mortality was 0.00 in all three replicates.

Table 28.2: Insecticidal Action of Plants Extract Against *Tribolium castaneum*

Concentration in (ppm)	Different Plant Extracts, Mortality in Percentage and Arc Sign Value	
	Annona squamosa	Argemone maxicana
2000	46 (6.85)	33 (5.83)
3000	60 (7.81)	46 (6.85)
4000	73 (8.60)	60 (7.81)
5000	86 (9.32)	73 (8.60)

Values in the paranthesis are arc $\sqrt{n+1}$ control mortality was 0.00 in all three replicates.

Table 28.3: Phagodeterrent Activity of *Annona squamosa* Nees (Methanol) against *Tribolium confusum*

Concentration in (ppm)	Total Food Consumed After 5 Days/by 10 Insect/ Three Replicates	Average Food Consumed by One Insect in One Day mean ± SE
200	280	5.6 ± 0.44
300	240	4.8 ± 0.36
400	220	4.4 ± 0.04
500	0.0	0.0 ± 0.0 *
Control	320	6.4 ± 1.24
Untreated	450	9.0 ± 3.84

* No consumption of food.

1ml + 1gm flour in each replicate (value are average of three replicates).

Table 28.3 reports the result for the alkaloids compound extracted from methanol extract of *Annona squamosa*. From the result, it is evident that this alkaloidal compound shows higher deterrent activity than the petroleum ether extract of the same plant against *Tribolium confusum*. The total food consumed by ten beetles varies from 200 mg to 220 mg. The 5 per cent concentration caused total deterrency to the adult beetle. The average percentage of food consumed by single beetle per day showed quite significant result at 3 and 4 per cent concentration. At 5 per cent there was no consumption of food at all. The beetles as soon as come in contact with the food mixed with 5 per cent concentration of the extract tried to avoid the touching of food and immediately rushed to the another site.

Discussion

Total alkaloids present in *Annona squamosa* and *Argemone maxicana* have been precipitated by using the method of Chopra *et al.* (1993). The further purification on TLC and column chromatographic techniques revealed total five fractions in *Annona* alkaloids and six fraction in *Argemone* out of which only FR2 of *Annona* and FR6 of *Argemone* have been analysed and purified further for detail phytochemical and spectral analysis for the active ingredient present in them.

The results indicated that as for as larvicidal activity are concerned, *Argemone maxicana* proved highly toxic than *Annona squamosa* methanol and *Annona squamosa* p.ether extracts.

Antifeedant or phagodeterrent are usually defined as factors that inhibit feeding or cause the cessation of feeding but don't directly kill the insect as indicated by Nawrot and Harmatha ((1994) in their recent review of natural product as antifeedant against stored product insects. The identification of antifeedant properties in natural products depends upon the food eating by the treated and untreated insects.

In the last few years various phagodeterrent of plant origin have been isolated including the most widely used neem antifeedant 'Azadirachtin'. Such reports are available from, Cunat *et al.* (1990); Isman *et al.* (1990); Singh and Upadhyay (1993); Deshpande and Sharma (1990); Koul and Isman (1990); Shrivastava *et al.* (1990); Gupta *et al.* (1993) and Duriraj *et al.* (1991).

Similar views regarding insecticidal and phagodeterrent activity have been expressed for stored grain pests by Saxena *et al.* (1992); Tiwari and Saxena (1993) and Diwan and Saxena (2008-10).

Conclusion

The pesticidal compound isolated from the two plants showed not only insecticidal activity to the stored grain pest but also showed feeding deterrent as well as growth inhibitory activity. Thus, the used of biopesticides to control stored grain pest can minimize the environmental pollution.

Acknowledgements

First three authors express their thanks to the Principal, Govt. P. G. College, MHOW. Gangaram Masar also thanks to the Principal, Govt. P.G. College, Khargone for giving him permission to carry out the research work. One author (N A) is highly thankful to UGC for RGNJRF.

References

Bowers, W.S., Fales, H.M., Thompson, M.J. and Uebel, E.C. (1966). Juvenile hormone: Identification of an active compound from Balsam fir. Science, 154: 1020-1021.

Bowers, W.S. and Nishida, R. (1980). Juvocimenes: Patent juvenile hormone mimics from sweet-basil. Sci., 209: 1030-1032.

Brooks, G.T., Pratt, G.E. and Jenning, R.C. (1979). The action of precocene in milk weeds bug, *Oncopetus fasciatus* and locust (*Locusta migratoria*). Nature, 291: 570-572.

Chopra, R.N., Gupta, J.C. and Mukherjee, B. (1983). The pharmacological action of an alkaloid obtained from *Rauwolfia serpentina* B. Ind. J. Med. Res., 21: 261-271.

Cunat, P., Primo, E., Snaz, I., Garcera, M.D., March, M.C., Bowers, W.S. and Pardo, R.M. (1990). Biocidal activity of some Spanish Mediterranean Plants. J. Agric. Food Chem., 38: 497-500.

Deshpande, S.G. and Sharma, R.N. (1990). Antifeedant action of eight selected forest seed oils against three lepidoptrain pests. Agric. Sci. Digest, 10(4): 217-219.

Diwan, R.K., and Saxena, R.C. (2008). Effect of sesquiterpene lectone on the control of *Callosobruchus chinensis* Linn. Biosciences, Biotechnology Research Asia, 5(2): 773-778.

Diwan, R.K. and Saxena, R.C. (2010). Insecticidal property of flavinoid isolated from *Tephrosia pupuria*. Int. J. Chem. Sci., 8(2): 777-782.

Duriraj, C., Soorianatha Sundaram, K. and Nambisan, K.M.P. (1991). Antifeedant effect of plant extracts on *Nodostoma pubecolle* (Coleoptera: Eumopidae), a pest on pear (*Pyrus communis*). Int. J. Agric. Sci., 61(12): 959-960.

Gupta, K.S., Radke, C.D. and Renwick, A.A. (1993). Antifeedant activity of cucurbitacins from *Iberis amara* against larvae of *Pieris rapae*. Phytochemistry, 33: 1305-1388.

Isman, M.B., Koul, O., Luczynski, A. and Kaminski, J. (1990). Insecticidal and antifeedant bioactivities of neem oils and their relationship to azadirachtin content. J. Agric. Food. Chem., 38(6): 1407-1411.

Koul, O. and Isman, M.B. (1990). Antifeedant and growth inhibitory effects of sweet flag *Acorus calamus* L. oil on *Peridoma saucia* (Lepidoptera: Noctuidae). Int. Sci. Appl., 11(1): 47-53.

Krishna, V.K., Uikey, J. and Saxena, R.C. (2008). Mosquito larvicidal and chemosterilant activity of flavonods of *Annona squamosa*. Life Science Bulletin, 5(1): 85-88.

Nawrot, J. and Harmatha, J. (1994). Natural products as antifeedants against stored products insect. Post harvest News and Information, 4(5): 217-221.

Saxena, R.C., Dixit, O.P. and Harshan, V. (1992). Insecticidal action of *Lantana camara* against *Callosobruchus chinensis* (Coleoptera: Bruchidae). J. Stored Prod. Res., 28(4): 279-281.

Schooneveld, H. (1979). Precocene induced necrosis and haemocyte mediated breakdown of corpara allata in nymph of the locust, *Locusta migtratoria*. Cell. Tissue. Res., 203: 25-33.

Shrivastava, R.P., Proksh, P. and Wray, V. (1990). Toxicity and antifeedant activity of a sesquiterpene lectone from Encella against *Spodoptera litloralis*. Phytochemistry, 29(11): 3445-3448.

Singh, G. and Upadhyay, R.C. (1993). Essential oils-A potent source of natural pesticides. J. Sci. and Industrial Research, 52(10): 676-683.

Slama, K. (1973). Simple peoplite derivatives with insect juvenile hormone activity N. Nesvadha, Peptides. Asnstardan: 286.

Stall, G.B. (1967). Plant as a source of insect hormone. Nederi Akad. Wererschappen of Ser. C, 409-419.

Tiwari, A. and Saxena, R.C. (1993). Repellent and feeding deterrent activity of *Sphaeranthus indecus* against *Tribolium castaneum* (Herbst). Bio. Sci. Res. Bull., 9(1-2): 57-60.

2013, Environmental Technology

Editors: **D.R. Khanna, A.K. Chopra, Gagan Matta, R. Bhutiani & Vikas Singh**

Published by: **DAYA PUBLISHING HOUSE, NEW DELHI**

Pages 245–257

Chapter 29

Textile Wet Processing Sector: Preventing Measures for Environmental Protection

Anita Rani[1] and Swati Pant[2]

Department of Clothing and Textiles, College of Home Science,
G.B.P.U.A. and T., Pantnagar

ABSTRACT

India has a large network of textile industries of varying capacity and have registered a quantum jump owing to its contribution to high economic growth during the past few decades. Accounted for 14 per cent of the total industrial production, 30 per cent of the total exports and being second largest employment generator after agriculture; it has also been major source of severe environmental pollution. The World Bank in its report stated that 17 to 20 per cent of industrial water pollution comes from textile processing sector. Total 72 toxic chemicals have been identified in our water solely from textile dyeing; 30 of which cannot be removed. Among various activities in textile industry as sizing, desizing, scouring, bleaching, mercerization, dyeing, finishing and ultimately washing contributes about 70 per cent of total pollution from textile sector.

The nature of the waste generated depends on the type of textile facility, the processes and technologies being operated, and the types of fibres and chemicals used. This represents an appalling environmental problem for the clothing designers and other textile manufacturers in terms of compliance to eco standards and acceptability among consumers. To some extent the pollution enforcement agencies followed a lenient attitude towards such industries on account of their socio-economic contributions and low investment capacity towards pollution control. Cleaner production is an attractive approach to tackle environmental problems associated with textile sector. Finding low cost pollution abatement technologies, move from pollution control to pollution prevention and international cooperation – these are most viable but not so easy solutions to environmental protection.

Keywords: Textile industry, Wet processing, Pollution, Clean production.

Introduction

Indian Textile Industry is among one of the industries which were earliest to come into existence in India. It has earned a unique place in our country owing to its contribution towards industrial production (14 per cent of the total), exports (nearly 30 per cent of the total) and is the second largest employment generator after agriculture. This industry fulfils one of the most basic needs of people and holds importance by maintaining sustained growth for improving quality of life. It has an image of self-reliant industry, from the production of raw materials to the delivery of finished products, with substantial value-addition at each stage of processing, contributing towards the country's economy. Indian textile industry is one of the leading in the world. Currently the Indian Textile Industry is estimated to be around US$ 52 billion and is projected to be around US$ 115 billion by the year 2012. The current Indian domestic market of textile is expected to be increased to US$ 60 billion by 2012 from the current US$ 34.6 billion.

Despite the large network of textile industry of varying capacity, it has been placed in the category of most polluting industries by the Ministry of Environment and Forests, Government of India. Textile industries in India were initially centred round big cities like Ahmedabad, Mumbai, Chennai, Coimbatore, Bangalore and Kanpur. Since independence, India has had strong policies to promote the small-scale industrial sector owing to various characteristics that it is labor intensive and thus creates more jobs; it contributes to decentralized industrial development and the units are flexible and are able to quickly reorient themselves to emerging demands. In these units Western technological systems are getting adopted far and wide which produce enormous gaseous, liquid and solid wastes. But the pollution control technologies developed in the West are not economically suitable for these small enterprises. The number of such enterprises is huge and so is the pollution from them. They contribute some percent of the total industrial wastewater in India. Clusters of textile dyeing industries have led to serious problems in towns situated on small rivers like Pali, Balotra and Jodphur in Rajasthan, Jetpur in Gujarat and Tiruppur in Tamil Nadu. Units producing dyes and dye intermediates have also become major sources of both groundwater and surface water pollution. The village of Bichhri has seen all its wells become black in one monsoon.

To some extent, the pollution enforcement agencies are responsible for these emerging situations who followed a lenient attitude to industries on account of its socio-economic contributions and low investment capacity towards pollution control. Unfortunately, this approach is further discouraged the industries to introduce successful pollution management strategies either through effective effluent treatment or through production process change by adopting cleaner production technologies. This paper discusses the materials consumed in different wet processes followed in textile industry; their impact on environment and preventive measure required thereof.

Textile Wet Processing

The textile industry is one of the most chemically intensive industries on earth, and the number one polluter of clean water (after agriculture). It takes about 500 gallons of water to produce 20 metres of fabric. The enormous amount of water required by textile production competes with the growing daily water requirements of the half billion people that live in drought-prone regions of the world. By 2025, the number of inhabitants of drought-prone areas is projected to increase to almost one-third of the world's population. If global consumption of fresh water continues to double every 20 years, the polluted waters resulting from textile production will pose a greater threat to human lives. The World Bank estimates that 17 to 20 per cent of industrial water pollution comes from textile dyeing and treatment. They've also identified 72 toxic chemicals in our water solely from textile dyeing, 30 of which are cannot be removed. This represents an appalling environmental problem for the clothing designers and other textile manufacturers.

The industry uses vegetable fibers such as cotton; animal fibers such as wool and silk; and a wide range of synthetic materials such as nylon, polyester, and acrylics. The production of natural fibers is approximately equal in amount to the production of synthetic fibers. Polyester accounts for about 50 per cent of synthetics. The stages of textile production are fiber production, fiber processing and spinning, yarn preparation, fabric production, bleaching, dyeing and printing, and finishing. Textile wet processing industry is characterized not only by consumption of the large volume of water for various unit operations but also by the variety of chemicals used for various processes. There is a long sequence of wet processing stages requiring inputs of water, chemical and energy and generating wastes at each stage (Tables 29.1 and 29.2).

The generation of pollution is significantly high in the preparatory and dyeing operations compared to the post dyeing operations. In fact, one third of the pollution caused by textile industries results from waste generated during desizing operations. The wastes from the dyeing houses are let out into the drains which in turn empty into the main sewerage causing hazard to those who use this water. The harmful effects of textile chemicals are given in Table 29.3.

Highly Toxic Chemicals and Dyestuff Used in Indian Textile Chemical Processing Industry

In short span of time, Indian textile industry has faced serious challenges such as German ban on penta chlorophenate, certain azo dyes, formaldehyde, various other chemicals etc. on one hand, and court order for compliance with environmental regulations on the other. There is a list of chemicals and dyestuffs given in Table 29.4 which are prospective source of pollution and health hazard among workers and consumers. These have been identified by the government agencies that have to be phased out.

Besides these harmful chemicals there are several routine chemicals which are excessively used than the required quantity which produces heavy effluent. To reduce effluent load so to save the environment, we have to examine our existing chemical treatments. The optimization of process treatments, quantity of chemicals process parameters and correct sequence is inevitable.

Table 29.1: Summary of Waste Generated and Nature of the Effluent during Textile Wet Processing

Process	Possible Pollutants	Nature of Effluents
Desizing	Starch, glucose, CMC, PVA, resins, fats and waxes not exert a high BOD	Very small volume, High BOD: (30 – 50 per cent of total) CMC and PVA
Kiering	Caustic soda, waxes and grease, soda ash, sodium silicate and fragments of cloth	Small volume, strongly alkaline, dark colour, high BOD: (30 per cent of total)
Bleaching	Hypochlorite, chlorine, caustic soda, hydrogen peroxide, acids	Small volume, strongly alkaline, low BOD: (5 per cent of total)
Mercerization	Caustic soda	Small volume, strongly alkaline, low BOD: (less than 1 per cent of total)
Dyeing	Dyestuff, mordant and reducing agents like sulphides, acetic acid and soap	Large volume, strongly coloured, fairly highBOD: (6 per cent of total)
Printing	Dyes, starch, gums oil, China clay, mordants, acids and metallic salts	Very small volume, oily appearance, fairly high BOD
Finishing	Traces of starch, tallow, salts, special finishes, etc.	Very small volume, less alkaline, low BOD

Source: http://www.fibre2fashion.com/industry-article/22/2195/pollution-and-its-control-in-textile-industry1.asp.

Table 29.2: The Pollution Load of different Chemical Processing Treatments

Parameters	Desizing	Kiering	Bleaching	Mercerising	Dyeing	Printing
pH	8.6-10	10.9-11.8	8.4-10.9	8.1-9.8	9.2-11.0	6.7-8,2
TDS mg/lit	5580-6250	12260-38500	2780-7900	2060-2600	3230-6180	1870-2360
SS mg/lit	2290-2670	1960-2080	200-340	160-430	360-370	250-390
BOD$_5$ days at 20°C, mg/lit	1000-1080	2500-3480	87-535	100-122	130-820	135-1380
COD mg/lit	1650-1750	12800-19600	1350-1575	246-381	465-1400	410-4270

Source:http://www.fibre2fashion.com/industry-article/19/1815/eco-friendly-chemicals-enzymes-and-dyes1.asp

Table 29.3: Impact of Harmful Chemicals on Human Health

Process Description	Chemicals Employed	Possible Health and Safety Related Problems
Bleaching	Hydrogen peroxidesodium hypochlorite	Respiratory and skin allergies
Optical brightening	Organic substances	Skin allergies
Dyeing	Diverse dyestuff Heavy metal carriers (benzidine, dyestuff carriers)	Skin allergies, carcinogenic
Finishing	Urea formaldehydemelamine formaldehyde	Skin allergies, carcinogenic tendency
Mercerization	Sodium hydroxide Ammonia	Corrosive
Softening	Quaternary ammonium siloxanes. Fatty acid-modified melamine resins	Skin allergies
Flame-proof finish	Halogenated hydrocarbons Phosphorous compounds	Carcinogenic tendency, allergenic substances

Table 29.4: List of Banned Dyestuffs and Other Processing Chemicals

Banned Dyestuffs	Banned Chemicals
• 4 amino diphenyl benzidine	• Chlorine bleaches
• 4- Chloro –o- toludine	• Acetic acid
• 2-Naphthylamine	• Starch based warp size
• o-aminiazo toluene	• Kerosene in pigment printing
• 2-Amino-4-nitro toloune	• Formaldehyde
• p- chloroaniline	• Pentachlorophenol
• 2,4 Diaminoanisol	• Halogenic carriers
• 4,4 –Diaminodiphenylmethane	• 1,2 dichloro Ethane
• 3,3- Dichloro benzidine	• Carbon tetra chloride
• 3,3- Diethoxy benzene	• Pentachloro biphenyl
• 3,3- Dimethoxy benzidine	• 1,1,1 Trichloro ethane
• 3,3 Dimethyl-4,4 diaminodiphenylmethane p- Kresidine	• Nonyl phenol ethoxylates
• 4,4- Methylene –bis-(2-chloraniline)	• Dibutyl phthalates
• 4,4-Oxydianiline	• Tri (2,3 dibromopropyl)-phosphate
• 4,4 –Thiodianiline	• Triphenyl phosphate
• o-Toluidine	
• 2,4- Toluyendiamine	
• 2,4,5–Trimethyl aniline	

Categorization of Waste Generated in Textile Industry and Approaches Required

Textile waste is broadly classified into four categories, each of having characteristics that demand different pollution prevention and treatment approaches. The attention of units is focused towards revamping the processing methods, recovery systems and effluent treatment techniques to make textile processing eco-friendly. Intensive efforts are being directed towards using a viable alternative technology for all processes. Such categories are discussed in Table 29.5.

Alternatives to Existing Processes and Chemicals

It is the large consumption of water and harmful chemicals by the textile industry that has fueled the creation of new technology and created a change in conventional chemical processes. The water used in textile processing is used in two ways, one as a solvent for the application of chemicals to the textile, and secondly as a washing or

Table 29.5: Classification of Waste Generated in Textile Industry

Waste Type	Source	Approaches required	
Hard to treat waste	Non-biodegradable organic or inorganic materials	• Colour and metal • Dyeing operation • Phosphates • Preparatory processes and dyeing • Surfactants	• Source reduction. • Chemical or process substitution • Process control and optimization • Recycle/reuse • End-of-pipe treatment
Hazardous or Toxic Wastes	Subgroup of hard to treat wastes	• Metals • Chlorinated solvents • Non-biodegradable or volatile organic materials	• Source reduction. • Chemical or process substitution • Process control and optimization • Recycle/reuse • End-of-pipe treatment
High Volume Wastes	High volume of waste water	• Wash water from preparation and continuous dyeing processes • Alkaline wastes from preparatory processes • Batch dye waste containing large amounts of salt, acid or alkali	• Recovery and Reuse of Process Chemicals • Process and equipment modification. • Reduction of waste water volume • End-of-pipe treatment
Dispersible Wastes	—	• Waste stream from continuous operation • Print paste • Lint • Foam from coating operations • Solvents from machine cleaning • Still bottoms from solvent recovery • Batch dumps of unused processing	• Reduction of waste water volume • Process modification • Source reduction • Recycle/reuse • End-of-pipe treatment

rinsing medium. In order to protect the environment, almost all of the countries in the world have regulations on the condition of effluents that can be disposed into sewer systems, lakes or rivers. These demands on the quality of effluents and water scarcity have lead to new technology and the use of new chemicals. New technology in textile manufacturing is focused on reducing the amount of water needed to perform processing techniques while maintaining the quality of finish on the product.

Source Reduction

☆ Audit the wastes from an operation, including both process and non-process items.

☆ Once all wastes have been listed, each should be viewed as a potential recycle product, and markets should be sought.

☆ Identify specific reduction strategies for each identified waste.

Some broadly applicable techniques include raw material control, conservation and optimization of chemicals, chemical substitution, process modification, equipment modification, maintenance procedures, housekeeping, waste recovery (for reuse and recycle), and segregation. Using a combination of the above techniques has produced documented annual savings in the millions of dollars.

Reducing Water Consumption

☆ *Reduce the number of process steps:* This involves a study of all the processes and determining where changes can be made. For example, fewer rinsing steps may be required if a dye with high exhaustion is used.

☆ *Optimize process water use:* Examples include using batch or stepwise rinsing rather than overflow rinsing, introducing counter-current washing in continuous ranges, and installing automatic shut-off valves.

☆ *Recycle cooling water:* Cooling water is relatively uncontaminated and can be reused as make-up or rinse water. This will also save energy as this water will not require as much heating.

☆ *Re-use process water:* This requires a study of the various processes and determining where water of lower quality can be used. For example, final rinse water from one process can be used for the first rinse of another process.

☆ *Using water efficient processes and equipment*: Although replacing outdated equipment with modern machines which operate at lower liquor ratios and are more water efficient requires capital investment, the savings that can be made ensure a relatively short pay-back period.

Reducing Chemical Consumption

☆ *Recipe optimisation:* Optimising the quantity of chemicals required will lead to more efficient chemical use and lower costs. Continual updating of recipes should be carried out when new dyestuffs enter the market as, in general, less of these chemicals are required.

☆ *Dosing control:* Overdosing and spillages can be reduced by mixing chemicals centrally and pumping them to the machines. Check that manual measuring and mixing is carried out efficiently and automatic dispensers are properly calibrated.

☆ *Pre-screen chemicals and raw materials:* Avoid dyestuffs containing heavy metals, solvent-based products and carriers containing chlorinated aromatics. Safety data sheets should be obtained from the chemical manufactures to obtain information such as toxicity, BOD and COD. Check that raw materials do not contain toxic substances. Check that companies will accept expired raw materials for disposal.

Chemical substitution

New chemicals are being used in textile processing that are replacing conventional chemicals in order to reduce pollution in the effluent. Alternative chemicals can be substituted for conventional chemicals that would produce a 50 percent reduction of pollution in the effluent. Review chemicals used in the factory and replace those hazardous to the environment with those that have less of an impact. Use dyes that have high exhaustion rates and require less salt. Specifically replace metal-containing dyes, use bi-reactive dyes in place of mono-reactive, avoid the use of APEO detergents and replace with more biodegradable alternatives, dye wool with dyes that do not require after-chroming. Using enzymes for desizing, bio-polishing, the change in hand of textile, the reduction of pilling, the removal of print thickeners used during printing textiles, and the production of a 'worn look' on a fabric. Using solvent instead of water as the main medium during the process cycle reduces the effluent of a washing system from 16 kilograms an hour to zero. This application process is being looked to as the future of textile wet processing.

Process Modification

The possible modifications in the process as well as its parameters is another way of eliminating unnecessary wastes and thus helps to reduced the water pollution for *e.g.*: substituting the standing baths for rinsing by running ones thus conserving water and thus reduced waste load. Process modification includes improved operating practices, improved housekeeping, change in raw materials, change in technology and in-house reuse or recycling. Examples of new processes and technology used in textile industry are:

☆ *Ultrasonic waves* are being researched in order to replace processes that require water or to decrease the amount of reagents in the effluent. The ultrasonic waves can produce effects on textiles that are similar to current physical and chemical techniques with the advantage of not using water. The waves also increase the dye absorption into a textile product so less dye will be in the effluent at the end of the process.

☆ *Foam Technology* is replacing the conventional padding system. In the padding system, the reagent is diluted in water which is then impregnated into the textile by running the textile through the diluted solution and then through two rollers that squeeze the excess solution off the textile. During the foam application method, air replaces water as the transport medium between the reagent and textile. This system does not require water to apply a dye or finish to textiles and less waste is produced when compared to the conventional padding method because not all of the reagent is impregnated into the textile during padding

End-of-Pipe Treatment Methods

Once waste minimization has been carried out in the factory, effluent will still be produced that will require some form of treatment prior to disposal to sewage, river or sea. Few effective treatments must follow essentially like:

☆ *Effluent segregation:* Prior to the installation of any end-of-pipe treatment method, it is essential to carry out segregation of the effluent streams to separate the contaminated streams from the relatively clean streams for treatment. This result in a more effective treatment system as a smaller volume of wastewater is treated (resulting in lower capital and operating costs) and it allows for the use of specific treatment methods rather than trying to find one method to treat a mixture of waste with different characteristics. The segregated clean streams can then be reused with little, or no, treatment elsewhere in the factory.

☆ *Treatment technologies:* There are two possible locations for treating the effluents, namely, at the textile factory or at the sewage works. The advantage of treatment at the factory is that it could allow for partial or full re-use of water. The following technologies have all been used: Coagulation and/or flocculation, membranes (microfiltration, nanofiltration, and reverse osmosis), adsorbents (granular activated carbon, silica, clays, fly ash, synthetic ion-exchange media, natural bioadsorbants, and synthetic bioadsorbants), oxidation (Fenton's reagent, photocatalyis, advanced oxidation processes, ozone) and biological treatment (aerobic and anaerobic).

☆ *Coagulation and/or Flocculation:* Chemicals are added that form a precipitate which, either during its formation or as it settles, collects other contaminants. This precipitate is then removed either through settling or by floating it to the surface and removing the sludge. This is a well-known method of purifying water. Both inorganic (alum, lime, magnesium and iron salts) and organic (polymers) coagulants have been used to treat dye effluent to remove colour, both individually and in combination with one another. With the changes in dyes and stricter discharge limits on colour, inorganic coagulants no longer give satisfactory results. They have the added disadvantage of producing large quantities of sludge.

Formulation and Implementation of Legislation

Various rules and regulations have been formed by the government agencies to curb the menace of textile pollution. Formulation and implementation of legislation plays a very important role in making the textile industry less polluted.

The ISO 14000

The ISO 14000 family addresses various aspects of environmental management. The very first two standards, **ISO 14001:2004** and **ISO 14004:2004** deal with environmental management systems (EMS). ISO 14001:2004 provides the **requirements** for an EMS and ISO 14004:2004 gives general EMS **guidelines.** An EMS meeting the requirements of ISO 14001:2004 is a management tool enabling an organization of any size or type to:

☆ Identify and control the environmental impact of its activities, products or services, and to

☆ Improve its environmental performance continually, and to

☆ Implement a systematic approach to setting environmental objectives and targets, to achieving these and to demonstrating that they have been achieved.

ISO 14001:2004 does not specify levels of environmental performance. If it specified levels of environmental performance, they would have to be specific to each business activity and this would require a specific EMS standard for each business.

Eco-labeling

Eco-labeling guarantees certain ecological criteria for all sorts of textile items and their manufacturing processes. It identifies:

☆ Products are environmentally safe

☆ Manufacturing by eco-friendly material and

☆ Avoid harmful chemicals

Oekotex 100 is one of its classified standard.

IS 15651:2006 TEXTILES

It covers quality as well as environmental requirements; denoted by earthen pot together with the standard mark of The Bureau of Indian Standards, is voluntary in nature. So far there are no certified producers of textiles with ecomark due to lack of demand in the internal market and absence of relevant legislation. The ecomark requirements of textiles and apparels are covered under is 15651:2006 textiles – requirements for Environmental labelling – specification.

It covers quality requirements as specified in the relevant product standard or as agreed between the buyer and the seller. The environmental requirements include limits for following as per international norms:

☆ Urea formaldehyde

☆ Insecticides and pesticides

☆ Chlorinated hydrocarbons

☆ Heavy metals

☆ Organo tin compounds

☆ PCP
☆ Banned aryl amines

Bluesign®

☆ Starts from the ground up; components and processes are selected in a manner to meet the specified criteria before production begins to guarantee maximum implementation of resource productivity and EHS – that's Environment, Health and Safety – in the most efficient and cost-effective way for everyone involved.

☆ Is practice-oriented; provides solutions for EHS issues on all levels, from raw material and chemical components suppliers to manufacturers and retailers.

☆ Is widely recognized within the complete textile production chain to assure efficient and economical solutions at all levels of the manufacturing chain.

It is recognized by leading chemical companies, numerous textile manufacturers as well as well-known retailer and brand companies.

Made in Green®

Made in green is a special symbol for all those who provide or who are seeking textile products manufactured with the guarantee that they are free from substances harmful to health. This certifies that the product, throughout its traceability chain has been manufactured in factories which respect the environment and the universal rights of workers. The object, therefore is, free from harmful substances, respect for the environment and respect for human rights.

Impact of REACH

Other initiatives include REACH (Registration, Evaluation and Authorisation and Restriction of Chemicals) legislation that aims to encourage safe and eco-friendly chemical production. In the USA the Toxic Substances Control Act (TCSA) enables the US Environmental Protection Agency (EPA) to track industrial chemicals produced in or imported into the country. Some man-made fibres, such as Lenzing's lyocell fibre 'Tencel', have a minimal impact on the environment. Also, organic cotton production is growing rapidly but still accounts for only a small fraction of global cotton output. Nonetheless, organic cotton is being adopted by high profile companies such as C&A, Coop, Nike, Wal-Mart, and Woolworths. A growing number of brand and manufacturing companies are also pursuing environmentally friendly strategies. For example, American Apparel, Gap, Interface, Patagonia, and Wal-Mart in the USA as well as Rohner Textil in Switzerland, and a small knitwear company in India, MaHan.

Conclusion

Cleaner production is an attractive approach to tackle environmental problems associated with industrial production and poor material efficiency. Although the solutions are in a way evident – find low cost pollution abatement technologies, move from pollution control to pollution prevention, and international cooperation – these are most viable but not so easy solutions to environmental protection. Resistance from the enterprises is very high. The integrated approach to environmental problems in textile industry will depend on the moves adopted by the dye/chemical manufacturer, textile processor and the environmental authorities towards safe production, secure waste management and proper implementation of legislations related to textile production, respectively. Effective education of the concerned population will surely reduce the amount of risk and will generate benefit from the required technology in sustainable manner.

References

Agarwal, A. 2010. Small-Scale Industries Drive India's Economy But Pollute Heavily: What Can Be Done? Retrieved November 11, 2010, from www.indiaenviromentportal.org.in/./Small per cent 20scale per cent 20industries.html

Anonymous. 2011. Water Pollution. Retrieved August 17, 2011 from http://eco360.me/what-is-eco360s-causes/water-pollution.

Anonymous. 1996. U.S. EPA/semarnap pollution prevention in Textile Industry. 49.168.34.62/ref/20/19041.pdf.

Arya, D. and Kohli, P. 2009. Environmental Impact of Textile Wet Processing. Retrieved September 22, 2011 from http://www.fibre2fashion.com/industry-article/21/2047/environmental-impact-of-textile1.asp

Bender, N. 2004. The Impact of Water Scarcity and Pollution on the Textile Industry: A Case Study from Turkey Textile Industry: A Case Study from Turkey. Retrieved August 30, 2011, from *www.airdye.com/downloads/20_Articles_112204_Impact.pdf*

Das, S. 2009. Textile Effluent Treatment – A Solution to the Environmental Pollution. Retrieved September 8, 2011 from *www.fibre2fashion.com/industry./Textile-Effluent-Treatment.pdf.*

Karthikeyan, N. and Alexander J. J. 2008. Waste minimisation in textile industry. The Indian Textile Journal. 118(12): 35-45

Khatri, Z. and Brohi, K. H. 2009. Environmental Friendly Textiles- a road to sustainability. Retrieved September 8, 2011, from http://knol.google.com/k/zeeshan-khatri/environmental-friendly-textiles-a-road/1gwk3g61mo0vs/2

Menezes, E. 2009. Eco-friendly Chemicals, Enzymes and Dyes. Retrieved September 22, 2011, from *http://www.fibre2fashion.com/industry-article/19/1815/eco-friendly-chemicals-enzymes-and-dyes1.asp*

Nelliyat, P. n.d. Socio-economic, environmental and clean technology aspects of textile industries in Tiruppur, South India. Retrieved September 22, 2011, from *www.sasnet.lu.se/EASASpapers/12PrakashNelliyat.pdf*

Parvathi, C., Maruthavanan, T. and Prakash, C. 2009. Environmental impacts of textile industries. The Indian Textile Journal. 119(11): 12-17

Shinde, L. M. 2009. Pollution and its Control in Textile Industry. Retrieved September 22, 2011, from http: //www.fibre2fashion.com/industry-article/22/2195/pollution-and-its-control-in-textile-industry1.asp

Turner, M. T. 1978. Wastewater Treatment and Re-use within the Textile Industry, Water Services Annual Technical Survey.

Vigneswaran, C., Anbumani, N. and Ananthasubramanian. 2011. Biovision in Textile Wet Processing Industry- Technological Challenges. Journal of Textile and Apparel, Technology and Management. 7(1): 3-13

2013, Environmental Technology
Editors: D.R. Khanna, A.K. Chopra, Gagan Matta, R. Bhutiani & Vikas Singh
Published by: DAYA PUBLISHING HOUSE, NEW DELHI

Pages 259–265

Chapter 30

Colored Cotton: Green Opportunity for Growing Fashion Industry

Tayyaba Fatma and Anita Rani

Department of Clothing and Textiles,
Govind Ballabh Pant University of Agriculture and Technology,
Pantnagar – 263 145

ABSTRACT

The Indian fashion industry is growing in importance as designers are becoming more experimental in term of clothing styles, western fashion along with use of naturally colored cotton fibre products. Cotton with natural colors *i.e.* orange, red, yellow, purple, brown, blue and black other than white is commonly referred as colored cotton. Colored cotton is also known as diploid cotton because of crossing the genes. Besides, use of germplasm, bio-genetic engineering and other high-tech techniques have opened the doors for producing cotton in different enchanting colors to emerge onto the world as "green movement".

Colored cotton is not a concept. It has a history of centuries, but has got momentum when the chemicals have been banned in many countries. Turning to the environment-friendly colored cotton as enthusiastic proposition has been readily accepted among apparel manufacturers having name in their field. These are one who shares concern about freedom form synthetic dyes and auxiliaries, human health and safety, cost of fabric production and threat to ecosystem thus helps in realizing "zero pollution" in fibre to cloth process. This form of cotton also feels softer to the skin and has a pleasant smell. So, the demand for naturally-colored cotton is increasing. Today's, naturally colored cotton is a hot topic in national and international green textiles and apparel research and development.

Keywords: *Colored cotton, Genetic engineering, Zero pollution and Green fashion.*

Introduction

During 18th century, the "natural" trend among consumers and the environment conscious social climate helped to create an initial demand and niche market for naturally colored cottons, organic fibers and other environment friendly textile products. Colored cotton is one among them and is looked upon as an enthusiastic proposition by all of us who share concern about health, safety and threat to ecosystem. Most people think that cotton only comes in the white color and needs to be dyed, but in reality cotton can be grown in a variety of colors lint other than white. Mostly, coloured cotton can be optained from varieties like *Gossypium hirsutum, Gossypium arboreum* and *Gossypium herbaceum* in our country. But the colored cotton fiber is also encountered in wide varieties of cotton species *i.e.* *G. nanking* Meyen, *G. purpurascens* Poir. and *G. punctatum* Schum, *G. sturtii* Mull., *G. schottii* Watt, *G. mustelinum* Miers, *G. darwinii* Watt, *G. harknessii* Brandegel (gray-brown fiber), *G. davidsonii* Kellogg (gold colored), *G. tomentosum* Nutt (dark rust color), *G. stocksii* Mast. (light brown) and *G. punctatum* Schum, which grows in the U. S. A., Jamaica, Costa Rico, Curacao, Africa and Nigeria.

Naturally colored cotton is cotton that has been bred to have colors other than the yellowish off-white typical of modern commercial cotton fibers. Colors grown include red, green, brown, orange, blue, black, purple and other several shades but all of them, colored cotton usually come in four standard colors - green, brown, red (a reddish brown) and mocha (similar to tan).

What Colored Cotton is?

Colored cotton refers to the cotton which is available in different natural colors *i.e.,* orange, red, yellow, purple, brown, blue and black other than white. This colored cotton is developed by gene transplantation and Crossing of genes. Besides, uses of germ plasm, bio-genetic engineering and other high-tech techniques have opened the doors for producing cotton in different enchanting colors.

Yield of naturally colored cotton is typically lower and fibre is shorter, coarse, weaker and uneven but has a softer feel than the more commonly available "white" cotton. Thus, this form of cotton feels softer to the skin and has a pleasant smell. Certain policies and legislative measures given by Cotton Corporation of India are used for safe and profitable cultivation of colored cotton. And application of latest technologies and modern farming techniques may give boost up result for exposure of colored cotton.

History

In nature, colored cotton is found from time immemorial. Exact origin of naturally colored cotton is around 5000 years ago in the Andes by Americans. Colored cotton agriculture began in Indo-Pakistan, Egypt, and Peru around 2700 B.C. It was the industrial revolution for colored cotton cultivation. Indians could be attributed with growing naturally colored cotton of myriad hues, which they maintained for over the last two millenniums. Colored cotton has been grown for years in Russia, India and South and Central America, and Israel. Colored cotton has been around from the beginning of sub-Himalayan regions of Asia and in ancient Indian.

Safe from Chemical Dyes in it

Naturally colored cotton is unique and exceptionally different from white cotton as it does not need to be dyed. Because dyeing is one of the most costly steps in fabric finishing due to water and energy use and waste production. Effluents after dyeing are high in toxicity because dyes containing traces of heavy elements such as arsenic, lead, cadmium, cobalt, zinc and chromium, which are not only affect the human beings but also affects surface water recourses and eco-system by environmental pollution. So, cost of colored cotton could be up to half of the value, environment friendly, eliminates disposal costs because it could not to be dyed.

Advantages

Coloured cotton is not a new concept. It has a history of centuries, but has got momentum now because of several advantages of naturally coloured cotton are briefly discussed below:

  ☆ *Safety concern of Human Health:* Artificial dyes done on Cotton fabrics, cause allergy aZnd itching on the skin and sometimes may cause skin cancer. It is a fact that most of dyes used in textile industries are carcinogenic. So, the fabric prepared from naturally coloured cotton lint is free from such adverse effects and also the fabric can be safely used even by those having sensitive skin and has been found to be the best for human health.

  ☆ *To reduce Environment pollution:* After dyeing, the chemical residues in the form of dyeing or finishing effluents are thrown in nearby river contaminating water and soil. This form is a major source of environmental pollution. When the fabric is manufactured from naturally coloured lint, there is no need of artificial dyes. Hence the residues of artificial dyes will not accumulate in the drains and which helps in reducing environmental pollution caused by artificial dyes.

  ☆ *Reduction of cost of fabric production:* The dyeing process is omitted when naturally coloured lint is used for manufacturing of the fabric. Thus the cost of production of fabric can be reduced to some extent through the use of naturally coloured cotton.

  ☆ *Less flammability:* There is experimental evidence to demonstrate that have found naturally colored cottons to have a higher oxygen index value than white cottons, which indicates less flammable.

  ☆ *Other benefits:* Naturally colored cottons are considered environment friendly because they have many insect and disease-resistant qualities, drought and salt tolerant and do not have to be dyed during fabric manufacturing. Naturally colored cotton knit fabrics has compared favourably with white cotton fabrics in evaluating abrasion resistance, dimensional stability and pilling resistance. Naturally colored cotton fabrics had less bursting strength than the white cotton fabrics.

Limitations

The naturally colored cotton has a small fibre and is not suitable for heavy machine spinning. Due to smaller fibre, it becomes unpractical to use the manufacturing of naturally colored cotton yarn. These are some other inherent drawbacks *i.e.* low yield potential, poor fibre properties, limited color, contamination, low market demand and lack of marketing facilities. But now, colored cotton is literally squeezed in with the conventional white cotton to make its fibre longer and stronger than other naturally colored cotton to be used in typical looms. Recently, the work on colored cotton has got momentum by incorporating the latest technology.

Up-gradation in Technology

Naturally colored cotton had smaller fibers which were not suitable for mechanical looms, therefore kept naturally colored cotton to enter in the commercial market. In 1982, Sally Fox, started researching on colored cotton introduced first long fiber of naturally colored cotton with the help of her knowledge and experience in technology. Sally Fox later started her company as a name of Natural Cotton Colors, Inc. It got patents in different shades including: green, Coyote brown, Buffalo brown, and Palo Verde green and registered it under different trademark likewise FoxFibre®. Later on the technology was further improved by a cotton breeder Raymond Bird in 1984. Bird began experimenting in Reedley, California, with red, green, and brown cotton to improve fiber quality. After that Raymond Bird along with his brother and C. Harvey Campbell Jr., a California agronomist formed BC Cotton Inc. to work with breeding of naturally colored cottons. Breeding objectives of colored cotton include improving color range, fiber quality and yield potential and can be used commercially. In many countries a number of researchers have studied on yield and quality improvement of naturally coloured cottons. Commercial production and processing remain a challenge predominantly due to the advances in technology and development of higher yielding cotton. Grown natural color cotton represents the best in man's efforts to work in concert with nature to produce beautifully colored fabrics without the use of dyes. However, color-grown fabrics may contain certain natural variations in color, light fastness and shrinkage. Today's Xinjiang in China is vital to producing colored cotton because of good ecological environment and fertile farmland.

Growing Fashion Industry in Reverence to Colored Cotton

The Indian fashion industry occupies a unique place in the Indian economy and has importance as designers are becoming more experimental in term of currently popular style or practice, especially in clothing, foot wear or accessories. Fashion means prevailing custom, usage, or style. Fashion references to anything that is the current trend in look and dress up of a person. Fashion always highly business, is in the process of becoming even more so as industry.

The traditional production and consumption of fashion has brought the humanity great material wealth, but at the same time, it also brought a lot of new problems like population explosion, uneven income, environment pollution, waste of resources, energy crisis, and worse ecology. Facing the new challenge, the humanity realizes they must change the original production and consumption way and seeks a new opportunities for production and consumption way. And these problems could be solved by green movement in term of development and production of green environmental protection clothing. In part to the Green movement, colored Cotton has risen in importance and economic viability. The movement for naturally coloured cotton has a long way to go. In India, colored cotton is also cultivated in small holdings in Andra Pradesh, Madhya Pradesh, Maharashtra and Orissa. Khadi and village industries commission have experimental producing dress material, particularly used in children clothes using colored cotton. Nature colored professional is producing natural colored cotton baby clothing brand, which is the brand name under Globaby Co. Ltd. As the fashion industry begins to embrace the green movement, how can consumers take decisions every day?

Through the garments and accessories, fashion designers within the fashion industry are responding to the challenge of creating more sustainable products with help of colored cotton. In addition, a variety of textile samples are on view those who are interested in sustainable fashion garment and textiles, can touch and examine. Fashion industries encourage the safe and healthy environment by including colored cotton in own collection. The fashion industries, which are manufactured using naturally coloured cotton garments, have the potential to compete successfully in the international market because the natural colours do not fade due to repeated washings but becomes darker and more intense. The main reason for liking the colour cotton garment is the no carcinogenic effect because the artificial dyes were not used as in the white cotton.

Naturally colored cotton products (garments and accessories) are also available to the consumer in different varieties. They are being utilized in casual clothing, home fashions and upholstery fabrics.

Future Thrust Areas

Today's naturally-colored cotton is a hot topic in national and international green textiles or green fashion related research and development. Colored cotton's future depends on continued improvement and development of appropriate manufacturing procedures and also signalizes future thrust area of research. These are as follows:

- ☆ Development of wide range of colours with different shades, uniformity and stability.
- ☆ Improvement of fibre properties such as length, strength and maturity.
- ☆ Development of high yielding, early maturing varieties and hybrids.
- ☆ Development of coloured cotton varieties and hybrids amenable to mechanical picking.
- ☆ Development of male sterility based hybrids in desi and upland coloured cottons.
- ☆ Development of coloured transgenic cottons.

Conclusion

Environmental issues influence all human activities. As fashion industries are more concerned about the environment and health of their customer, green marketing of garment enterprise. It ensures the interests of the organization and all its consumers are protected too. Both the buyer and seller mutually benefit. Fashion industry as the main part of all industries plays an active role in the operating actively and must establish new green marketing thinking, give up the temporary benefit and pursue the persistent environmental protection profit which helps in realizing "zero pollution" from fiber to cloth. It means coloured cotton not only the elimination of textile printing and dyeing in the course of harmful carcinogenic substances such as formaldehyde, carcinogenic aromatic amines of azo dyes and harmful heavy metal ions with a reduction of discharge of sewage pollution on the environment, protecting the living environment of mankind. Currently, the demand for naturally-colored cotton is increasing because of limited resources, high prices, and the huge potential market. Some businessmen use illegal means to deceive consumers and obtain profits, which impedes the development of naturally-colored cotton products. So, user of naturally-colored cotton fashion products can be identified with high accuracy by using a qualitative analysis method.

References

What is Color-Grown Cotton? Retrieved 2011, September 10 from http: // www.supercomfort.com/help/learn-about-organic-cotton.html\

Colored Cotton. Retrieved 2011, September 17 from http: //www.farminfo.org/ othercrops/cotton.htm

Recent Developments in Cotton. Retrieved 2011, September 17 from http: // sustainablefashion.blogspot.com/

Natural colored cotton. Retrieved 2011, September 17 from http: //en.wikipedia.org/ wiki/Naturally_colored_cotton

Ecological Textile Products. Retrieved 2011, September 17 from http: // www.perunaturtex.com/scientif.htm

Innovative life: colored cotton. Retrieved 2011, September 17 from http: // invention.smithsonian.org/centerpieces/ilives/lecture12.html

SEWA members grow coloured Cotton for their Hansiba Brand. Retrieved 2011, September 17 from http: //manvenrakesh.blogspot.com/2010/11/sewa-members-grow-coloured-cotton-for.html

Move over white, now Gujarat is growing coloured cotton. Retrieved 2011, September 17 from http: //www.dnaindia.com/money/report_move-over-white-now-gujarat-is-growing-coloured-cotton_1466896

Natural Cotton Color Retrieved 2011, September 17 from http: // www.coolhunting.com/style/brazil-natural.php

Gujarat growing coloured cotton. Retrieved 2011, September 17 from http: // daily.bhaskar.com/article/move-over-white-now-guj-is-growing-coloured-cotton-1550912.html

Natural Cotton Color. Retrieved 2011, September 17 from http: // www.indianjournals.com/ijor.aspx?target=ijor: ijgpb&volume= 58&issue=1&article=001

Dodamani M.T. and Kunnal L.B., "Value Addition to Organically Produced Naturally-Coloured Cotton under Contract Farming" Agricultural Economics Research Review, Vol. 20 (Conference Issue) 2007 pp 521-528

Singh P., Singh V.V. and Waghmare V.N., "NATURALLY COLOURED COTTON", Central Institute for Cotton Research, Nagpur.

Murthy M. S. S., Never Say Dye: The Story of Colored Cotton, General Artical, December, 2001.

Dickerson Dianne K., Lane Eric F. and Rodriguez Dolores F., "NATURALLY COLORED COTTON: RESISTANCE TO CHANGES IN COLOR AND DURABILITY WHEN REFURBISHED WITH SELECTED LAUNDRY AIDS", California Agricultural Technology Institute Design and layout by Steve Olson, October 1999.

2013, Environmental Technology   Pages 267–273
Editors: D.R. Khanna, A.K. Chopra, Gagan Matta, R. Bhutiani & Vikas Singh
Published by: DAYA PUBLISHING HOUSE, NEW DELHI

Chapter 31

Control of Mosquito Vector by Ecofriendly Methods

K.S. Baghel[1], R.K. Kaurav[1], Bhansingh Barela[2], Surjeet Jat[2], Prashant Meena[2] and R.C. Saxena[2]

[1]*Department of Zoology, Bherulal Patidar Govt. P.G. College, Mhow*
[2]*Pest Control and Ayurvedic Drug Research Laboratory,
S.S.L. Jain P.G. College, Vidisha – 464 001, M.P.*

Introduction

To tackle mosquito vector is a cumbersome task, because they have a high fecundity, short and complex life cycle strategies, easy availability of breeding ground, ability to recover population loss, capacity to develop resistance against chemical insecticides and various type of resting –hiding places. They are found in most parts of the world except Antarctica and few Islands. The anthropophilic behavior of some mosquitoes is responsible for the transmission of disease in human beings.

In the last century, mosquito control programmers were mainly based on chemical insecticides. Insecticides are very effective in mosquito control. D.D.T. (Dichloro diphenyl trichloroethane) was discovered in 1874 by Austrian chemist. Othemer zeitler and is resynthesised in 1939 by Paul Muller, who revealed its insecticidal value for which he got the Nobel Prize in 1948. With the discovery of D.D.T. an unprecedented interest was developed for finding insecticides. After that BHC (Benzene hexa chloride), Dieldrin, Malathion, Fenitrothion, Cchlordane, Diazinon Baytex and abate were discovered for mosquito and pest control. These insecticides have proved highly toxic to non target organisms and distractive to environment. Rao R. (1988)

The proposed study examined the following important aspects;

1. Larvivorous potential of local fishes.
2. Larvivorous efficacy of local insects.

Review of Literature

Miura and Takahshi (1988) carried out laboratory experiments to evaluate.the potential larviovorous behavior of *Enallagama civile* (Damsel fly) nymph. It was found that E.civile nymph is very capable of preying on Mosquito larvae as a good source of food. Hati (1988) studies four predaceous arthropods for biological control of Mosquitoes.The water bugs (Heteroptera; belostomatidae) are reported to be well known predators of different immature stages of Mosquitoes. Paul &Ghosh (1988) examined predatory habit of *Sphaerodema annulatum*, a water bug.on Mosquito larva and its prospects as biocontrol agents of Mosquito larvae.Phong *et al.* (2008) evaluaed fecundity and survivorship of six copepod species, in relation to selection of candidate biological control agent against Aedes aegypti. Mesocyclops as pericornis has been recommended as good biological agent for Mosquito larvae control in domestic water containers. Fischer and Nicolas (2008) have reported about the mosquito larvae control by predatory genera of insect such as Tropisternas, Rhantus Liodessus and Belostomain temporary rain pools of Buenos Aires Argentina.

Materials and Methods

Colonization of Mosquitoes

To understand the life cycle of mosquitoes, breeding behavior of adult. Larval development and fulfillment of larvae for experimental purpose, a viable insectary was maintained. Colonization of tree health important mosquito *Anopheles stephensi, Culex quinquefasciatus* and *Aedes aegypti* were done in the insectary. An insectary of 10 x10 sq.ft.in sizes was maintained in the laboratory equipped with thermostat, head converter, humidifier fluorescent tube light, incandescent lamps with regular dimmer, cooler, exhaust fan, RH meter and minimum-maximum thermometer to maintain the favorable environment for mosquitoes. The insectary was maintained at 27 ±2°C and RH 75±5 per cent. The photoperiod was maintained at L: D; 14:10.

Experimental Fishes

Three indigenous fishes, *Anabas testudineus, Ambasis nama* (Chanda nama) and *Punctius ticto* widely found in Satak dam and River Kunda at Khargone M.P.,the larvivorous potential of all the fishes were seen against mosquito larvae.

Experimental Bioassay on Larvivorous Potential of Indigenous Fishes

Larvivorous potential of experimental fish species were recorded by taking them separately and individually in 1000ml. glass beakers with larvae of an. Stephens. These experimental were conducted without fish food only IV instar larvae of An. Stephensi were given as feed total consumed larvae in 24 Hours duration were noted. The difference between the number of larvae released in the beginning of the

Table 31.1: Ranking of Larvicidal Potential of Fishes Based on Body Weight

Sl.No.	Name of Fishes	Body Weight of Fishes (gm)	Fourth Instar Larvae of Anopheles ste Phensi Consumed by Fish Phensi	Ranking Based on per gm. Body Weight	Fourth Instar Larvae of Culex Consumed by Fish	Ranking Based on per gm. Body Weight	Fourth Instar Larvae of Aedes ageypti Consumed by fish	Ranking Based on per gm. Body Weight	Ranking based on per gm. Body weight	Final Ranking of Larvivorous Potential of Fishes Based on per gm. Body Weight
1.	Ambassis nama (Chanda nama)	01	8.2	38.4	IV (4.68 larvae)	38.0	IV (4.63 larvae)	35.2	IV (4.29 larvae)	IV (4.53 larvae)
2.	Anabas testudineus	01	16.2	88.0	III (5.30 larvae)	84.4	III (5.08 larvae)	80.0	III (4.81 larvae)	III (5.06 larvae)
3.	Punctius ticto	01	10.2	71.8	II (7.03 larvae)	73.0	II (7.15 larvae)	71.0	I (6.96 larvae)	I (7.04 larvae)

Experiments were performed in 3 replicates.

Experiments were observed for 24 hours duration.

Table 31.2: Larval Performance of Fishes on Larval Instar of Mosquito Species

Sl.No.	Name of Fishes	Number of Fishes	Number of Larval Instar Consumed by Fishes																
			Larval Stage of *An. stephensi*				Larval Stage of *Cx.quinquelaciatus*				Larval Stage of *Aedes aegypti*			
			I	II	III	IV	I	II	III	IV	I	II	III	IV
1.	*Ambasis nama* (chanda nama)	01	+	+	++	+++	+	+	++	+++	+	+	++	+++
2.	*Anabas Iseludineus*	01	+	+	++	+++	+	+	++	+++	+	+	++	+++
3.	*Punctius ticto*	01	+	+	+++	+++	+	+	+++	+++	+	+	+++	+++

+: Less preferred; ++: More preferred; +++: Most preferred.

Experiments were conducted in three replicates.

Experiments were observed for 24 hours duration.

experiment and found alive after 24hours was considered as the fish. A control was also maintained by taking 50 larvae of mosquitoes in a beaker without fish.

Observation and Result

Average body weight of fishes was mentioned in the Table 31.1.

Table 31.1

Report ranking of Larvivorous potential of experimental fishes based on their body weight. Larvivorous potential of *Punctius ticto* was the highest among other fishes in relation to their body weight. It was recorded for *Puntius ticto*, 7.03, 7.15 and 6.96 fourth instar larvae per gram body weight per 24 hours for *Anopheles stiphensi*, *Culex quinquefasciatus* and *Aedes aegypti* respectively. *Anabas testudineus* was in the second which consumed 5.03, 5.08 and 4.81 fourth instar larvae/gm body weight/ 24 hours of *Anopheles* stephensi,*Culex quinquefasciatus* and *Aedes aegypti* respectively.*Ambassis nama* (Chanda nama) came in third which consumed 4.68, 4.63, and 4.29, fourth instar larvae/gm. Body weight/24 hours of *Anopheles stephensi, Culex quinquefasciatus* and *Aedes aegypti respectively.*

Table 31.2

The result in table II showed the larval performance of fishes. It was noticed that *Amabassis nama* was less preferred first instar of Anopheles first and second star of Culex and first and second instar of Ades larvae. Similarly *Anabas testudineus* was preferred maximum in fourth instar larvae of all the 3 stages of mosquito. As regards the choice of *Punctius ticto* it mostly preferred III and IV instar larvae of all the 3 species of mosquito. The reasum being large amount of fucsin found in the cuticles of these instar.

Discussion

As alternative to the chemical control of vector speices was worked out in the present study so as to minimize environmental pollution and to provide sustainable development. The present paper reported the biological control of vector species by *brvaevorous* fishes which are quite suitable according to the environmental condition and different eco ones, the study which was carried out at government college MHOW reports the larviovorous potential of 3 indigenous fishes collected form satak dam from Khargone district of M.P. the maximum number of larval consumption was noted in *Punctius ticto*. Similar result was reported by Soni *et al.*, 2006 and Arora 1992 but as regards the larviovorous potential as well as edible value and commercial potential *Anabas testudineus* was found to be of great commercial value. Dua have also reported the larviovorous potential of poecilia reticulate and observed a great larviovorous potential. Krishan *et al.*, 2008 have also expressed the similar views of bioenvironmental control of vector Mosquito by predatory fishes and insect.

Conclusion

The present paper report the larviovorous potential of three speices of indigenous fishes as an alternative to the chemical pesticide to minimize the human health hazards due to chemical pesticides and to save the ecosystem. Degradation due to chemical pesticides among the three species *Punctius ticto* was found to be highly efficacious as for as its predatory behavior is concerned but as far as commercial value is concerned *Anabas testudineus* is superior than others. The results showed that the fishes generally proffered in 3rd and 4th Instar larvae may be due to the presence of lycofuchin in their cuticle. For the bioenvironmental control using predatory fishes the following are pre exquisite to undertake the experimental protocol.

1. The fishes must be easily label around that particular area.
2. They should be cultured by the farmers in small reservoirs.
3. The fishes must have high edible value.

Acknowledgments

The Authors are very much thankful to the Principal, Govt. P.G. College, Mhow (M.P.) and the Principal, Govt. P.G. College, Khargone (M.P.) for encouraging them in the said work.

References

Arora, M.P. (1992). Animal behavior. Edit.-Mrs. Chandrakanta.Himalaya pub.House.Delhi, pp.351

Dua, V. k., A.C> Pandey, Swapnil Rai and A.P. Dash (2007). Larviovorous activity of poecilia reticulate against Culex quinquefasciatus larvae in a polluted water drain in Hardwar India.J.Amer. Mosq. Con. Ass., 23(4): 481-483

Fisher Sylvia and Nicolos schweigmann (2008).Association of Immature Mosquito and predatory insects in urban rain pools, J. of vect. Ecol., 33(1): 46-55

Hati, A.K. (1988). Studies on four predacious arthropods for biological control of Mosquitoes, Bicovas, 1: 25-40

Krishna V.K. (2008). A study on the control of vector Mosquito by eco-friendly methods, Ph.D. thesis, Barkutulla University Bhopal (M.P.) India, , pp 1-208

Krishna V.K., Jyoti Uikey and R.C. Saxena (2008). Mosquito Larvicidal and chemosterilant Activity of flavonoids of squamosa life science bulletin, Satana (M.P.) India, : 5(1) 85-88.

Miura T. and R.M. Takahashi (1988). A laboratory study of predation by damsel fly nymph Enallagama civil upon Mosquito Larvae Culex tarsalis, j. Amer. Mosq. Contr.Asso, 4(3), 129-131

Paul T. K. and K.K. Ghosh (1988). Predatory habit of sphaerodemia Annulatum (Hemiptera) on Mosquito larvae and its prospects as biocontrol agent, Bicovas., 2: 163: 172

Phong Tarun vu, Nabuko Tuno, Histohi Kawada and Mosahiro Takagi (2008). Comparative evolution of fecundity and survivorship of six copepod (copepod, cyclopoidae) species in relation to selection of candidate. Biological control agents against *Ades aegypti*, J. Amer. Mosq. Con. Asso., 24 (1), 61-69

Rao R. (1981). The Anopheles of India, Rev. Edn. New Delhi MRCI (IMRC) Delhi India.pp 1-538

Soni, K., O.P. Saxena, H. N. Khare and R.C. Saxena (2006). Study of some larvicidal indigenous fishes with species reference to biological control of malaria, National J. of Life Science., 3(supp): 537-540

2013, Environmental Technology
Editors: **D.R. Khanna, A.K. Chopra, Gagan Matta, R. Bhutiani & Vikas Singh**
Published by: **DAYA PUBLISHING HOUSE, NEW DELHI**

Pages 275–301

Chapter 32

Environmental Conservation and Management of Water for Future Perspectives

Bina Rani[1], Upma Singh[2] and Raaz Maheshwari[3]

[1] *Department of Engineering Chemistry and Environmental Engineering,
PCE, Sitapura, Jaipur, Rajasthan*
[2] *School of Applied Sciences, Gautam Buddha University,
Greater Noida, U.P.*
[3] *Department of Chemistry, SBDTC, Lakshman Garh – 332 311, Rajasthan*

ABSTRACT

Water is one of the five basic elements from which creation emanates. The evolution of human culture and civilization has evolved around river systems. Water has been described as elixir of life and cleaner of sins. In other words, mankind can't do without water. Only 2.5 per cent of all water in the globe is fresh water, the rest being sea/salt water of which, as per WHO estimates – only 0.007 per cent is readily available for human consumption. The remaining is frozen in glaciers or polar ice caps or is deep within the earth beyond our reach. This indicates that fresh water on the Earth is finite and also unevenly distributed. This is when 70 per cent of our body is comprised of water and water is essential to most our activities. Unfortunately, with a galloping population, urbanization and ever-increasing demand on it, water sources world over are fast depleting. The need for conserving water has therefore become imperative. Global water consumption has raised almost 10 fold since 1900, and many parts of the world are now reaching the limits of their supply. World population is expected to increase by 45 per cent in the next 30 years, whilst fresh water runoff is expected to increase by 10 per cent. UNESCO has predicted that by 2020 water shortage will be serious worldwide problem. 1/3rd of the world's population is already facing crisis due to water shortage and poor portable water quality. Effects include massive outbreak of disease, malnourishment and crop failure. Furthermore excessive use of water has seen the degradation of the environment costing the world billions of dollars. Many of our rivers, wet lands and bays are degraded partly because of water extracted, polluted surface run-off and storm water flushed into them. "Be under no illusions: the impact of general water shortage is going to hit our cities. In the 21st century, wars will be fought over water." – says former UN Secretary General Kofi Anan. According to the International Water Management Institute (IWMI) about 250km3 of water are extracted for irrigation each year in India, of which the rain put back around 100km3 only, resulting in gradual depletion of the aquifers. To cope up with global water scarcity, UN's General Assembly in 1992, declared March 22nd as World Water Day to create awareness of the dilemma amongst individuals and communities. Access to safe water should be a basic human right and it can be ensured by adopting various measures like watershed management through rehabilitation of existing systems like tanks, construction of check dams and Rainwater Harvesting (RWH). Among the different water management techniques, RWH is gaining favour but it's yet to become an essential part of day to day living for which mass awareness and group action necessary can be promoted by different committees, societies, NGOs, self help groups etc. It's true that if each of us uses the water judiciously and augment the water resources by becoming the custodian rather than the absolute owner, the water resources can well be protected for the future mankind. The traditional as well as new strategies of water management and conservation must be implemented wholeheartedly to save our BLUE PLANET – THE MOTHER EARTH.

Keywords: RWH, Artificial recharge, Erosion, Urbanization, Health hazards, Sustainable development, Check dams, Percolation tanks, Nadi-Talab-Johad-Baori-Kund.

Introduction

The importance of water as a vital resource to the life system and an essential component of societal development cannot be overemphasized. Recognising the importance of water resource development many ancient civilizations emphasized on various mechanisms of water appropriation, collection and distribution. In earlier times the state took care of the water supply by developing and maintaining several ingenious and indigenous ways of string rain and floodwaters. The maintenance of water quality and the means of regenerating the natural resource were crucial factors for sustainability, especially in the dry areas. The ancient man relied on water structures like ponds, lakes, tanks, wells, *baodies*, small *kutcha bunds, tankas, kund, khadins, gulls,* and *ahars.* Rainwater was meticulously conserved and stored at various places for irrigation purposes in these water bodies. As early as 300 BC Megasthenes, the Greek ambassador in court of Chandragupta Maurya, mentioned in his memoirs – the whole country is under irrigation and very prosperous because of the double harvests which they are able to reap each year because of irrigation.

The remnants of many of these age-old structures are architectural and engineering marvels and show that these water bodies were carefully developed, maintained and sustained over the ages. These traditional water-harvesting systems remained environmentally viable and sustainable until they were subjected to large-scale abuses as in the recent times. With recurring droughts year after year especially in Rajasthan, AP, Gujarat, MP and Odisha, environmentalists are emphasising the need to revive and revert back to the water harvesting systems which existed in earlier times. Ironically, the States currently experiencing drought have been endowed with several water harvesting systems which have now fallen into disuse due to neglect over the decades. In Rajasthan, tankas, talabs and baodies traditionally performed the jobs of collecting and storing runoff water. The tankas (underground tanks for drinking water) were one of the most reliable methods of water harvesting in desert towns. Its water was used judiciously to avoid shortage in summer. In the event of scarce rainfall, water from nearby talabs, nadis or village ponds was used to fill up the tankas. Rooftop harvesting was a common feature in villages and towns across the Thar Desert. The technique of rooftop harvesting involves collecting rainwater that falls on the sloping hose roofs through a pipe into an underground tanker built in courtyard. The locals created numerous other water bodies, these included johad, bandha, sagar, samand and sarovar. Wells including kua, kohar (owned by a community) and stepwells (baodis or jhalaras) were also important water sources. Stepwell is a unique form of underground well architecture very common in Rajasthan and Gujarat. Recent reports have indicated that most of the houses in Dwarka, Gujarat, are practising rooftop rainwater harvesting and thus the city is in far better condition than the rest of the drought-prone areas in the State,

In the history of irrigation of South India, tanks played a prominent role. At the end of the first 5-year plan AP had 58,527 tanks with an irrigated area of over 26 lakh acres. Many of these tanks were part of a system that helped store and conserve the rain runoff in such a manner that the water overflow from the tanks in the head was collected in lower ends so that no wastage of water took place. The storage tanks enriched the water table through percolation. Tank irrigation suffered a steady decline in the decades following the 1950s. This was mainly due to the British policy of considering small tanks as being un-remunerative and an unnecessary burden on the State. As the first FiveYear Plan emphasised on 'Grow-more-Food', tank beds were given to individual to cultivate more food thereby causing irreparable damage to these water bodies. Traditional water harvesting systems practised in other Stats include the Phad system in MS, Haveli system in MP, Khadin in Rajasthan, and Ahar-pyne in Bihar.

Over the years rising populations, growing industrialisation, and expanding agriculture have pushed up the demand for water. Efforts have been made to collect water by building dams and reservoirs and digging wells. Water conservation has become the need of the day. The idea of groundwater recharging by harvesting rainwater is gaining importance in many cities.

In the forests, water seeps gently into the ground as vegetation breaks the fall. This groundwater in turn feeds wells, lakes, and rivers. Protecting forests means protecting water 'catchments'. In ancient India, people believed that forests were the 'mothers' of rivers and worshiped the sources of these water bodies. The Indus Valley Civilization, that flourished along the banks of the river Indus and other parts of western and northern India about 5,000 years ago, had one of the sophisticated urban water supply and sewage systems in the world. The fact that the people were well acquainted with hygiene can be seen from the covered drains running beneath the streets of the ruins at both Mohenjodaro and Harappa. Another very good example is the well-planned city of Dholavira, on Khadir Bet, a low plateau in the Rann in

Gujarat. One of the oldest water harvesting systems is found about 130km from Pune along Naneghat in the Western Ghats. A large number of tanks were cut in the rocks to provide drinking water to tradesmen who used to travel along this ancient trade route. Each fort in the area had its own water harvesting and storage system in the form of rock-cut cisterns, ponds, tanks and wells that are still in use today. A large number of forts like Raigad had tanks that supplied water.

In urban areas, the construction of houses, footpaths and roads has left little exposed earth for water to seep in. In parts of the rural area of India, floodwater quickly flows to the rivers, which then dries soon after it stops raining. If this water can be held back, it can seep into the ground and recharge the groundwater supply. This has become a very popular method of conserving water especially in the urban areas. Rainwater harvesting essentially means collecting rainwater on the roofs of the building and storing it underground for later use. Not only does this recharging arrest groundwater depletion, it also raises the declining water table and can help augment water supply. Rainwater harvesting and artificial recharging are becoming very important issues. It is essential to stop the decline in groundwater levels, arrest seawater ingress, _i.e._ prevent seawater from moving landward, and conserve surface water run-off during the rainy season.

Town planners and civic authority in many cities in India are introducing bylaws making rainwater harvesting compulsory in all new structures. No water or sewage connection would be given if a new building did not have provisions for rainwater harvesting. Such rules should be implemented in all the other cities t ensure a rise in the groundwater levels. Realizing the importance of recharging groundwater, the CGWB (Central Groundwater Board) is taking steps to encourage it through rainwater harvesting in the capital and elsewhere. A number of government buildings have been asked to go in for water harvesting in Delhi and other cities of India. All we need for a water harvesting system is rain, and a place to collect it. Typically, rain is collected on rooftops and other surfaces, and the water is carried out to where it can be used immediately or stored. We can direct water run-off from this surface to plants, trees or lawns or even to the aquifer.

Some of the benefits of rainwater harvesting are as follows:

- ☆ Increases water availability
- ☆ Checks the declining water table
- ☆ Is environmentally friendly
- ☆ Improves the quality of groundwater through the dilution of fluoride, nitrate, and salinity
- ☆ Prevents soil erosion and flooding especially in urban areas

The most important step in the direction of finding solutions to issues of water and environmental conservation is to change people's attitudes and habits – this includes each one of us. Conserve water because it is the right thing to do. We can follow some of the simple things that have been listed below and contribute to water conservation.

☆ Try to do thing each day that will result in saving water. Don't worry if the savings are minimal – every drop counts! We can make a difference.

☆ Remember to use only the amount of water we actually need.

☆ Encourage the family to keep looking for new ways to conserve water, in and around home.

☆ Make sure that our home is leak-free. Many homes have leaking pipes that go unnoticed.

☆ Do not leave the tap running while we are brushing our teeth or soaping our face.

☆ See that there are no leaks in the toilet tank. We can check this by adding colour to the tank. If there is a leak, colour will appear in the toilet bowl within 30 minutes (Flush as soon as the test is done, since colouring may stain the tank.)

☆ Avoid flushing the toilet unnecessary.

☆ For a group of water-conscious people and encourage friends and neighbours to be a part of this group. Promote water conservation in community newsletters and on bulletin boards. Encourage friends, neighbours and co-workers to contribute.

Extracting the Elixir–Groundwater Development in India

During the past five decades, there has been phenomenal growth of groundwater abstraction structures in India. Their number has increased from 4 million in 1951 to about 18 million in 1998-99 while in the same period irrigation potential created from groundwater has increased from 6 to 30 million hectares. Groundwater extraction however, varies over space. It is intensive in the alluvial Indo-Gangetic plains of Punjab, Haryana, UP, UK and in parts of hard rock terrain of southern states. Although over-exploitation of this resource in pockets of the country has created serious problems, yet a large portion of the available groundwater still remains untapped, particularly in the north-eastern areas, where precipitation is high and the demand for irrigation is low. This is also true of the eastern states where the fragmented nature of land holdings has been a major factor in low development of groundwater usage.

How Much Groundwater Does India Have?

The annually 'replenishable' groundwater resources of the country have been assessed as 432 billion cubic meter (BCM). Keeping aside a basic provision of 71 BCM per year for domestic and industrial use, 361 BCM per year is available for irrigation. The Ganga basin has the highest potential followed by Godavari basin and Brahmaputra basin. In fact the Indo-Gangetic alluvial plain with an area of around 25,000 km^3 is one of the largest groundwater reservoirs in the world. In addition to the 'replenishable' resources we have the 'in storage' groundwater resources. These resources which lie below the lowest level of groundwater fluctuation can be used in extreme situations. The in-storage groundwater resources up to depth of 450 meters in hard rock terrain have been estimated as 10812 BCM.

What Contaminates our Groundwater?

It was way back in mid eighties that a systematic study of groundwater quality and identification of 'problematic zones' was under taken by Central Pollution Control Board (CPCB). Over the years the problem areas that emerged witnessed excessive exploitation of groundwater for domestic and industrial uses. As pollution control enforcement activities gained momentum there were observed cases of indiscriminate waste disposal, subsurface discharge of effluent and inappropriate wastewater management by industries. Today, as a result of these malpractices there is a palpable stress on groundwater, in terms of quantitative imbalances as well as qualities deterioration.

Fluoride in Groundwater

In the world, around 200 million people from 25 nations have great health risks, with high fluoride in the drinking water. in the country (India), almost 60-65 million people drink fluoride contaminated groundwater and the number affected by fluorosis is estimated at 2.5 to 3 million in many states, especially, Rajasthan, AP, Punjab, TN and UP. In India, safe limit of fluoride in potable water is between 0.6 and 1.2ppm (mg/l)[BIS 2003]. Lower limit of fluoride (<0.6ppm) than that of the prescribed limit (0.6ppm) causes dental caries, while higher limit of fluoride (>1.2ppm) than those of the recommended limit (1.2ppm) results in fluorosis. The source of fluoride in groundwater is not only because of a wide spread occurrence of fluoride rich soil in India but also because of the excessive use of phosphatic fertilizer, mining (copper and iron) and allied industries. Fluorides are released into the environment naturally through the weathering and dissolution of minerals, in emissions from volcanoes and in marine aerosols. Fluorides are also released into the environment via coal combustion and process waters and wastes from various industrial processes, including steel manufacture, primary aluminium, copper and nickel production and use, glass, brick and ceramic manufacturing, and glue and adhesive production. The use of fluoride containing pesticides as well as the controlled fluoridation of drinking-water supplies also contributes to the release of fluoride from anthropogenic sources. Based on available data, phosphate ore production and use as well as aluminium manufacture are the major industrial sources of fluoride release into the environment. Other minerals containing fluoride are sellaite (MgF_2), villianmite (NaF), fluorite or fluorospar (CaF_2), cryolite (Na_3AlF_6), bastnaesite [(Ce, La)(CO_3)F] and fluorapatite [$Ca_3(PO_4)_2F$].

Fluoride ingested with water goes on accumulating in bones up to age of 55 years. At high doses fluoride can interfere with carbohydrate, lipid, protein, vitamins, enzymes and mineral metabolism. Long term consumption of water containing 1ppm of fluoride leads to dental fluorosis. White and yellow glistening patches on the teeth are seen which may eventually turn brown. The yellow and white, patches when turned brown present itself has horizontal streaks. The brown streaks may turn black and affect the whole tooth and may get pitted, perforated and chipped off at the final stage. Skeletal problems: Fluoride can also damage the foetus-if the mother consumes water and food with a high concentration of fluoride during pregnancy/breast feeding, infant mortality due to calcification of blood vessels can also occur. The others are: (1) Severe pain in the backbone, hip region and in the joints, (2) Stiffness of the backbone, (3) Immobile/stiff joints, (4) Increased density of bones, besides calcification of ligaments, (5) Construction of vertebral canal and vertebral foramen-pressure on nerves, and (6) Paralysis. Non-skeletal problems: (1) Neurological manifestations: Nervousness, depression, tingling sensation in fingers and toes, excessive thrust. Tendency to urinate frequently (polydypsia and polyurea are controlled by brain-appears to be adversely affected). (2) Muscular manifestations: Muscle weakness, stiffness, pain the muscle and loss of muscle power. (3) Allergic manifestations: Very painful skin rashes, which are perivascular inflammation – present in women and children, pinkish, red or bluish red spots on the skin that fade and clear up in 7-10 days, they are round or oval shape. (4) Gastero-intestinal problems, acute abdominal pain, diarrhea, constipation, blood in stools, bloated feeling (gas) tenderness in stomach, feeling of nausea and mouth sores (5) Headache and (6) Loss of teeth (edentate) at an early age.

Treatment of high fluoride groundwater: treatment with lime and alum is the most common method practised both at the community and the household level but, more recently, activated alumina has come into use. In situ treatment has received much less attention. Alkaline soils can be remediated through the application of gypsum, pyrite or sulphuric acid. On a long-term basis, the planting of trees like *Acacia nilotica, Prosopis juliflora, Albizia lebbek* and *Polpulus deltoids* may alleviate sodicity in soils. Treatment of alkaline soils is usually initiated as a result of problems with soil structure and permeability. As is indicated by the relationship between soil, pH and fluoride in groundwater, lowering alkalinity may also decrease the mobility of fluoride. Gypsum treatment is the classical method of alleviating soil alkalinity. It has advantages in being cheap as gypsum is abundant in India, even in hard rock areas. The gypsum treatment will give harder water. this may be an advantage as that means a higher intake of Ca which, as mentioned earlier, mitigates the effect of fluoride.

GROUNDWATER IN INDIAN CITIES

With twenty five per cent of India's population living in urban areas, Indian cities are posed with the problem of over-exploitation of groundwater. The poor urban infrastructure has no systematic provision of sewage or solid waste management. Unplanned growth, unorganised land-use and poor drainage system further compound the groundwater quality concerns. In fact the very process of urbanization in Indian cities, has led to phenomenal decrease of natural (groundwater) recharge due to paved roads and soil compaction, thus promoting imbalance in the overall groundwater budget.

Arsenic in Groundwater

Arsenic contamination of groundwater is a natural occurring high concentration of arsenic in deeper levels of groundwater, which became a high-profile problem in recent years due to the use of deep tubewells for water supply in Ganges Delta, causing serious arsenic poisoning in large number of people. A 2007 study found that over 137 million people in more than 70 countries are probably affected by arsenic poisoning of drinking water. Arsenic contamination of groundwater is found in many countries throughout the world, including the USA. Approximately 20 incidents of groundwater arsenic contamination have been reported from all over the world. Of these, four major incidents were in Asia, including locations in Thailand, Taiwan, and Mainland China. South American countries like Argentina and Chile have also been affected. There are also many locations in the United States where the groundwater contains arsenic concentrations in excess of the Environmental Protection agency standard 10ppb adopted in 2001. According o a recent film funded by the US Superfund, "In Small Doses", millions of private wells have unknown arsenic levels, and in some areas of the US, over 20 per cent of wells may contain levels that are not safe.

Arsenic is a carcinogen which causes many cancers including skin, lungs, and bladder as well as cardiovascular disease. Some research concludes that even at the lower concentrations, there is still a risk of arsenic contamination leading to major causes of death. A study conducted in a contiguous six-country area of southeastern Michigan investigated the relationship between moderate arsenic levels and 23 selected disease outcomes. Disease outcomes included several types of cancer, diseases of the circulatory and respiratory system, diabetes mellitus, and kidney and liver diseases. Elevated mortality rates were observed for all diseases of the circulatory system. The researchers acknowledged a need to replicate their findings. A preliminary study shows a relationship between arsenic exposure measured in urine and type II diabetes the resulted supported the hypothesis that low levels of exposure to inorganic arsenic in drinking water may play a role in diabetes prevalence. Arsenic in Drinking water may also compromise function "Scientist link influenza A (H_1N_1) susceptibility to common levels of arsenic exposure".

The story of arsenic contamination of groundwater in Bangladesh is a tragic one. Many people have died from this contamination. Diarrhoeal diseases have long plagued the developing world as a major cause of death, especially in children. Prior to the 1970s, Bangladesh had one of the highest infant mortality rates in the world. Ineffective water purification and sewage systems as well as periodic monsoons and flooding exacerbated these problems. As a solution UNICEF and the World Bank advocated the use of wells to tap into deeper groundwater for a quick and inexpensive solution. Millions of wells were constructed as a result. Because of this action, infant mortality and diarrheal illness were reduced by fifty per cent. However, with over 8 million wells constructed, it has been found the last two decades that approximately one in five of these wells is now contaminated with arsenic above government's water standard.

In the Ganges Delta, the affected wells are typically more than 20 m and less than 100 m deep. Groundwater closer to the surface typically has spent a shorter time in the ground, therefore likely absorbing a lower concentration of arsenic; water deeper than 100 m is exposed to much older sediments which have already been depleted of arsenic. Dipankar Chakraborty from West Bengal brought the crises to international attention in 1995. Beginning his investigation in west Bengal in 1988, he eventually published, in 2000, the results of a study conducted in Bangladesh, which involved the analysis of thousands of water samples as well as hair, nail and urine samples. They found 900 villages with arsenic above the government limit. Chakraborty has criticized aid agencies, saying that they denied the problem during the 1990s while millions of tube wells were sunk. The aid agencies later hired foreign experts, who recommended treatment plants which were not appropriate to the conditions, were regularly breaking down, or were removing the arsenic.

Chakraborty says that the arsenic situation in Bangladesh and West Bengal is due to the negligence. He also adds that in West Bengal water is mostly supplied from rivers. Groundwater comes from deep tubewells, which are few in number in the state. Because of the low quantity of deep tubewells, the risk of arsenic patients in West Bengal is comparatively less. According to the WHO, "In Bangladesh, West Bengal (India) and some other areas, most drinking-water used to be collected from open dug wells and ponds with little or no arsenic, but with contaminated water transmitting diseases such as diarrhoea, dysentery, typhoid, cholera and hepatitis. Programmes to provide 'safe' drinking water over the past 30 years have helped to control these diseases, but in some areas they have had the unexpected side-effect of exposing the population to another health problem – arsenic." The acceptable level as defined by WHO for maximum concentrations of arsenic in safe drinking water is 0.01 mg/l. the Bangladesh government's standard is at a slightly higher rate, at 0.05 mg/l being considered safe. WHO has defined the ares under threat: Seven of the nineteen districts of West Bengal have been reported to have groundwater arsenic concentrations above 0.05 mg/l. the local population in these seven districts is over 34 million, with the number using arsenic-rich water is more than 1 million (0.05 mg/l). The number increases to 1.3 million when the concentration is above 0.01 mg/l. according to British Geological Survey (BGS) study in 1998 on shallow tube-wells in 61 of the 64 districts in Bangladesh, 46 per cent of the samples were above 0.01 mg/l and 27 per cent were above 0.05 mg/l. When combined with the estimated 1999 population, it was estimated that the number of people exposed to arsenic concentrations above 0.05 mg/l and the number of those exposed to more than 0.01 mg/l is 46-57 million (BGS 2000).

Throughout Bangladesh, as tubewells get tested for concentrations of arsenic, ones which are found to have arsenic concentrations over the amount considered safe are painted red to warn residents that the water is not safe for drink. The solution according to Chakraborty is "By using surface water and instituting effective withdrawal regulation West Bengal and Bangladesh are flooded with surface water. We should first regulate proper watershed management. Treat and use available surface water, rain-water and others. The way we're doing at present is not advisable."

There are many locations across the United States where groundwater contains naturally high concentrations of arsenic. Cases of groundwater-caused acute arsenic toxicity, such as those found in Bangladesh, are unknown in the United States where the concern has focussed on the role of arsenic as a carcinogen. The problem of high arsenic concentrations has been subject to greater scrutiny in recent years because of changing government standards for arsenic in drinking water. Some locations in the United States, such as Fallon, Nevada, have long been known to have groundwater with relatively high arsenic concentrations (in excess of 0.08 mg/l). Even some surface waters, such as Verde River in Arizona, sometimes exceed 0.01 mg/l arsenic, especially during low-flow periods when the river is dominated by groundwater discharge. A drinking water standard of 0.05 mg/l (equal to 50ppb) arsenic was originally established in the United States by the Public Health service in 1942. The Environmental Protection Agency (EPA) studied the pros and cons of lowering the arsenic Maximum Contaminant Level (MCL) for years in the late 1980s and 1990s. A study of private water wells in the Appalachian mountains found that 6 per cent of the wells had arsenic above the US MCL of 0.010 mg/l.

In Nepal there is a serious problem with arsenic contamination particularly in Terai region, worst being near Nawakparasi District, where 26 per cent of shallow wells failed to meet WHO standard of 10ppb. A study by Japan International Cooperation Agency and the Environment in the Kathmandu Valley, particularly in deep wells, of which 71.6 per cent failed to meet the WHO standard, and 11.9 per cent failed to meet the Nepali standard of ppb.

The first signs of arsenic contamination in India were detected in West Bengal as early as 1988. Today symptoms of arsenicosis are being observed in more and more states. Recently, the groundwater in UP's Ballia district was found to be contaminated with arsenic. The UN's estimate is that currently 35 million people in Bangladesh and India are in danger of drinking arsenic contaminated water. In January 2004, as many as 2,404 samples of water drawn from handpumps in 55 villages in Ballia were tested. More than half the samples had arsenic level above the Indian guideline of 10µg per litre; eight per cent had arsenic levels above 500. The samples of water were tested and analysed under the guidance of Dr Dipankar Chakraborty, Director, School of Environmental Studies (SOES), Jadhavpur University, Kolkata, Dr. Chakraborty is often offered as the 'arsenic hunter'.

The report on arsenic contamination in Ballia district was released by Sunita Narain, Director of Delhi-based Centre for Science and Environment (CSE), at a conference in Delhi recently. Narain said the local administration had promised to send Dr Chakraborty's report to the Industrial Toxicology Research Centre (ITRC) in Lucknow. ITRC had reported in July 2004 that the samples of water they tested from the district were safe, and so the villagers in the area had gone back to drinking water drawn up by handpumps. Having seen obvious symptoms of arsenicosis in the villagers of Ballia, CSE sponsored their own tests on handpump water, as well as blood, nail and hair samples from the area in August 2004. Analysed at the Shri Ram Institute for Industrial Research Laboratory (Delhi), the water samples were found to contain levels of arsenic ranging from 15 to 129 µg per litre – dangerously high considering what the Indian guideline level suggests as safe. Arsenic in groundwater can enter the body by drinking arsenic laced water or by eating food cooked in the water. arsenic does not evaporate in to the air and is not easily absorbed through the skin. Most foods, including vegetables, fish and seafood also contain some arsenic. Studies have shown that drinking water containing elevated levels of arsenic can cause the following health effects:

☆ Thickening and discolouration of the skin, sometimes leading to skin cancer, which may be curable at an early stage.

☆ Digestive problems such as stomach pain, nausea, vomiting and diarrhoea.

☆ Numbness in the hands and feet.

Arsenic laced water is mostly found at an intermediate water depth of 20-100 meter below ground level. Occurrence of 'arseno-pyrite' in the region and the change of geo-chemical environment due to over-exploitation of groundwater and excessive fluctuations of groundwater table, has introduced this dreaded substance into fresh water.

Small Scale Water Treatment

A simpler and less expensive form of arsenic removal is known as the Sono arsenic filter, using 3 pitchers containing cast iron turnings and sand in the fist pitcher and wood activated carbon and sand in the second. It is claimed that thousands of these systems are in use can last for years while avoiding the toxic waste disposal problem inherent to conventional arsenic removal plants. Although novel, this filter has not been certified by any sanitary standards such as NSF, ANSI, and WQA, and does not avoid toxic waste disposal similar to any other iron removal process. In the United States small "under the sink" units have been used to remove arsenic from drinking water. this option is called "point of use" treatment. The most common types of domestic treatment use the technologies of adsorption (using media such as Bayoxide E33, GFH, or titanium dioxide) or reverse osmosis. Ion exchange and activated alumina have been considered but not commonly used.

Coagulation/filtration removes arsenic by coprecipitation and adsorption using iron coagulants. Coagulation/filtration using alum is already by some utilities to remove suspended solids and may be adjusted to remove arsenic. But problem of this type of filtration system is that it gets clogged very easily, mostly within two or three months. The toxic arsenic sludge are disposed of by concrete stabilization, but there is no guarantee that they won/t leach out in future.

Iron oxide adsorption filters the water through a granular medium containing ferric oxide, which has a high affinity for adsorbing dissolved arsenic. The iron oxide medium eventually becomes saturated, and must be replaced. The sludge disposal is a problem here too.

Activated alumina is another filter medium known to effectively remove dissolved arsenic. It has also been used to remove undesirably high concentration of fluoride.

Ion exchange has long been used as a water-softening process, although usually on a single-home basis. It can also be effective in removing arsenic with a net ionic charge. (Note that arsenic oxide, As_2O_3, is a common form of arsenic in groundwater that is soluble, but has no net charge.) But the main advantage is that, the media is pretty much expensive.

Both Reverse osmosis and electrodialysis (also called *electrodialysis reversal*) can remove arsenic with a net ionic charge. Some utilities presently use one of these methods to reduce total dissolved solids and therefore improve taste. A problem with both methods is the production of high-salinity wastewater, called brine, or concentrate, which then must be disposed of.

Subterranean Arsenic Removal (SAR) Technology

In SAR, aerated groundwater is recharged back into the aquifer to create an oxidation zone which can coprecipitate iron and arsenic. The oxidation zone created by aerated water boosts the activity of the arsenic-oxidizing microorganisms which can oxidize arsenic from +3 to +5 state SAR technology. no chemicals are used and almost no sludge is produced during operational stage since iron and arsenic are trapped under the earth. Thus toxic waste disposal and risk of its future mobolizatuion is prevented by this technology. Also, it has very long operational life similar to the long lasting tube wells drawing water from the shallow aquifers. The first community water treatment plant based on SAR technology was set up at Kashinpore near Kolkata in 2004 by a team of European and Indian engineers. Researchers from Bangladesh and the united kingdom have recently claimed that dietary intake of arsenic adds a significant amount of total intake, where contaminated water is used for irrigation.

Arsenic Fixer Identified

Biologists have identified genes in plants that control arsenic accumulation. The study promises to prevent the metalloid from entering the food web and also clean contaminated sites through bioengineered plants. Researchers had been looking for the genes for 25 years. "Identifying them is a crucial step in keeping arsenic from accumulating inthe edible parts of the plants, like rice grains and fruits," stated Julian Shroeder, biologist at Columbia University, USA. As the plant absorbs water, arsenic binds with a plant molecule, phytochelatin, which transports it to vacuoles, a storage structure in plant cells. Eating plant parts that contain arsenic filled vacuoles is harmful. Shroeder's team analysed the genetic material of a yeast, *Schizosaccharomyces pombe,* and found its ABC2 gene controls the activity of the phytochelatin and plays a role in arsenic accumulation in the yeast, the findings were reported in the November, 10, issue of *Journal of Biological Chemistry.*

Shroeder later teamed up with researchers of University of Zurich in Switzerland and found two genes in *Arabidopsis* plant that control transport and storage. Knocking them off will prevent the uptake and accumulation of arsenic in plants. Researchers at the US Geological survey, stated the findings suggest non-food plants can be bioengineered to absorb arsenic from contaminated sites. Rhay can be later burnt to eliminate the metalloid.

Nitrate in Groundwater

Nitrate is a naturally occurring form of nitrogen in soil essential to all forms of life. Most crop plants require large quantities to sustain high yields. The formation of nitrates is an integral part of the nitrogen cycle in our environment. in moderate amounts, nitrate is harmless constituents of food and water. due to its high mobility, nitrate also can leach into groundwater. Although nitrates occurs naturally in some groundwater, in most cases higher levels are thought to result from human activities. Common sources of nitrate include:

- ☆ Fertilizer and manure,
- ☆ Animal feed,
- ☆ Municipal wastewater and sludge,
- ☆ Septic tanks, and
- ☆ N-fixation from atmosphere by legumes, bacteria and lightning

Health Effects of Nitrates

High nitrate levels in water can cause methemoglobinemia or blue baby syndrome, a condition found especially in infants less than six months. The stomach acid of an infant is not as strong as in older children and adults. This causes an increase in intestinal bacteria that can readily convert nitrate (NO_3^-) to nitrite (NO_2^-). Nitrate is absorbed in the blood, and haemoglobin (the oxygen carrying component of blood) is converted to methemoglobin, which does not carry oxygen efficiently. This results in a reduced oxygen supply to vital tissues such as brain. Methemoglobin in infant blood cannot change back to haemoglobin, which normally occurs in adults. Severe methemoglobinemia can result in brain damage and death. Pregnant women, adults with reduced stomach acidity, and people deficient in the enzyme that changes methemoglobin back to normal haemoglobin are all susceptible to nitrite-induced methemoglobinemia. The most obvious symptom of methemoglobinemia is a bluish colour of the skin, particularly around the eyes and mouth. Other symptoms include headache, dizziness, weakness or difficulty in breathing.

Healthy adults can consume fairly large amounts of nitrate with few known health effects. In fact, most of nitrate we consume is from our diets, particularly from raw or cooked vegetables. This nitrate is readily absorbed and excreted in the urine. However, prolonged intake of high levels of nitrate are linked to gastric problems due to the formation of nitrosamines. N-nitrosamine compounds have been shown to cause cancer in test animals. Clinical infant methemoglobinemia was first recognized in 1945. About 2,000 cases were reported in North America and Europe by 1971.fatality rates were reported to be approximately 7 to 8 per cent.

Protecting our Drinking Water

Protecting our drinking water supply from contamination is important for health and to protect property values and minimise potential liability. High nitrate levels often are associated with poorly constructed or improperly located wells. Locate new wells uphill and at least 100 feet away from septic systems, barn yards and chemical storage facilities. Properly seal or cap abandoned wells. Manage non-point sources of water pollution (fields, lawns) to limit the loss of excess water and plant nutrients. Match fertilizer and irrigation applications to precise crop uptake needs in order to minimise groundwater contamination. A recent report by the Central Ground Water Board (CGWB) has revealed nitrate contamination in excess of the admissible limit (45ppm) in water in various parts of country. The report says that nitrate content in excess of the desirable limit from groundwater sources has been noticed in several localised pockets of the country. Actions have been initiated to provide safe water in the affected areas, with allocation of funds for the purpose.

Purification of Contaminated Water

While it may be technically possible to treat contaminated groundwater, it can be difficult, expensive and not totally effective. For this reason, prevention is the best way to ensure clean water. water treatment include distillation, reverse osmosis, ion exchange or blending.

☆ Distillation boils the water, catches the resulting steam, and condenses the steam on a cold surface (a condenser). Nitrates and other minerals remain behind in the boiling tank.

☆ Reverse osmosis forces water under pressure through a membrane that filters out minerals and nitrates. One-half to two-thirds of the water remains behind the membrane as reject water. Higher-yield systems use higher water pressures.

☆ Ion-exchange takes another substance, such as chloride, and trades places with nitrate. An ion exchange unit is filled with special resin beads that are charged with chloride. As water passes over the beads, the resin takes up nitrate in exchange for chloride. As more water pass over the resin, all the chloride is exchanged for nitrate. The resin is recharged by back washing with sodium chloride solution. The back wash solution, which is high in nitrate, must be properly disposed of.

☆ Blending is another method to reduce nitrates in drinking water. mix contaminated water with clean water from another source to lower overall nitrate concentration. Blended water is not safe for infants but id=s acceptable fpr livestock and healthy adults.

Charcoal filters and water softeners do not adequately remove nitrates from water. Boiling nitrate-contaminated water does not make it safe to drink and actually increases the concentration of nitrates. In many cases, the most effective alternative is to use bottled water for drinking and cooking.

Indiscriminate Disposal of Sewage and Garbage

With increasing urbanization, the groundwater pollution due to indiscriminate disposal of untreated sewage and garbage has also required alarming proportions. With 70 to 80 per cent of water supply getting converted into wastewater and limited facility of only 26 per cent for its treatment, further compound the problem, and in many instances led to outbreak of water borne diseases apart from microbial contamination of groundwater.

Salt Water Intrusion

Along about 7000 km long Indian coast line coastal aquifers form a vital source of fresh water. On the other hand, the aquifers being in hydraulic contact with sea are equally vulnerable to contamination due to intrusion of salt water from sea. The intrusion in these areas is caused by concentrated withdrawal of groundwater and reversal of natural hydraulic gradient. The problem has been reported in areas of Saurashtra, TN, AP and WB. It must be clearly understood that less than even 2 per cent of sea water can diminishes water portability. The recommended remedial methods for salt water intrusion include modification of pumping pattern, artificial recharge, physical barrier and hydraulic barrier.

PROTECTING GROUNDWATER

☆ Mapping of vulnerable areas of groundwater depletion and pollution

☆ Notification of critical areas of groundwater

☆ Notification for banning commercial sale of groundwater

☆ Special studies on areas of high concentration of carcinogenic elements in groundwater

☆ Directives in industries/mining/commercial establishments for regulation over withdrawal of groundwater

☆ Environment impact study for groundwater

☆ Campaigns to create public awareness for judicious use and conservation of groundwater

---

Groundwater Management Perspectives

India's groundwater is not in a very good state. The annual recharge o water is far less than what is consumed. The situation is more alarming in urban areas due to population pressure and industrial growth. In spite of this, groundwater constitutes one vital component of water resource system and shall continue to play a key role in meeting the water needs. Hence if requires an integrated approach for groundwater management necessary needs to incorporate following asects.

☆ Water conservation

☆ Watershed management

☆ Conjunctive use surface and groundwater

☆ Augmentation of groundwater by artificial recharge

Apart from above, site specific practice can also be highly purposeful and yield good results. Especially in coastal areas, increasing the direct use of brackish water combined with suitable crop pattern can reduce growing stress on fresh water. In Israel, a similar exercise has yielded impressive results in case of cotton cultivation using brackish water for irrigation.

Rapid industrialization and urbanization coupled with continuous decline in per capita water availability is putting a lot of pressure on the available water resources in the country. As per report of standing sub-committee for assessment of availability and requirement of water for diverse uses in the country (August, 2000) the future water requirement for meeting the demands of various sections in the country for the year 2025 and 2050 have been estimated to be 1093BCM and 1447BCM respectively. The increasing gap between water availability and demand highlights the need for conservation of water. The National Policy 2002 also lays stress on conservation of water. It has been stipulated that efficiency of utilization in all the diverse uses of water should be optimized and an awareness of water as a scarce resource should be fostered. There is need of water conservation. Not only to restore the fast deteriorating eco-system of the country but also to meet the inevitable emergency of the shortage even for domestic and drinking water in near future. The following points are to be pondered upon to plan strategies to meet the crisis:

1. Water is a finite resource and cannot be replaced/duplicated.

2. Water resources are theoretically "renewable" through hydrological cycle. However, what is renewable is only the quality, but pollution, contamination climate change, temporal and seasonal variations have affected the water quality and reduced the amount of, "usable water".

3. As per Ministry of Rural Development, 182 districts (972 blocks) comprising an area of 7,45,914sqkm have been covered under 'Drought Prone Areas Programmes'.

4. About 310 blocks in the country are over-exploited where groundwater is withdrawn more than its replenishment from rainfall.

5. The groundwater levels have declined by more than 4 meters in 40 districts of 16 states in the country during last decade.

6. Rainfall is highly unevenly distributed over time and space in various parts of the country.

7. About 87.2 BCM of surplus monsoon run-off is available in 20 river basins of the country. Out of which 21.4BCM can be re charged to groundwater reservoirs.

8. Increased demand in coastal areas is threatening the fresh water aquifers with sea water intrusion.

9. In inland saline areas, the freshwater is becoming saline due to excessive withdrawal of groundwater.

10. Water conservation practices in urban areas can reduce the demand as much as by one third, in addition to minimizing pollution of surface and groundwater resources.

11. Watershed programmes tended to concentrate on harvesting rainwater through surface structures. There is a need to look at surface and groundwater holistically and prepare a conjunctive use plan.

Action Plan for Water Conservation

Conservation of Surface Water Resources

A large number of dams have been constructed in the country to store rainwater. At the end of IX Plan, 4050 large dams creating live storage capacity of 213BCM have been constructed and 475 large projects are on ongoing, which will add another 76BCM on completion. Projects under consideration will add another 108BCM of storage. All efforts have to be made to fully utilize the monsoon run-off and store rainwater at all probable storage sites. In addition to creating new storages it is essential to renovate the existing tanks and water bodies by desilting and repairs. The revival of traditional water structures should also be given due priority.

Conservation of Groundwater Resources

Groundwater is an important component of hydrological cycle. It supports the springs in hilly regions and the river flow of all Peninsular rivers during the non-monsoon period. For sustainability of groundwater resources it is necessary to arrest the groundwater outflows by

1. Construction of sub-surface dams
2. Watershed management
3. Treatment of upstream areas for development of springs
4. Skimming of freshwater outflows in coastal areas and islands

Rainwater Harvesting

Rainwater harvesting is the technique of collection and storage of rainwater at the surface or in subsurface aquifers, before it is lost as surface run-off. Groundwater augmentation through diversion of rainfall to subsurface reservoirs, by various artificial recharge techniques, has special relevance in India where due to terrain conditions most of the rainwater is lost as flash floods and local streams remain dry for most of the part of the year. Central Groundwater Board (CGWB) has identified an area of about 4.5lakhs sq km in the country, which shows a declining trend in groundwater levels and needs urgent attention to meet the growing needs for irrigation, industry and domestic purpose. It is estimated that in these identified areas of water scarcity, about 36.1BCM of surplus monsoon surface run-off is available which can be fruitfully utilized to augment the groundwater resources. A twin strategy of adopting simple artificial recharge in rural areas like percolation tanks, check dams, recharge shafts, dug well recharge and sub-surface dykes and adopting roof-top rainwater harvesting in urban areas, can go a long way in redeeming the worsening situation of groundwater.

About 2.25 lakhs artificial recharge structures in rural areas and about 37 lakhs Rooftop rainwater harvesting structures in the cities are feasible. The design and viability of various low cost structures have been demonstrated by CGWB by undertaking 174 schemes throughout the country during Ninth Five Year Plan under the Central Sector Scheme "Study of Groundwater Recharge". Rainwater harvesting is to be taken up in a big way to solve the crisis of water scarcity. Uncovered areas, particularly in urban and semi-urban localities, are continuously diminishing due to phenomenal pace of industrialization and urbanization and massive use of concrete all around in the country. This phenomenon is constantly causing reduced scope for percolation of rain waters to the ground during monsoon and thus perpetual reduction in groundwater recharge year after year. With a view to offset this loss in recharge of groundwater there is apparent need for making roof rainwater harvesting mandatory, either through legislation or by promulgating ordinance, for every public as well as private new and existing buildings in urban and semi-urban areas within specified time frame. Apart from this, harvesting of surface run-off in open areas, both public and private, may also need to be encouraged.

Protection of Water Quality

The rapid in density of human population in certain pockets of the country as a result of urbanization and industrialization is making adverse impact on the quality of both surface and groundwater. Demand for water is increasing on one hand and on the other hand the quantity of "utilizable eater resources" is decreasing due to human intervention in the form of pollution of fresh water. Thus the protection of existing water resources from pollution is a very vital aspect of water conservation.

Cleaning Up of Polluted Rivers, Lakes and Water Bodies

Rivers, lakes, ponds and water bodies are the main sources of water on which a civilization grow and develop. Water bodies get polluted as a result of human interference and unplanned development activities. The main reason for pollution is discharge of untreated domestic and municipal waste and also the industrial waste. The cleaning up of these water bodies is of utmost importance to provide water supply to the population on the one hand and on the other hand to maintain the environment to the desired level. The action points in this regards are as follows:

1. To control and check the flow of pollution to the rivers, lakes and ponds through appropriate measures/action.
2. Treatment of effluent up to the appropriate standard before discharging the river.
3. Proper maintenance and uninterrupted operation of the sewage treatment plant.
4. System of incentive and disincentive for discharging pollutants/untreated waste into the rivers.
5. Adopting remedial measures in the particular river stretch where the problem is acute.
6. Adopting appropriate technology for removal of pollution from lakes and reservoirs.
7. Declaring particular site/location as water heritage site and adoption by different organizations/departments for maintaining the same to the desired standard.

On account of continuous discharge of industrial effluents in water bodies like rivers, canals, lakes, ponds, etc. and contamination of groundwater aquifers with polluted waters, these water bodies at places have become polluted to an enormous extent and apparently huge financial resources are needed for decontaminating them. This suggests for taking stringent measures like imposition of huge penalty for abusing such water bodies, cancellation of license or permission for operation of water polluting industrial units. Pollution control boards at Central and State levels may be provided legal powers through legislation to deal with such delinquent agencies and industrial units. Sensitizing general public and involvement of non-governmental organizations with requisite experience and interest in implementation of legislation for control of pollution of water bodies may also prove useful and effective. Media has also a very vital role to enact by way of highlighting lapses on the part of individuals and industrial units. Traditionally, in India, rivers are revered as Goddess. With time, such a feeling has started diluting. People particularly young generation, may be inculcated to bestow respect to rivers and other water bodies to strengthen this traditional belief of sacred status of rivers and streams and maintain their aesthetic values through mass awareness,

Groundwater Protection

Groundwater resources are getting polluted at an alarming pace due to lack of proper wastewater and sewerage disposal system in urban areas. The application of excessive fertilizers in agriculture sector and disposal of hazardous effluents from the industries are putting great strain on availability of fresh water. the action points to safeguard the water bodies may be as follows:

1. Use of organic fertilizers should be encouraged to protect groundwater from pollution due to excessive use of chemical fertilizers. Groundwater vulnerable zones may be identified by preparing vulnerability maps for physical, chemical and biological contaminants for the whole country.
2. Notification on banning industries, landfills and disposal sites of industrial effluents and sewerage, which are hazardous to groundwater aquifer systems.
3. Devising groundwater solute transport model for contaminants plum migration studies.
4. Research and development studies for corrective action techniques on polluted aquifers.

Recent Attempts for Water Conservation

1. The MP Government implemented watershed development under Rajiv Ghandi Watershed Mission. In the initial phase beginning with 1994, the objective of the programme was to arrest degradation of resources that were critical to peoples' livelihood. The programme evolved over a period of time and culminated in the year 2001 as "Pani Roko Abhiyan". It was a peoples' movement, which was backed by financial commitment and technical support of the Government. The resources available from the Government for drought relief were placed at the disposal of the community, which took up programmes for water harvesting and water conservation in a decentralized manner. The programme was so successful that 14 districts, which were not covered under the drought relief programme in 2001, were also enabled through the banking channels to take up "Pani Roko Abhiyan".
2. PRADHAN, a NGO has adopted this strategy in Jharkhand. Under the Indo–German Bilateral Watershed Project, it seeks to promote livelihood improvement through water harvesting. It is an innovative and simple technique of collecting rainwater in a two-meter deep pit in 5 per cent of the total area of the plot. PRADHAN provided assistance and guidance to the villagers for construction of farm tanks. The farmers have been able to

harvest two crops from the same land due to availability of the water in the field tank. It would appear that harvesting of water in every plot could be a possible option for rainwater conservation. This, however, ignores the close link between stabilization of water channels, even in the plain areas. This problem would be more acute in areas with higher slopes as without treating the upper reaches the soil erosion would continue to be rampant and would require de-silting of ponds every second or third year.

3. The Government of Gujarat has adopted rainfall harvesting technique by constructing small dams in water deficit areas. Farmers are also forward to adopt this methodology. Other states may undertake similar water conservation measures.

Action Points for Water Conservation

An important of component of water conservation involves minimizing water losses, prevention of water wastage and increasing efficiency in water use. "Resource saved is resource created" should be kept uppermost in mind. The action points towards water conservation in different sectors of water use are listed below:

Irrigation Sector

Important action points towards water conservation in the irrigation sector are as follows:

☆ Performance improvement of irrigation system and water utilization;

☆ Proper and timely system maintenance;

☆ Rehabilitation and restoration of damaged/and silted canal system to enable them to carry designed discharge;

☆ Rehabilitation and restoration of damaged/and silted canal systems to enable them to carry designed discharge;

☆ Selective lining of the canal and distribution systems, on technoeconomic consideration, to reduce seepage losses;

☆ Registration/provision of appropriate control structures in the canal system with efficient and reliable mechanism;

☆ Conjunctive use of surface and groundwater to be restored to, specially in the areas where there is threat to water logging;

☆ Adopting drip and sprinkler systems of irrigation fof crops, where such systems are suitable;

☆ Adopting low-cost innovative water saving technology;

☆ Renovation and modernisation of existing irrigation systems;

☆ Preparation of realistic and scientific system operation plan keeping in view the availability of water and crop water requirements;

☆ Execution of operation plan with reliable and adequate water measurement structures;

☆ Revision of cropping pattern in the event of change in water availability;

☆ Utilization of return flow of irrigation water through appropriate planning;

☆ Impairing training to farmers about consequences of using excess water for irrigation;

☆ Rationalization of water rate to make the system self-sustainable;

☆ Formation of Water Users Associations and transfer of management to them;

☆ Promoting multiple use of water;

☆ Introducing night irrigation practice to minimize evaporation loss;

☆ In arid regions crops having longer root such as linseed, berseem, lucerne guar, gini grass, etc. may be grown as they can sustain in dry hot weather;

☆ Assuming timing an optimum irrigation for minimizing water loss and water logging;

☆ Introducing rotational cropping pattern for balancing fertility of soil and natural control of pests;

☆ Modern effective and reliable communication systems may be installed at all strategic locations in the irrigation command and mobile communication systems may also be provided to personal involved with running and maintenance of systems. Such an arrangement will help in quick transmission of messages and this in turn will help in great deal in effecting saving of water by way of taking timely action in plugging canal breaches., undertaking repair of systems and also in canal operation particularly when water supply is needed to be stopped due to sudden adequate rainfall in the particular areas of the command;

☆ With a view to control over irrigation to the field on account of un-gated water delivery systems, all important outlets should be equipped with flow control mechanism to optimize irrigation water supply;

☆ As far as possible with a view to make best use of soil nutrients and water holding capacity of soils, mixed cropping such as cotton with groundnut, sugarcane with black gram or green gram or soya bean may be practiced;

☆ It has been experienced that with scientific use of mulching in irrigated agriculture, moisture retention capacity of soil can be increased to the extent of 50 per cent and this in turn may increase yield up to 75 per cent.

Domestic and Municipal Sector

Important action points for water conservation in domestic and municipal sector are as under:

☆ Action towards reduction of losses in conveyance;

☆ Management of supply through proper meter as per rational demand;

☆ Intermittent domestic water supply may be adopted to check its wasteful use;

☆ Realization of appropriate water charges so that the system can be sustainable and wastage is reduce;

☆ Creation of awareness to make attitudinal changes;

☆ Evolving norms for water use for various activities and designing of optimum water supply system accordingly;

☆ (a) Modification in design of accessories such as flushing system, tap, etc. to reduce water requirement to optimal level;

(b) Whenever necessary, BIS code may be revised;

☆ (a) Possibility for recycling and reuse of water for purpose like gardening, flushing to toilets, etc. may be explored;

(b) Wastewater of certain categories can be reused for other activities as per feasibility;

☆ Optimum quantity of water required for waste disposal to be worked out;

☆ In public building the taps, etc. can be fitted with sensors to reduce water losses.

Industrial Sector

Important action points for water conservation in industrial sector are given below;

☆ Setting-up norms for water budgeting;

☆ Modernization of industrial process to reduce water requirement;

Table 32.1: Typical Use of Water

Drinking	4 per cent
Cooking and other kitchen uses	8 per cent
Personal hygienic	29 per cent
Washing cloths	10 per cent
Toilet flushing	39 per cent
House cleaning/gardening, etc.	10 per cent

☆ Recycling water with a re-circulating cooling system can greatly reduce water use by using the same water to perform several cooling operations;

☆ Three cooling water conservation approaches are evaporative cooling, ozonation and air heat exchange. The ozonation cooling water approach can result in a five-fold reduction in blow down when compared to traditional chemical treatment and should be considered as an option for increasing water saving in a cooling tower;

☆ The use of de-ionized water in reusing can be reduced without affecting production quantity by eliminating some plenum flushes, converting from a continuous flow to an intermittent flow system and improving control of the use.

☆ The reuse of deionised water may also be considered for other uses because it may still be better than supplied municipal water;

☆ The wastewater should be considered for use for gardening, etc.

☆ Proper processing of effluents by industrial units to adhere to the norms for disposal;

☆ Rational pricing of industrial water requirement to ensure consciousness/ action for adopting water saving technologies.

Table 32.2: Water Usage and its Saving

What we Do?	What Should be Done?	Saving of Water
Bathing with shower: 100 litre	Bathing with bucket: 18 litre	82 litre
Bathing with running water:40 litre	Bathing with bucket: 18 litre	22 litre
Using old style flush: 20 litre	Using new style flush: 6 litre	14 litre
Shaving with running water: 10 litre	Shaving by taking water in a mug: 1 litre	9 litre
Brushing teeth with running water: 10 litre	Brushing teeth by taking water in mug: 1 litre	9 litre
Washing clothes with running water: 116 litre	Washing clothes with bucket: 36 litre	80 litre
Washing car with running water: 100 litre	Washing car with wet cloth: 18 litre	82 litre
Washing clothe with running water (15'x10') 50 litre	Washing floor with wet cloth: 10 litre	40 litre (per 150 aq kt area)
Washing hands with running water tap: 10 litre	Washing hands with mug: 0.5 litre	9.5 litre

Regulatory Mechanism for Water Conservation

Groundwater is an unregulated resource in our country with no price tag. The cost of construction of a groundwater abstraction structure is the only investment. Unrestricted withdrawal in many areas has resulted in decline of groundwater levels. Supply side management of water resources is very important for conserving this vital resource for a balanced use. An effective way is through energy pricing restriction on supply and proving incentives to help in conservation of water. Action plan, in this regard, may include the following:

☆ Rationalizing pricing policy of water in urban and rural area. Industries should be discouraged to exploit groundwater with high price slabs;

☆ Restriction on new construction of groundwater structures in all the over exploited and dark blocks of the country;

☆ Metering of all groundwater abstraction structures;

☆ Controlled supply of electricity and downsizing of pump capacity in rural areas;

☆ Regulating the water trading or selling;

☆ Providing incentives for adorption of rainwater harvesting;

☆ Modification in building bye-laws in urban areas to make it mandatory to adopt rainwater harvesting;

☆ Action has been initiated by Delhi, AP, Gujarat, Haryana, Karnataka, Kerala, MS, Rajasthan, TN and UP in this respect. Other States are required to take initiatives in this regard.

Mass Awareness

Water conservation is a key challenge, which requires public participation. Mass awareness on the need for water conservation and providing common tips to effectively participate in this important mission is need of the time. The simple domestic purpose and how to save water under this sector, as given below, may help in creating awareness. Electronic and print media, posters, stickers, handbills, etc. may be used liberally to inculcate sense of responsibility and belongingness for precious natural resource water among various sectors of society. Small documentary films, in regional languages, on importance of water and techniques adopted for water saving and water conservation may be telecast periodically from regional television channels to create awareness among countrymen particularly people living in rural areas.

Tips for Conserving Water for Domestic and Municipal Use

☆ Timely detection and repair of all leaks;

☆ Turning off water tap while brushing teeth;

☆ Use of mug rather than running water for saving;

☆ Avoiding/minimizing use of shower/bath tub in bath room;

☆ Turning off faucets while soaping and rinsing clothes;

☆ Avoiding use of extra detergent in washing clothes;

☆ Using automatic washing machine only when it is fully loaded;

☆ Avoiding use of running water while hand-washing;

☆ Avoiding use of running water for releasing ice tray ahead of time from freezer;

☆ Using smaller drinking glasses to avoid wastage;

☆ Using overflow stop valve in the overhead tanks to check over flow of water;

☆ Turning off the main valve of water while going outdoor;

☆ Avoiding use of hose for washing floors; use of broom may be preferred;

☆ Minimizing water used in cooling equipment by following manufacturer's recommendations;

☆ Watering of lawn or garden during the coolest part of the day (early morning or late evening hours) when temperature and wind speed are the lowest. This reduces losses from evaporation;

☆ Avoiding use of excess fertilizers for lawns in view of the fact that application of fertilizer increases the requirement of water in addition to polluting the groundwater;

☆ Planting of native/or drought tolerant grasses, ground covers, shrubs and trees. Once established, they do not need to be watered as frequently and they usually survive a dry period without much watering;

☆ Grouping of plants based on water needs while planting them;

☆ Turning off water tap so as to use full water available in hose;

☆ Avoiding over watering of lawns. A good rain eliminates the need for watering for more than a week;

☆ Setting sprinklers to water the lawn or garden only, not the street or sidewalk;

☆ Avoiding installation or use of ornamental water features unless they recycle the water and avoiding running them during drought or hot weather;

☆ Installation of high-pressure, low-volume nozzles on spray washers;

☆ Replacement of high-volume hoses with high-pressure, low volume cleaning systems;

☆ Equipment spring loaded shutoff nozzles on hoses;

☆ Installation of float-controlled valve on the makeup line, closing filling line during operation, provision of surge tanks for each system to avoid overflow;

☆ Adjusting flow in sprays and other lines to meet minimum requirements;

☆ Washing vehicles less often, or using commercial car wash that recycles water.

In case of big establishments like hotels, large offices and industrial complexes, community centres, etc. dual piped water supply may be insisted upon. Under such an arrangement one supply may carry fresh water for drinking, bathing and other human consumptions whereas recycled supply from second line may be utilized for flushing out human solid wastes. Similarly, water harvesting through storming of water run-off including rainwater harvesting in all new buildings on plots of 100 sq m and above may be made mandatory.

Tips for Conserving Water for Industrial Use

☆ Using fogging nozzles to cool product;

☆ Installing in-line strainers on all spray headers; regular inspection of nozzles for clogging;

☆ Adjusting pump cooling and water flushing to the minimum required level;

☆ Determining whether discharge from any one operation can be substituted for fresh water supply to another operation;

☆ Choosing conveying systems that use water efficiently;

☆ Handling waste materials in a dry mode whenever possible;

☆ Replacing high-volume hoses with high-pressure, low volume cleaning systems;

☆ Replacing worn-out equipments with water-saving models;

☆ Equipping all hoses with spring loaded shutoff nozzles – it should be ensured that these nozzles are not removed;

☆ Instructing employees to use hoses sparingly and only when necessary;

☆ Turning off all flows during shutdowns unless flows are essential for cleanup – using solenoid valves to stop the flow of water when production stops (the valves could be activated by trying them to drive motor controls);

☆ Adjusting flows in sprays and other lines to meet minimum requirements;

☆ Instead of hosing, sweeping and shovelling may be practiced;

☆ Making an inventory of all cleaning equipments, such as hoses in the plant – determining how often equipment are used and whether they are water-efficient;

☆ Washing cars, trucks and bus fleets less often;

☆ Driveways, loading docks, parking areas or sidewalks with water may be avoided – using sweepers and vacuum may be considered;

☆ Avoiding run-off and making sure that sprinklers cover just the lawn or garden, not sidewalks, driveways, or gutters;

☆ Watering on windy days may be avoided as far as practicable.

It is imperative that users from all sectors of water use, stakeholders including state and central governments, agencies, institutions, organizations, NGOs, municipalities, village panchayats, public sector undertakings and other such bodies directly or indirectly involved in planning, development and maintenance of water resources projects and providing services to the users, may need to be involved for making integrated and continuous efforts for creating mass awareness towards importance of saving and conservation of water, and duties and responsibilities of individuals as well as of organizations and institutions towards judicious and optimal use of water.

Water Users' Association (WUA) and Legal Empowerment

Water Users' Association, though relatively a new concept in the country but is prevalent in some states in irrigation sector. It is considered that involvement of farmers in water management will facilitate equitable and judicious allocation of irrigation waters among farmers of head, middle and tail reach and improve collection of water charges, irrigation projects may not languish for maintenance for want of funds and in this way overall efficiency of irrigation system will improve. This will help saving of water and optimum utilization of water. Such a concept *i.e.*, involvement of users in the distribution and management process may also be extended in domestic and industrial sectors of water use. It has been observed that adverse situation of water supply to domestic sector, when supply is not adequate to meet demand, some residents use water pumps in water supply lines to boost supplies in their dwellings and thereby causing hardship to other residents of the locality. Illegal tapping of water from supply lines or lifting water of canals are also prevalent at places. It has also been observed that inhabitants, in general, are less sensitive to leakage or water loss from the system. Similarly in case of industrial sector, it is not very uncommon to discharge untreated or partially treated industrial effluents in water bodies like rivers, lakes, ponds, canals, etc. including groundwater aquifers. WUA in domestic and industrial sectors of water use may address these issues and may help in conservation of water and control pollution of water bodies from industrial pollutants. WUA may be duly empowered through legislation or promulgation of ordinance to punish errant water users.

(*Source*: Central Water Commission: http://cwc.gov.in/
Acts_laws_rules_guidelines.htm)

2013, Environmental Technology   *Pages 303–309*
Editors: D.R. Khanna, A.K. Chopra, Gagan Matta, R. Bhutiani & Vikas Singh
Published by: DAYA PUBLISHING HOUSE, NEW DELHI

Chapter 33

Application of Nanotechnology in Medical Fields

Akhilesh Kumar

Department of Physics,
Govt. P.G. College, Rishikesh – 249 201, Dehradun, Uttarakhand

ABSTRACT

Nanotechnology is an extremely powerful emerging technology, which is expected to have a substantial impact in medical fields now and in the future. The potential novel applications on disease diagnosis, therapy, and prevention is foreseen to change health care in a fundamental way. Furthermore, therapeutic selection can increasingly be tailored to each patient's profile. This paper presents the promising nanotechnology approaches for medical technology. In particular, in surgery, cancer diagnosis and therapy, biodetection of disease markers, molecular imaging, implant technology, tissue engineering, and devices for drug, protein, gene and radionuclide delivery. Many medical nanotechnology applications are still in their initial stage. However, number of products is currently under clinical investigation and some products are already commercially available, such as surgical blades and suture needles, contrast-enhancing agents for magnetic resonance imaging, bone replacement materials, wound dressings, anti-microbial textiles, etc.

Introduction

Nanoparticles, a unique subset of the broad field of nanotechnology, include any type of particle with at least one dimension of less than 100 nm [1]. The word nano comes from the Greek word "nanos" meaning dwarf. The compound term "nano" is the factor 10^{-9} meter or one billionth. In case of semiconductor, the artificially made semiconductor structure shows a surprising variety of new interesting properties, which are completely different from the solid state bulk material after reduce its dimension. These size effects [2] can be observed, when the average size of the crystalline grains does not exceed 100 nm and evident most vividly for grain sizes smaller than 10 nm. Polycrystalline materials with an average grain size below 100 nm are called nanocrystalline materials. These materials represent a special state of matter, which consists of matter ensembles of extremely small particles with size of few nanometers. In nanomaterials with grains sizes ranging from 100 down to 10 nm, the surface contains 10 per cent to 50 per cent of all atoms of the nano crystalline solid [3]. When, we reduce size of material from bulk to nano, its physical, chemical, structural, optical, biological, thermal and electrical properties changes.

Nobel Laureate Physicist Richard P. Feynman in 1959 [1] delivered an inspirational lecture in American Physical Society. He emphasized the importance of the materials and devices on nanoscale ranges for future applicability. He categorically emarked *that the physical laws, in principle, are not against the possibility of manipulating things atom by atom but it is not done, in practice, because we are too big.* This lecture received most attention by the investigators in physical sciences and stimulated the research activities in Physics, Material Science and Microelectronics disciplines. Interestingly, in 1974, NorioTaniguchi [4] Tokyo Science University Professor, coined the term Nanotechnology, while manufacturing materials with nanometric tolerances.

Nanotechnology and Nanoscience has encompassed all walks of life such as science, society, engineering and industry. Eric Drexler [5, 6] used the term nanotechnology from Taniguchi's work and wrote a book in 1986 entitled *'Engines of Creation: The coming era of Nanotechnology'*. Richard-E Smalley discovered and characterize C_{60} (Bucky Ball) having shape of soccer ball. *Fullerene is general symbolic name of any cage –like, hollow molecules composed of hexagonal and pentagonal group of carbon atoms. C_{60} along with other fullerenes such as C_{70}, constitute the third elemental form of carbon, after graphite and diamond.* Smalley's work triggered researches on buckytubes, elongated tubes, commonly known as Carbon Nano Tubes (CNTs). These CNTs are finding applications in nearly every technology, where electron conduction dynamics are used. In Feb. 2000, findings on buckytubes led to the start up of a new company, *Carbon Nanotechnologies, INC.*

Since, design and fabrication of devices in micrometer range led to the microelectronics revolution in 20th century, *it is hoped that the design and fabrication of the same in nanometric range would revolutionize the 21st century.* The scientific knowledge acquired by investigators are extensively applied to the areas like Electronics, Bioinformatics, Photonics, Biology and Molecular Electronics.

The interaction of nanotechnology with different field of sciences has generated terms such as Nano-Electronics, Nano-Photonics, Nano-Biotechnology, Nano-Medicine and Nano-Bioengineering etc.

Methodology

Nanotechnology is an interdisciplinary research field, several methods are used in fabrication of nanomaterials such as *Physical* (mechanical and vapor deposition techniques), *Chemical* (Colloids, Sol-gel, L-B films etc.), *Biological* (Biomembranes, DNA etc) and *Hybrid* (Electrochemical, CVD etc). *The guiding principle for the technique, to be employed for synthesis, depends largely upon,* (a) material of interest, (b) type of nanostructure (oD, 1D, 2D, 3D), and (c) size and quantity etc.There are two basic approaches for manufacturing nanomaterials, (i) Top down approach (ii) Bottom up approach. In top-down approach, we reduce the size of material upto nano size, where as in bottom up approach, we start from atomic size and collecting atoms to nano size.

Among various methods for fabrication of nanomaterials the 'WET CHEMICAL METHOD is the simplest one because of following considerations:

1. Inexpensive, least instrumentation required as compared to Physical Methods,
2. Moderate temperature (< 350°C) is required for most of nanomaterials,
3. Dopant can be easily inserted during synthesis process,
4. Nanomaterials of different size and shapes are possible. Materials so obtained can be dried to obtain powder and thin films.

Medical Applications of Nanomaterials

In Surgery

The performance of surgical blades can be enhanced significantly when microstructured hard metal is coated with diamond and processed. Major advantages of the diamond nano-layers in this application are low physical adhesion to materials or tissues and chemical/biological inertness [6]. New suture needles for ophthalmic and plastic surgery are made of stainless steel incorporating nano-size particles (1-10 nm quasi crystals) by using thermal ageing techniques. Such needles coated with nanoparticles have good ductility, exceptional strength, and corrosion resistance. Nanotechnology will provide new tools for medicine. It could radically change the way surgery is done. It will make it possible to do molecular scale surgery to replace defective cells, repair and rearrange cells. Since disease is the result of physical disorder, misarranged molecules and cells, medicine at this level should be able to cure most diseases. Mutations in DNA could be repaired and cancer cells [7], toxic chemicals and viruses could be destroyed through the use of medical Nano devices. Ultra-short pulse laser can perform extremely precise surgery and cut nano-sized cell structures as has been shown in nerve cells. Usually, conventional lasers first heat the target area, and then cut it, but this increases the risk for tissue damage. The advantage of "nanoscissors" is its ability to cut cell organelles without harming surrounding tissue. The technique uses a series of low-energy femto second (femto = 10^{-15}) near-infrared laser pulses. Nanotechnological tools such as optical tweezers used for cell manipulation and immobilisation, "nanoscissors" for cell compartment restructuring, and nano needles used as external communicators to deliver/withdraw substances have been proposed to enable a novel "lab-in-a-cell" concept that could eventually lead to cell surgery.

Nanoparticles for Cancer Therapy

Nanotechnology has the power to radically change the way cancer is diagnosed, imaged and treated [8]. Currently, there is a lot of research going on to design novel

Nano devices capable of detecting cancer at its earliest stages, pinpointing its location within the body and delivering Anticancer drugs specifically to malignant cells. The fundamental challenges in cancer chemotherapy are its toxicity to healthy cells and drug resistance by cancer cells. In cancer therapy anti-angiogenesis therapy is an elegant concept based on the starvation of tumour cell by impairment of blood supply. However, lack of oxygen prompt tumour cells to release a cell signaling molecule known as hypoxia-inducible factor-1α which triggers metastasis and the development of resistance to further chemotherapy [9]. Nanoscale devices smaller than 50 nanometers can easily enter most cells, while those smaller than 20 nanometers can transit out of blood vessels. As a result Nanoscale devices can readily interact with Biomolecules on both the cell surface and within the cell [10].

Biopharmaceuticals Coated with Nanoparticles

Biopharmaceuticals are peptides or protein molecules that trigger multiple reactions in the human body. They are extensively used in the treatment of life – threatening diseases like cancer. The effectiveness [11] of Biopharmaceuticals can be increased many times by coating them with nano particles, which will proficiently deliver the peptides or proteins at the tumor site and in this manner cure cancer without causing extensive damage to the adjacent tissues and organs.

Neuro-surgery

Neuro–electronic devices are unique machines based on nanotechnology that connect the nervous system with the computer. These devices detect and interpret the signals from the nervous system, also control and respond to them. They can be used in the treatment of diseases that slowly and steadily decay the nervous system like multiple sclerosis.

Pacemakers and Hearing Aids

New pacemakers and hearing aids will be equipped with nano sensors that uses nanomaterials to sense, also employ nano electronics technology, *i.e.* Spintronics. Application of Spintronics in pacemakers will enable non-invasive high-speed communication. For hearing aids these tiny sensors will automatically adjust to accommodate the source of sounds.

Nano Nephrology

This branch of nanomedicine is concerned with the detection and treatment of kidney diseases. In this various devices based on nanotechnology are being used for studying various kidney processes and detecting their disorder. After that, nanoparticles and drug delivery system are being used for curing the disorder.

Bone Replacement Materials

Hydroxyapatite nanoparticles used as bone replacement material, *i.e.* bone cement with improved mechanical properties, is commercially available. Indications for bone replacement materials are bone fractures, periprosthetic fractures during hip prosthesis revision surgery, acetabulum reconstruction, osteotomies, filling cages in spinal column surgery, and filling in defects in children.

In Orthopaedics

Various nanotechnology based instruments and nanomaterials are being used in orthopedics such as carbon nanofibres are used for bone replacement which is stronger than steel. Another way to improve the performance of orthopaedic/dental implants can be achieved by modification of the surface roughness, specifically by creating nanometer-scale roughness. Cell responses might be triggered by changes in surface roughness, *i.e.* in horizontal as well as vertical direction, in the nanometer domain (<100 nm) rather than on submicron scale (>100 nm).

Nanorobots

The entry of Nanorobots will literally revolutionize the World of Medicine. These miniature devices are only be capable of entering into the body detecting the diseases and infection, but also they will be capable of repairing internal injuries and wounds. Nanotechnology is the science of maneuvering the structure and properties of matter at an atomic and molecular scale. As a result of this maneuvering, the properties of matter change dramatically *i.e.* the inert elements start to function as catalyst and insoluble matter develop unique solubility capacity. Likewise, non-colloids begin exhibiting excellent colloidal properties and electrical non-conductors start conducting electricity. Owing to their size and properties, nanomaterials are extensively used for the treatment of a number of diseases. Cancer is such a disease where Nanotechnology can play a significant role.

Biosensors/Biodetection

A nano biosensor or nano sensor is a biosensor that has dimensions on the nanometer size scale. Nano sensors could provide the tools to investigate important biological processes at the cellular level. Two types of nano sensors with medical application possibilities are cantilever array sensors and nano tube/nano wire sensors. In micro fabricated cantilever array sensors are used as ultra-sensitive mechanical sensors converting (bio) chemical or physical processes into a recordable (electrical) signal in micro electromechanical systems (MEMS) or nano electromechanical systems (NEMS). Cantilevers are typically rectangular-shaped silicon bars. The unique feature of micro cantilevers is their ability to undergo bending due to molecular adsorption or binding induced changes in surface tension. The major advantages of such miniaturized sensors are their small size, fast response times, high sensitivity, and direct transduction without the need for any labels. Nanotube based biosensors can be used for blood glucose monitoring, DNA detection, etc.

In Magnetic Resonance Imaging

Various magnetic nanoparticles [12] as magnetite (Fe_3O_4), magnetite (γ-Fe_2O_3) or other insoluble ferrites, iron oxides etc are being used in MRI for fast response and better contrast.

Microchip-Based Drug Delivery Systems

Microchip-based drug delivery systems are devices incorporating micrometer-scale pumps, valves, and flow channels and allow controlled release of single or multiple drugs on demand. Micro- and nanotechnology-based methods (*e.g.*, UV-photolithography, reactive ion etching, chemical vapor deposition, electron beam evaporation) can be used for the fabrication of these silicon-based chips.

Nanoshells

Destruction of solid tumors using high heat has been in investigation for some time. Some thermal therapies include the use of laser light, focused ultrasound and microwaves. The advantage of using this is that most procedures are non-invasive, relatively simple and have the potential to treat tumors where surgery is impossible. However to reach underlying tumors, the energy sources has to penetrate healthy tissues, often destroying healthy tissue [13]. Nanoshell–Assisted Photo Thermal therapy (NAPT) is a simple, non invasive procedure for selective photo-thermal tumour removal. It makes use of nanoshells that absorb light in the Near InfraRed (NIR) region. Nanoshells [14-15] have a core of silica coated with an ultra-thin metallic layer, normally gold. By adjustment of core and shell thickness, nanoshells can be designed to absorb and scatter light at a desired wavelength. Nanoshells for cancer therapeutic purposes have been designed to have a peak optical absorption in the NIR region, as this is the wavelength that optimally penetrates tissue. The metal shell converts the absorbed light into heat with great efficacy and stability. Due to their small size, nanoshells are preferentially concentrated in cancer cells by EPR or enhanced permeation retention. By supplying a NIR light from a laser, the particle heats up and kill the tissue. It was found that the temperature within the nanoshell-treated tumors rose by about 40°C compared to a rise in 10°C in tissues that was treated with NIR light alone. Thus using a NIR laser, cancer tissue can be destroyed by local thermal heating around the nanoshells.

Conclusion

Nanotechnology will change the way to diagnose, treat and prevent cancer to meet the goal of eliminating suffering and death from cancer. Nanotechnology can provide the technical power and tools that enables development of new diagnostics, therapeutics and preventives. With Nanomedicine, we will be able to stop cancer even before it develops. Nanomedicine has the ability to improve health care by leaps and bounds. It has positive impact to people from all walks of life. Nanotechnology improves cancer treatment in terms of efficiency and quality and also helps in the process of understanding cancer as a disease process. Nanomedicine is a powerful and revolutionary development that has significant impact on society, the economy and life in general. Nanotechnology is a very important tool for medical sciences.

References

1. `Feynman R. There's plenty of room at the bottom. Science. 1991, **254**: 1300–1301.

2. R Tripathi, A Kumar and T P Sinha, *Pramana J. Phys.* Vol. 72, No.6, June 2009, pp. 969-978.

3. Ramna Tripathi, Akhilesh Kumar, Chandrahas Bhati and T. P. Sinha, *J. Current Appl. Phys.*, Vol. 10, 2010, pp. 676-681.

4. Stirling '*Opportunities for Industry in the Application of Nanotechnology, Institute of Nanotechnology*': 2000.

5. K.E.Drexler, 1991, MIT, USA, Ph.D. Thesis entitled '*Nanosystems: Molecular Machinery, Manufacturing and Computation*'.

6. K.E. Drexler, 1992, Nanosystems: Molecular Machinery, Manufacturing and Computation, John Wiley.

7. Robert A. Freitas Jr., *Introduction to Nanomedicine*, 2000.

8. National Cancer Institute, Cancer Nanotechnology: Going small for big advances, *NIH publication*, Jan 2004.

9. I. Brigger, C. Bubernet and P. Couvreur, " *Advanced Drug Delivery Reviews*, 2002, 54: 631-651.

10. Roy I, Ohulchanskyy TY, Pudavar HE, Bergey EJ, Oseroff AR, Morgan J, Dougherty TJ, Prasad PN. J Am Chem Soc. 2003, **125**: 7860–7865.

11. Reich DH, Tanase M, Hultgren A, Bauer LA, Chen CS, Meyer GJ. J Appl Phys. 2003, **93**: 7275–7280.

12. Weissleder R, Elizondo G, Wittenburg J, Rabito CA, Bengele HH, Josephson L. Radiology. 1990, **175**: 489–493.

13. D.P. O'Neal, L.R. Hirsch, N.J. Halas, J.D. Payne and J.L. West, *Cancer Letters*, 2004, 209, 171-176.

14. J. Brabury, Lancet Oncology, 2003 4, 711.

15. M.L. Brongersma, Nature Materials, 2 2003, 296-297.

2013, Environmental Technology   Pages 311–316

Editors: D.R. Khanna, A.K. Chopra, Gagan Matta, R. Bhutiani & Vikas Singh

Published by: DAYA PUBLISHING HOUSE, NEW DELHI

Chapter 34

The Goal of Data Mining with some Novel Softwares Used in Biotechnology

Najiya Sultana and S.S. Sarangdevat

Department of CS and IT,
Rajasthan Vidyapeeth University, Udaipur

ABSTRACT

In this article one can find the routinely used protocols in biotechnology and Molecular Biology Laboratories. The students and researchers can find the easy to use protocols with productive results. This article highlights some of the basic concepts of Biotechnology in data mining. The application of data mining in the domain of bioinformatics is explained. It shows some of the current challenges and opportunities of data mining in Biotechnology. In this paper we have shown that data mining techniques can be applied successfully in the field of Biotechnology.

Keywords: *Data mining, Biotechnology, Protein sequences analysis, Biotechnology tools, DNA analysis.*

Introduction

Here we can find the routinely used protocols in Biotechnology. The new students and young researchers working in molecular biology labs can find the easy to use protocols with productive results. In recent years, rapid developments in genomics and proteomics have generated a large amount of biological data. Drawing conclusions from these data requires sophisticated computational analyses. Bioinformatics, or computational biology, is the interdisciplinary science of interpreting biological data using information technology and computer science. The importance of this new field of inquiry will grow as we continue to generate and integrate large quantities of genomic, proteomic, and other data.

A particular active area of research in Biotechnology is the application and development of data mining techniques to solve biological problems. Analyzing large biological data sets requires making sense of the data by inferring structure or generalizations from the data. Examples of this type of analysis include protein structure prediction, gene classification, cancer classification based on microarray data, clustering of gene expression data, statistical modeling of protein-protein interaction, etc. Therefore, we see a great potential to increase the interaction between data mining and bioinformatics.

Bioinformatics

The term bioinformatics was coined by Paulien Hogeweg in 1979 for the study of informatics processes in biotic systems. It was primary used since late 1980s has been in genomics and genetics, particularly in those areas of genomics involving large-scale DNA sequencing. Bioinformatics can be defined as the application of computer technology to the management of biological information. Bioinformatics is the science of storing, extracting, organizing, analyzing, interpreting and utilizing information from biological sequences and molecules. It has been mainly fueled by advances in DNA sequencing and mapping techniques [1]. The primary goal of bioinformatics is to increase the understanding of biological processes.

Biotechnology

Some of the area of research in Biotechnology includes:

Sequence Analysis

Sequence analysis is the most primitive operation in computational biology. This operation consists of finding which part of the biological sequences are alike and which part differs during medical analysis and genome mapping processes. The sequence analysis implies subjecting a DNA or peptide sequence to sequence alignment [2], sequence databases, repeated sequence searches, or other bioinformatics methods on a computer.

Genome Annotation

In the context of genomics, annotation is the process of marking the genes and other biological features in a DNA sequence [3]. The first genome annotation software system was designed in 1995 by Dr. Owen White. Computational approaches to genome comparison have recently become a common research topic in computer science.

Analysis of Gene Expression

The expression of many genes can be determined by measuring mRNA levels with various techniques such as microarrays, expressed cDNA sequence tag (EST) sequencing, serial analysis of gene expression (SAGE) tag sequencing, massively parallel signature sequencing (MPSS) [4], or various applications of multiplexed in-situ hybridization etc. All of these techniques are extremely noise-prone and subject to bias in the biological measurement. Here the major research area involves developing statistical tools to separate signal from noise in high-throughput gene expression studies [5].

Modeling Biological Systems

Modeling biological systems is a significant task of systems biology and mathematical biology. Computational systems biology aims to develop and use efficient algorithms [6], data structures, visualization and communication tools for the integration of large quantities of biological data with the goal of computer modeling. It involves the use of computer simulations [7] of biological systems, like cellular subsystems such as the networks of metabolites and enzymes, signal transduction pathways and gene regulatory networks to both analyze and visualize the complex connections of these cellular processes. Artificial life is an attempt to understand evolutionary processes via the computer simulation of simple life forms

High-Throughput Image Analysis

Computational technologies are used to accelerate or fully automate the processing, quantification and analysis of large amounts of high-information-content biomedical images [8]. Modern image analysis systems augment an observer's ability to make measurements from a large or complex set of images. A fully developed analysis system may completely replace the observer [14]. Biomedical imaging is becoming more important for both diagnostics and research. Some of the examples of research in this area are: clinical image analysis and visualization, inferring clone overlaps in DNA mapping, Bioimage informatics, etc.

Data Mining

Data mining, or knowledge discovery, has become an indispensable technology for businesses and researchers in many fields. Drawing on work in such areas as statistics, machine learning, pattern recognition, databases, and high performance computing [13], data mining extracts useful information from the large data sets now available to industry and science. This collection surveys the most recent advances in the field and charts directions for future research [5].

Data Mining Tasks

The two "high-level" primary goals of data mining, in practice, are prediction and description. The main tasks well suited for data mining, all of which involves mining meaningful new patterns from the data, are:

Classification

Classification is learning a function that maps (classifies) a data item into one of several predefined classes.

Estimation

Given some input data, coming up with a value for some unknown continuous variable.

Prediction

Same as classification and estimation except that the records are classified according to some future behavior or estimated future value).

Association Rules

Determining which things go together, also called dependency modeling.

Clustering

Segmenting a population into a number of subgroups or clusters.

Description and Visualization

Representing the data using visualization techniques.

Learning from data falls into two categories: directed ("supervised") and undirected ("unsupervised") learning [12]. The first three tasks – classification, estimation and prediction – are examples of supervised learning. The next three tasks – association rules, clustering and description and visualization – are examples of unsupervised learning. In unsupervised learning, no variable is singled out as the target; the goal is to establish some relationship among all the variables. Unsupervised learning attempts to find patterns without the use of a particular target field.

The development of new data mining and knowledge discovery tools is a subject of active research. One motivation behind the development of these tools is their potential application in modern biology [9].

Bioinformatics Tools

Some of the important tools for biotechnology given in the Table 34.1.

Application of Data Mining

Applications of data mining to bioinformatics and Biotechnology include gene finding, protein function domain detection, function motif detection, protein function inference[15], disease diagnosis, disease prognosis, disease treatment optimization, protein and gene interaction network reconstruction, data cleansing, and protein sub-cellular location prediction. For example, microarray technologies are used to predict a patient's outcome. On the basis of patients' genotypic microarray data, their survival time and risk of tumor metastasis or recurrence can be estimated [10]. Machine learning can be used for peptide identification through mass spectroscopy [16]. Correlation among fragmentations in a tandem mass spectrum is crucial in reducing stochastic mismatches for peptide identification by database searching [11]. An efficient scoring algorithm that considers the correlative information in a tunable and comprehensive manner is highly desirable.

Table 34.1: Some of the Software Tools for Biotechnology [DNA Analysis Software]

jambw 1.1	The Java based Molecular Biologist's Workbench.
AnnHyb v.4.942	A tool for working with and managing nucleotide sequences in multiple formats.
ChromasPro 2.33	a software for chromatogram files created by sequencers
DnaSP 5.10.01	DNA Sequence Polymorphism is a software package for the analysis of nucleotide polymorphism from aligned DNA sequence data.
DFW 2.51 trail	A compact, easy to use DNA analysis program, ideal for small-scale sequencing projects.
Artemis R12	A free genome viewer and annotation tool that allows visualization of sequence features and the results of analyses within the context of the sequence, and its six-frame translation.
ACT R9.05	(Artemis Comparison Tool) is a DNA sequence comparison viewer based on Artemis.
GDA 1.1	Genetic Data Analysis software
Sequencher 4.9 Demo	The industry standard software for DNA sequence analysis.
MeltCalc	The ultimate thermodynamic modelling spreadsheet for Excel
GenomePixelizer 2003.10.1	Be useful in the detection of duplication events in genomes, tracking the "footprints" of evolution, as well as displaying the genetic maps and other aspects of comparative genetics.
GenescanView1.2 4	Allows to visualize genescan files (.fsa format from ABI PRISM sequencers) and to view the exact peak size.
Sequin 9.50	A stand-alone software tool developed by the NCBI for submitting and updating entries to the GenBank, EMBL, or DDBJ sequence databases.
DNAuser 1.0	An integrated DNA/Protein analysis tool package for molecular biology/ biotech researchers.
Gene Construction Kit 3.0 Demo	Allows graphic manipulation of DNA sequences and sophisticated plasmid drawing options.
GenomeComp 1.3	a DNA sequence comparison tool and graphical user interface (GUI) viewer implemented in Perl/Tk.

Conclusion and Challenges

Data mining is the exploration and analysis, by automatic or semiautomatic means, of large quantities of data in order to discover meaningful patterns and rules. So, data mining is defined as the process of extracting interesting and previously unknown information from data, and it is widely accepted to be a single phase in a complex process known as Knowledge Discovery in Databases (KDD). Bioinformatics, Biotechnology and data mining are developing as interdisciplinary science. Data mining approaches seem ideally suited for bioinformatics, since bioinformatics is data-rich but lacks a comprehensive theory of life's organization at the molecular level. However, data mining in bioinformatics is hampered by many facets of biological databases, including their size, number, diversity and the lack of a standard ontology to aid the querying of them as well as the heterogeneous data of the quality and provenance information they contain. Another problem is the range of levels the domains of expertise present amongst potential users, so it can be difficult for the database curators to provide access mechanism appropriate to all. The integration of biological databases is also a problem. It is important to examine what are the important research issues in bioinformatics and develop new data mining methods for scalable and effective analysis.

References

1. Li, J., Wong, L. and Yang, Q. (2005). *Data Mining in Bioinformatics*, IEEE Intelligent System, IEEE Computer Society.

2. Liu, H., Li, J. and Wong, L. (2005). *Use of Extreme Patient Samples for Outcome Prediction from Gene Expression Data*, Bioinformatics, vol. 21, no. 16, pp. 3377–3384

3. Yang, Qiang. *Data Mining and Bioinformatics: Some Challenges*, http://www.cse.ust.hk/~qyang

4. Kuonen, Diego. *Challenges in Bioinformatics for Statistical Data Miner*, Bulletin of the Swiss Statistical Society, 46, 10-17.

5. Richard, R.J. A. and Sriraam, N. (2005). *A Feasibility Study of Challenges and Opportunities in Computational Biology: A Malaysian Perspective*, American Journal of Applied Sciences 2 (9): 1296-1300.

6. Nayeem, Akbar, Sitkoff, Doree, and Krystek, Jr., Stanley. (2006). *A comparative study of available software for highaccuracy homology modeling: From sequence alignments to structural models*, Protein Sci. April, 15(4): 808–824

7. N., Cristianini and M., Hahn. (2006). *Introduction to Computational Genomics*, Cambridge University Press. ISBN 0-5216-7191-4.

8. Mewes, H.W., Frishman, D., X.Mayer, K. F., Munsterkotter, M., Noubibou, O., Pagel, P. and Rattei, T. (2006). *Nucleic Acids Research*, 34, D169.

9. Hirschman, Lynette, C. Park, Jong, T., Junichi, Wong, L. and H. Wu., Cathy (2002). *Accomplishments and challenges in literature data mining for biology*, BIOINFORMATICS REVIEW, Vol. 18 no. 12, 1553–1561

10. Han and Kamber (2006). *Data Mining concepts and techniques*, Morgan Kaufmann Publishers.

11. Hand, D. J., Mannila, H. and Smyth, P. *Principles of Data Mining*, MIT Press.

12. Baxevanis, A.D., Petsko, G.A., Stein, L.D. and Stormo, G.D., eds. (2007). *Current Protocols in Bioinformatics*. Wiley.

13. Mount, D. W. (2002). *Bioinformatics: Sequence and Genome Analysis* Spring Harbor Press.

14. Gilbert, D. (2004). *Bioinformatics software resources. Briefings in Bioinformatics*, Briefings in Bioinformatics.

15. Jiong, Lei Liu, Yang, A. and Tung, K. H. *Data Mining Techniques for Microarray Datasets*, Proceedings of the 21st International Conference on Data Engineering (ICDE 2005).

16. Pevzner, P. A. (2000). *Computational Molecular Biology: An Algorithmic Approach* The MIT Press.

Index